Weed Ecology and Diversity

Weed Ecology and Diversity

Editor

Ilias Travlos

MDPI

Basel • Beijing • Wuhan • Barcelona • Belgrade • Novi Sad • Cluj • Manchester

Editor
Ilias Travlos
Agricultural University
of Athens
Athens
Greece

Editorial Office
MDPI
St. Alban-Anlage 66
4052 Basel, Switzerland

This is a reprint of articles from the Special Issue published online in the open access journal *Diversity* (ISSN 1424-2818) (available at: https://www.mdpi.com/si/diversity/weed_diversity).

For citation purposes, cite each article independently as indicated on the article page online and as indicated below:

Lastname, A.A.; Lastname, B.B. Article Title. *Journal Name* **Year**, *Volume Number*, Page Range.

ISBN 978-3-7258-0123-7 (Hbk)
ISBN 978-3-7258-0124-4 (PDF)
doi.org/10.3390/books978-3-7258-0124-4

Contents

Ioannis Gazoulis, Nikolaos Antonopoulos, Panagiotis Kanatas, Nikolas Karavas, Irena Bertoncelj and Ilias Travlos
Invasive Alien Plant Species—Raising Awareness of a Threat to Biodiversity and Ecological Connectivity (EC) in the Adriatic-Ionian Region
Reprinted from: *Diversity* **2022**, *14*, 387, doi:10.3390/d14050387 . . . . **1**

Philipp Köllmann and Rainer Waldhardt
Farming Intensity Affects Soil Seedbank Composition and Spontaneous Vegetation of Arable Weeds
Reprinted from: *Diversity* **2022**, *14*, 111, doi:10.3390/d14020111 . . . . **28**

Heba Sallam, Mashail Nasser Alzain, Amani Omar Abuzaid, Naglaa Loutfy, Mohamed O. Badry, Ahmed K. Osman and Sabah A. Hammad
Wild Plant Diversity and Soil Characteristics of Desert Roadside Vegetation in the Eastern Desert
Reprinted from: *Diversity* **2023**, *15*, 874, doi:10.3390/d15070874 . . . . **47**

Swati Shrestha, Gourav Sharma, Shandrea Stallworth, Edilberto D. Redona and Te Ming Tseng
Exploring the Genetic Diversity among Weedy Rice Accessions Differing in Herbicide Tolerance and Allelopathic Potential
Reprinted from: *Diversity* **2022**, *14*, 44, doi:10.3390/d14010044 . . . . **64**

Ioannis Gazoulis, Panagiotis Kanatas and Nikolaos Antonopoulos
Cultural Practices and Mechanical Weed Control for the Management of a Low-Diversity Weed Community in Spinach
Reprinted from: *Diversity* **2021**, *13*, 616, doi:10.3390/d13120616 . . . . **74**

Wei-Hsun Hsieh, Yen-Chiun Chen, Hsien-Chun Liao, Yann-Rong Lin and Chih-Hui Chen
High Differentiation among Populations of Green Foxtail, *Setaria viridis*, in Taiwan and Adjacent Islands Revealed by Microsatellite Markers
Reprinted from: *Diversity* **2021**, *13*, 159, doi:10.3390/d13040159 . . . . **90**

Barbara Kutasy, Zoltán Farkas, Balázs Kolics, Kincső Decsi, Géza Hegedűs, Judit Kovács, et al.
Detection of Target-Site Herbicide Resistance in the Common Ragweed: Nucleotide Polymorphism Genotyping by Targeted Amplicon Sequencing
Reprinted from: *Diversity* **2021**, *13*, 118, doi:10.3390/d13030118 . . . . **107**

Kaitlyn E. Trepanier, Brea Burton and Bradley D. Pinno
Do Weeds Hinder the Establishment of Native Plants on a Reclaimed North American Boreal Mine Site?
Reprinted from: *Diversity* **2021**, *13*, 76, doi:10.3390/d13020076 . . . . **123**

Sami Asir Al-Robai
Ecology and Diversity of Angiosperm Parasites and Their Host Plants along Elevation Gradient in Al-Baha Region, Saudi Arabia
Reprinted from: *Diversity* **2023**, *15*, 1065, doi:10.3390/ d15101065 . . . . **133**

Yessica P. Duque, Carlos E. Giraldo-Sánchez, Mario A. Quijano-Abril and Jose M. Rojas
Ecology and Diversity of Weed Communities in the Northern Andes under Different Anthropogenic Pressures
Reprinted from: *Diversity* **2023**, *15*, 936, doi:10.3390/d15080936 . . . . **155**

diversity

MDPI

Review

Invasive Alien Plant Species—Raising Awareness of a Threat to Biodiversity and Ecological Connectivity (EC) in the Adriatic-Ionian Region

Ioannis Gazoulis [1,*], Nikolaos Antonopoulos [1], Panagiotis Kanatas [2], Nikolas Karavas [3], Irena Bertoncelj [4] and Ilias Travlos [1]

1 Laboratory of Agronomy, Agricultural University of Athens, 75, Iera Odos Str., 11855 Athens, Greece; nikolasantwno@gmail.com (N.A.); travlos@aua.gr (I.T.)
2 Department of Crop Science, University of Patras, P.D. 407/80, 30200 Mesolonghi, Greece; pakanatas@gmail.com
3 Department of Botany, Faculty of Biology, University of Athens, 15784 Athens, Greece; nik.karavas@gmail.com
4 Department of Agricultural Ecology and Natural Resources, Agricultural Institute of Slovenia, Hacquetova ulica 17, SI-1000 Ljubljana, Slovenia; irena.bertoncelj@kis.si
* Correspondence: giangazoulis@gmail.com

Abstract: Improving ecological connectivity (EC) within landscapes by establishing corridors and ecological networks has been proposed to counteract the negative effects of habitat fragmentation and climate change on biodiversity. To be functional, ecological networks should be kept free of opportunistic invasive species that can disrupt EC between protected areas and cause biodiversity loss. The present study focused on perennial herbaceous species whose occurrence in the Adriatic-Ionian region has increased in the last two decades, namely common milkweed (*Asclepias syriaca*), Jerusalem artichoke (*Helianthus tuberosus*), Japanese knotweed (*Reynoutria japonica*), Bohemian knotweed (*Reynoutria* × *bohemica*), giant hogweed (*Heracleum mantegazzianum*), giant goldenrod (*Solidago gigantea*), Canadian goldenrod (*Solidago canadensis*), and Bermuda buttercup (*Oxalis pes-caprae*). All species have a high potential to spread in grasslands, abandoned agricultural fields, forest edges, and riparian areas and pose a significant threat to native plant communities and biodiversity. Restoring heavily infested sites is a major challenge because these perennial invaders are very persistent and tend to alter the soil environment in invaded habitats and prevent the recolonization of native plant communities. Therefore, early action should be taken to prevent the spread of these environmental weeds in ecological networks and protected areas with high conservation value.

Keywords: environmental weeds; perennial weeds; dispersal; grasslands; riparian areas; abandoned agricultural fields; Balkan Peninsula; eradication

Citation: Gazoulis, I.; Antonopoulos, N.; Kanatas, P.; Karavas, N.; Bertoncelj, I.; Travlos, I. Invasive Alien Plant Species—Raising Awareness of a Threat to Biodiversity and Ecological Connectivity (EC) in the Adriatic-Ionian Region. *Diversity* **2022**, *14*, 387. https://doi.org/10.3390/d14050387

Academic Editor: Michael Wink

Received: 19 April 2022
Accepted: 12 May 2022
Published: 13 May 2022

1. Introduction

Improving ecological connectivity (EC) within landscapes through the establishment of corridors and ecological networks has been proposed to counteract the negative effects of habitat fragmentation and climate change on biodiversity [1]. At a time when biodiversity loss has become a serious issue in Europe, the conservation of EC has become a major challenge [2]. This is also true for a large part of the Balkan Peninsula, where political and economic circumstances have created transboundary barriers between associated countries, posing a major challenge for the conservation of EC and biodiversity [3]. In this context, the "Interreg Adrion DINALPCONNECT" project was launched in 2020 with the aim of improving EC by combating environmental vulnerability and habitat fragmentation and securing ecosystem services in the Adriatic-Ionian region. Seven countries in the Adriatic-Ionian region, namely Italy, Slovenia, Croatia, Bosnia and Herzegovina, Montenegro, Albania, and Greece, are included in the broader project area. Four pilot regions have been established as linkages between Natura 2000 sites and protected areas on the borders between Italy and Slovenia, Slovenia and Croatia, Croatia and Bosnia and Herzegovina, and Albania and Greece (Figure 1).

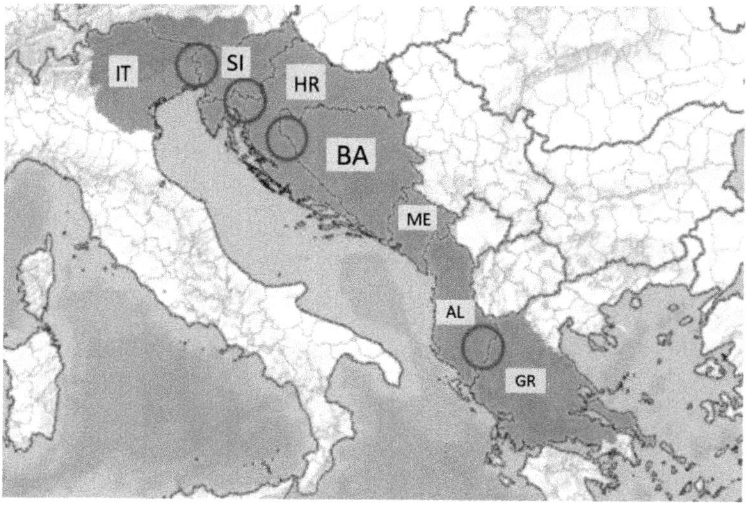

Figure 1. Countries of the Adriatic-Ionian region where ecological connectivity (EC) will be explored and consolidated. The four circles represent pilot regions selected as linkages between Natura 2000 sites and protected areas by the "Interreg Adrion DINALPCONNECT" project team. Abbreviated names of associated countries: IT; Italy, SI; Slovenia, HR; Croatia, BA; Bosnia and Herzegovina, ME; Montenegro, AL; Albania, GR; Greece.

By enabling the movement of organisms, corridors can reduce the likelihood of species extinction by increasing genetic exchange between populations and promoting recolonization of empty habitat patches in spatially structured populations [4,5]. However, to be sustainable and functional, ecological networks should be kept free of opportunistic invasive species that can disrupt EC between protected areas and lead to biodiversity loss [6]. To achieve this goal, the status of biological invasions in larger areas should not be overlooked but continuously monitored, and eradication programs should be initiated in a timely manner [7]. According to the Driver, Pressure, State, Impact, and Response (DPSIR) models developed in [8], extensive urbanization and improper management practices in agricultural areas, forests, grasslands, and riparian zones generate serious threats to biodiversity and EC in the Balkan Peninsula. Among others, the risk of further spread of exotic non-native plant species has been identified as one of the major threats to biodiversity in the Adriatic-Ionian region [8].

Biological invasions by alien plant species are among the main causes of biodiversity loss, as the presence of invaders leads to the displacement of native vegetation and can have negative impacts on the soil environment and associated fauna taxa [9–12]. In addition to alien invasive trees and herbaceous species with an annual life cycle, there is much evidence that the occurrence of perennial herbaceous species with high invasive potential has increased in the Adriatic-Ionian region in the last two decades [13–19]. Perennial herbaceous species have a strong invasive potential, as most of them can reproduce both by seeds and by underground vegetative organs. Vegetative propagation allows for the rapid establishment of dense, monospecific, clonal populations in invaded habitats, which makes eradication a very challenging goal because it requires sufficient control of the belowground vegetative organs that give rise to new clonal plants. If only the aboveground vegetative parts are removed, plants regrow from their belowground vegetative organs [20–24]. In addition, seed production and dispersal facilitate effective colonization of new areas for species that can reproduce both by seed and vegetatively [15,21,25]. Therefore, measures to prevent the spread of these species and eradication programs should be implemented in the wider Adriatic-Ionian region before invasions become problematic in ecological networks and protected areas of high conservation value.

The present study focused on common milkweed (*Asclepias syriaca* L.), Jerusalem artichoke (*Helianthus tuberosus* L.), Japanese knotweed (*Reynoutria japonica* Houtt.), Bohemian knotweed (*Reynoutria* × *bohemica* Chrtek et Chrtková), giant hogweed [*Heracleum mantegazzianum* Sommier & Levier], giant goldenrod (*Solidago gigantea* Aiton), Canadian goldenrod (*Solidago canadensis* L.), and Bermuda buttercup (*Oxalis pes-caprae* L.). Our first objective was to raise public awareness of the spread status of these perennial herbaceous alien plant species in the countries of the Adriatic-Ionian region, as described in the recent literature. Another objective was to focus on the main ecological features contributing to the success of these species as invaders and to present the impact of their invasions on biodiversity, based on data from case studies conducted in the wider European region. The importance of raising public awareness and taking preventive measures to prevent further spread of these species is emphasized. Available eradication methods that can be applied to infested sites are also presented.

2. Spread of the Selected Alien Plant Species in the Adriatic-Ionian Region

The alien plant species included in the current study are perennial herbaceous species that have been classified by the European and Mediterranean Plant Protection Organization (EPPO) as having a high potential to spread and are capable of altering the environment in invaded habitats, thus posing a significant threat to native plant communities, the environment, and biodiversity in the EPPO region. All species are already present in at least four countries in the Adriatic-Ionian region (Table 1).

Table 1. Summary of the presence of the selected alien plant species in the countries of the Adriatic-Ionian region.

Binomial Name	Botanical Family	Common Name	Origin	Infested Countries [1]
Asclepias syriaca	Apocynaceae	Common milkweed	North America	IT, SI, HR, BA, ME
Helianthus tuberosus	Asteraceae	Jerusalem artichoke	North America	IT, SI, HR, BA, ME, AL, GR
Reynoutria japonica	Polygonaceae	Japanese knotweed	East Asia	IT, SI, HR, BA
Reynoutria × *bohemica*	Polygonaceae	Japanese knotweed	East Asia	IT, SI, HR, BA
Heracleum mantegazzianum	Apiaceae	Giant hogweed	Eastern Europe	IT, SI, HR, BA
Solidago gigantea	Asteraceae	Giant goldenrod	North America	IT, SI, HR, BA, ME
Solidago canadensis	Asteraceae	Canadian goldenrod	North America	IT, SI, HR, BA, ME
Oxalis pes-caprae L.	Oxalidaceae	Bermuda buttercup	South Africa	IT, HR, AL, GR

[1] Country names are represented by three-code abbreviations—e.g., IT; Italy, SI; Slovenia, HR; Croatia, BA; Bosnia and Herzegovina, ME; Montenegro, AL; Albania, GR; Greece.

Common milkweed has been naturalized in Italy and occurs in seven administrative regions; it spreads in all northern regions of the country, except in Liguria and Aosta Valley [17]. In Slovenia, this species has been recorded in the central part of the country along roadsides, near Ljubljana, east and northeast of it; there are also reports from Brežice, Podravje, and Pomurje regions, and two other cases of occurrence in the northeastern part of the country, near watercourses along the Mura River on the border with Croatia [26]. In Croatia, Boršić et al. [16] found that common milkweed occurs mainly in the northwestern and eastern parts of the country along roadsides and railways, riverbanks, pastures, and open forest areas. The presence of local populations has been confirmed in wet meadows in the Posavina region of Bosnia and Herzegovina [27]. According to Stešević and Petrović [19], common milkweed is one of the alien plants detected in Montenegro, where only a few populations have been reported infesting wet meadows outside the city of Podgorica.

Jerusalem artichoke is one of the most noxious invaders in field crops in northern Italy, especially where it has been grown in the past as an energy or industrial crop, or when allowed to spread uncontrollably on abandoned agricultural land [28]. This species is also reported as invasive in Tuscany, where it occurs mainly on riverbanks [29]. This species is also introduced in Greece and Albania but is not yet invasive in these countries [13,14]. In Montenegro, Jerusalem artichoke is widely distributed on abandoned agricultural lands, ruderal areas, and riverbanks [19]. In addition, it is considered invasive in northwestern Bosnia and Herzegovina and also in the central part, including the Sarajevo Canton area, along roadsides, railways, riverbanks, and streams [15,27]. Jerusalem artichoke is a serious invader in riparian zones of streams in Slovenia [30]. Küzmič and Šilc [18] reported its occurrence in six habitats of the European Nature Information System (EUNIS). Follak et al. [31] also mentioned that this weed has a high potential to spread across roadsides and railways in the country and colonize new areas. This is also supported by the recent studies in which dense populations were discovered along a roadside near Zbilje, a settlement in the Upper Carniola region of Slovenia [32]. In Croatia, escaped Jerusalem artichoke populations have spread aggressively along forest edges, drainage systems, and riverbanks in the continental part of the country, as well as in flooded woodlands on the riverbanks of Sava, Sutla, Krapina, and Drava [33]. Its occurrence on forest edges was also confirmed in Medvednica Nature Park by Vuković et al. [34].

According to Mincheva et al. [35], Japanese knotweed has been well distributed in grasslands in the Piedmont region of northwestern Italy during the last 30 years. Populations have also been observed in the western Italian Alps [36]. In Slovenia, this species is also widespread in urban areas, along watercourses, riparian corridors, roadsides, and railroad lines [31]. In particular, the southeastern part of the municipality of Ljubljana in central Slovenia is known as the most infested area in the country [37]. According to Vuković et al. [38], Japanese knotweed populations have infested at least 25 sites in Croatia. Most detections are from roadsides and railways, urban areas and areas along watercourses at several sites around the city of Zagreb in the north and sites in the Alpine biogeographical region on the Dinaric mountain of Velebit. There are also reports of grasslands infested with this species between the settlements of Krstinja and Brusovačanear near the border with Bosnia and Herzegovina. Dense populations of Japanese knotweed have also been found along roadsides and railways, watercourses, urban areas, and grasslands in several locations in Bosnia and Herzegovina, including the larger areas of Sarajevo, Zenica, and Žepce, as well as outside cities in the northeastern part of the country (Banja Luka, Derventa and Doboj) [15,27]. It should be noted at this point that Japanese knotweed may be confused with Bohemian knotweed, which is the hybrid of Japanese knotweed and giant knotweed (*Reynoutria sachalinensis* (F. Schmidt) Nakai) [39].

Jovanović et al. [40] recorded Bohemian knotweed populations in ruderal and urban areas in the cities of Celje and Ljubljana, in Slovenia. Galasso et al. [17] reported that Bohemian knotweed is invasive and widespread in Italy. Giulani et al. [41] collected samples from three populations of Bohemian knotweed growing along artificial banks of the Ombrone River in Podere della Chiesa, along stream shores and artificial banks under woody riparian vegetation along the stream Vincio in Pontelungo, and also along the Arno River in Subbiano; in Subbiano, Bohemian knotweed was also widespread along cultivated areas and habitats disturbed by anthropogenic activities. According to Lazzaro et al. [42], the hybrid occurs in at least nine different regions and is classified as one of the most important invasive alien plants in the country. In Croatia, Vuković et al. [38] reported large populations growing along the wider area of the Kašina stream in the settlement of Kašina, eastwards of Medvednica and in a grassland located between the settlements of Dekanovec and Domašinec. The same authors detected Bohemian knotweed populations in urban areas, in the city of Zagreb, and also across roadsides in several localities throughout the country. In Bosnia and Herzegovina, the hybrid is found in riparian habitats near the towns of Višegrad, Zvornik, Nova Kasaba, Dobrun, and Sarajevo; this weed is also found in ruderal sites in Dobrun, Rogatica, and in urban areas in Uvac [40]. In Montenegro, these researchers have detected Bohemian knotweed occurrences on riparian and ruderal sites near Pljevlja and in gardens in the small town of Žabljak.

In the Sava River basin, populations of giant hogweed have been found only in Slovenia near Ljubljana, probably as escapees from the Ljubljana Botanical Garden [43]. Although active eradication measures are implemented every year to control the spread of this invasive species, there is recent evidence that roadsides serve as corridors for further spread in Slovenia [31]. In Croatia, the occurrence of this species has been confirmed in disturbed habitats in Međimurje County and in Gornja Šemnica in Krapina-Zagorje County near roadsides [16]. As for its occurrence in Bosnia and Herzegovina, populations have been reported in the central part of the country and especially near Hadžići, southwest of Sarajevo, along the railway between Sarajevo and Mostar, and also near Lokve between Hadžići and Pazarić [44]. In Italy, populations of giant hogweed have been classified as naturalized in some areas and invasive in others [17]. According to maps provided by Jahodová et al. [45], this species also occurs in the northeastern part of the country, while Siniscalco and Barni [36] reported infestations in mountainous areas in the Aosta Valley and Alpine Botanical Gardens.

Giant goldenrod and Canadian goldenrod are among the most widespread invasive alien plant species in Slovenia. Zelnik [30] reported that extensive populations of both species infest the riparian zones along streams, disturbed habitats with standing waters, and the edges of floodplain forests. In the study by Marinšek and Kutnar [46], their occurrence was found in 37 out of 130 plots near the Mura River in the northeastern part of Slovenia, in the sub-Pannonian region. These species are also common in shrublands and afforestation areas. According to Küzmič and Šilc [18], giant goldenrod has spread in 16 EUNIS habitats, while Canadian goldenrod is already present in 7 different EUNIS habitats. Šajna [47] found that giant goldenrod occurred in 65% and 91% of mowed and abandoned forest meadows, respectively, within the natural forests that developed along the riverbanks of the Drava river. Canadian goldenrod has a high potential to spread across roadsides and railways in Slovenia [31]. In Croatia, both species are invasive [48], but giant goldenrod seems to be more common. In the continental part of the country, it has been found on many different ruderal areas, usually disturbed due to human activities; the highest infestation rates are observed on abandoned agricultural land, riverbanks, and forest edges [49]. Novak and Novak [49] reported that at least 10 ha were infested by the giant goldenrod at such sites in Zagreb, Varaždin, Koprivnica-Križevci, Krapina-Zagorje, Međimurje, Bjelovar-Bilogora, and Karlovac counties. Maslo et al. [27] mentioned that both species were found along rivers and forest edges, on roadsides, and in gardens in Bosnia and Herzegovina; they also indicated that giant goldenrod populations have invaded the wider area of Trebević mountain. In Montenegro, two distinct occurrences of giant goldenrod were found near roadsides in the suburban areas of Nikšić and Mojkovac, near gardens where it is grown as an ornamental plant [50]. A population of Canadian goldenrod was found on the roadside in the village of Vir, near the town of Nikšić [51]. The giant goldenrod is the most common of the two species in northern Italy [52].

Bermuda buttercup occurs in Greece in all 18 habitat groups, namely marine habitats, coastal habitats, inland water bodies, fens and fogs, grasslands, shrublands, forests, rocky areas and screes, cultivated arable fields, cultivated areas in gardens and parks, perennial orchards and vineyards, man-made urban areas, transportation networks (roadsides and railways), artificial constructed boundaries, and abandoned places [13]. It is common in urban and suburban ruderal areas, olive groves, meadows, and forest edges, where it forms dense populations that prevent the growth of other species, especially during the growing season from autumn to early spring [13]. This species is considered invasive in similar habitats in Italy [17]. It was also introduced in Croatia in the last two decades [48]. In all mentioned countries, islands are highly susceptible to invasions of this species [13,17,48]. In addition, Barina et al. [14] reported that Bermuda butter cup is well established in Albania, but not yet invasive.

3. Species Description and Key Ecological Characteristics

3.1. Common Milkweed

Common milkweed is a perennial broadleaved weed species; it is diploid ($2n = 22$) and belongs to the family Apocynaceae, subfamily Asclepiadoideae [53]. It is native to the northeastern, north central, and southeastern United States and adjacent areas of Canada [54]. The plants have a C_3 photosynthetic pathway and grow up to 1.5 m tall. They grow in clusters of stout stems with short downy hairs; all plant parts contain a milky sap [55]. The conspicuously veined leaves are smooth on the upper surface, pubescent on the lower surface, 10–20 cm long and 5–11 cm wide, and appear in pairs on opposite sides of the stem [56]. Inflorescences are large spherical clusters (umbels) that arise in the upper leaf axils and at the tips of the stem [57]. The number of inflorescences per plant ranges from 2 to 5; the number of flowers per inflorescence ranges from 30 to 108 [58]. The flowers are fragrant, pink to white, and considered an excellent source of nectar for butterflies, bees, and other insects [59].

Reproduction is by seed as well as by underground vegetative organs. The flowers are self-sterile and are pollinated by insects [60]. Only 2% of the flowers produce mature pods. On average, each plant produces 4–6 pods containing 150–425 seeds [55]. Consequently, the presence of 1–6 stems m^{-2} can lead to a production of up to 87 million seeds ha^{-1}. Csontos et al. [61] reported that at densities of 14 and 18.1 stems m^{-2}, seed production reached 7820 and 10597 seeds m^{-2}, respectively. In addition to seed production, seeds have inherent dormancy; they require at least a post-maturation period of 1 year to germinate [62]. Although seed survival is influenced by environmental conditions, the high seed production capacity and seed dormancy allow the formation of persistent seed banks in infested areas [63]. The seeds are pinnate, and their weight is very low (e.g., the weight of 1000 seeds is usually 7–8 g). Therefore, they can disperse over long distances by wind. For example, early studies by Platt and Weis [64] reported that the diameter of the plume was 5.59–5.73 cm and the seeds were dispersed by the wind at a distance of 13.3–14.3 m with a velocity (terminal velocity) of 0.0249 m s^{-1} with a wind dispersion of 2.78–4.18 m s^{-1}. In more recent studies, Moravcová et al. [65] reported that seeds were dispersed by wind at a velocity of 0.284–0.422 m s^{-1}.

Seed dispersal is the main mechanism for milkweed to spread over long distances and colonize new areas, while establishment of dense stands over shorter distances occurs by vegetative propagation with rhizomes [21]. Seed germination occurs between April and May. Three weeks after germination, the formation of underground rhizomes occurs. Flowering does not occur during the first growing season [63]. The rhizomes contain buds that develop the next spring. New shoots develop from these buds under favorable conditions. The shoots are viable for one year but are formed in the same place each year [55]. Most rhizomes normally grow horizontally 10 to 40 cm below the soil surface at a growth rate of 0.01 to 0.25 m per year [63]. Some rhizomes penetrate deeper soil profiles (1–1.5 m depth) and reach groundwater, which provides water and nutrients to the shoots; new shoots appear from mid-spring to early summer, and plants flower from June to August [63]. Plants grow as single shoots or as groups of 2–5 shoots connected by rhizomes for many years after establishment [21]. Cultivation for ornamental purposes, cultivation for fiber production, cultivation as a nectar source by beekeepers, improper cleaning of agricultural machinery used near infested sites, transport of soil contaminated with rhizomes to new areas, abandonment of agricultural land, and other anthropogenic activities also contribute significantly to its spread [21,66–68]. In addition, this species spreads along roadsides and railways [31].

3.2. *Jerusalem Artichoke*

Jerusalem artichoke is a hexaploid (2*n* = 102), erect, rhizomatous, perennial, broadleaved weed species originally from North America [69]. The plants have a C3 photosynthetic pathway, but also possess enzymes typical of the C_4 type of photosynthesis [70]. Stems are light green or green-purple in color, hairy, sparsely to moderately branched in the upper half, woody in the years after establishment, and can reach a height of up to 3–4 m with a stem diameter of 1.6–3 cm [15,71]. Leaves are 10–25 cm long and 4–12 cm wide, broad near the base, and glabrous on the upper surface. In the lower part of the stem, they are opposite or arranged in whorls of three leaflets, while in the upper part of the stem, the leaves are alternate, simple, winged above, the blade ovate to ovate-lanceolate, with three prominent veins and a petiole 2–8 cm long [72]. Inflorescences are sunflower-like heads with bright yellow disk flowers; they are formed as single heads or in corymbs at the ends of the main stems and axillary branches and are 4–11 cm in diameter, while disk diameters range from 1.5 to 2.3 cm [71,72]. The bracts are lanceolate, green, finely pubescent dorsally, and hairy at the margin; the petals are oblong-spatulate, acute, and hairy in the upper part [73]. The flowers are insect-pollinated and are an excellent nectar source for honeybees, wasps, flies, and butterflies [74]. It should be noted that this species can be confused with wild and weedy forms of cultivated sunflower (*Helianthus annuus* L.) [75]. Fewer than five seeds are produced per flower head, forming glabrous and hairy achenes. The seeds are wedge-shaped and smooth achenes, usually gray or brown in color, 6–8 mm long, and 2 mm wide [15].

The root system is adventitious and fibrous and develops cord-like rhizomes that are 1 m long with a swollen apical portion and axillary buds. Tubers are formed by the thickening of the rhizomes [76]. Reproduction is vegetative, most commonly by tubers or tuber pieces, but also by rhizomes or rhizome pieces. The majority of seeds are infertile, but their production is an important means of maintaining the genetic diversity of the species. Seed dormancy is also a feature that contributes to persistence [72]. Growth begins from tubers or seeds in April or May and is followed by a period of rapid vegetative growth. Rhizome sprouting occurs at the beginning of the flowering period, from early July to late September. At this time, tubers are initiated [77].

It should be highlighted here that, although Jerusalem artichoke is a C_3 species, the high aboveground biomass acts as a reservoir of carbohydrates that are stored in the tubers and rhizomes to maintain the clone throughout the winter and provide a ready energy source for rapid spring regrowth [78,79]. The main type of carbohydrates includes homologous series of polyfructofuranose units, comprised of linear chains of *D*-fructose molecules terminated by a *D*-glucose molecule; the dominant term used to describe all such polysaccharides consisting largely of fructose units is often referred to as inulin [77]. It should be also noted that plants are photoperiod-sensitive, requiring alternate cycles of long light followed by periods of shorter light to stimulate tuber formation and further development [80]. In any case, Jerusalem artichoke exhibits rapid vegetative growth and tuber development in a single growing season [78]. A single plant is alive for up to five generations, but stands are persistent after establishment since new ramets emerge from tubers continuously over years [72].

The plant thrives on riverbanks; water is an important means of dispersal over long distances. There is evidence that rivers, streams, and creeks can transport rhizomes and tubers to new locations [81]. Seed dispersal may also contribute to further colonization of new areas at a given time [15]. Dispersal by animals is an important aid in dispersal. Rodents feeding on tubers, rhizomes, and seeds of Jerusalem artichoke in late summer carry the invader's reproductive organs through their digestive system and subsequently release them in new locations—a phenomenon known as endozoochory [29,82]. An example is the study by Mori et al. [29], who reported that the initial presence of a pair of adult Crested porcupine (*Hystrix cristata* L., 1758) increased Jerusalem artichoke density by 41% in five years, which was observed in a total area of 212 hectares. In any environment, Jerusalem artichoke can spread rapidly through vegetative propagation and form dense, monospecific stands once established. Thus, in the study by Žgančíková et al. [81], its density increased by 30–80% from 2010 to 2011 on ruderal areas, railroad lines, and riverbanks. The same authors recorded a population density of up to 265 plants m^{-2}. However, it should be noted that this species is self-limiting at high densities, and fewer tubers are produced per plant under crowded conditions [72]. In any case, human activities are also one of the most important reasons for the invasiveness of this species in non-native areas. In particular, Jerusalem artichoke is widely cultivated in Europe as a multipurpose crop for feed production and industrial, energy, and medicinal purposes [83,84]. Therefore, the plants can escape from cultivation and grow in the wild [85].

3.3. *Japanese Knotweed and Bohemian Knotweed*

Japanese knotweed is a rhizomatous, perennial, broadleaved weed species with a C_3 photosynthetic pathway, native to Japan, Taiwan, and Korea [15]. In Europe, populations with a chromosome number of $2n = 88$ are predominant [86]. Rhizomes are dark brown, nodular with annular structures spaced about 2 to 4 cm apart, thick and woody when old, up to 8 cm in diameter, and often bright orange inside [87]. Stems grow more than 3 m tall, branch, become woody with age, and often have reddish spots [88]. Leaves are 5–12 cm × 5–8 cm, broadly ovate, pointed at the apex and truncate at the base, coarse in texture, and glabrous with petioles 1–3 cm long [88]. Inflorescences are membranous ochreates that are initially erect and fall off at maturity. Flowers are cream-colored and appear in clusters of three to six on terminal and axillary inflorescences [15]. In male sterile plants, flowers have five tepals, the outer three of which are keeled, and eight stamens in the perianth [88]. The fruits are trigonous achenes, 2–4 mm long and 2 mm wide, dark brown in color and shiny [89]. Although most European populations reproduce vegetatively, propagation by seed can also be observed [90]. Seedlings emerge from late March to May. To survive, plants from seed should form at least five true leaves in the year of establishment (BBCH: 15) [91]. The potential of the seed bank is negligible because seeds do not exhibit significant dormancy levels [92].

As for vegetative propagation, a clonal patch of Japanese knotweed can arise from rhizome fragments as small as 1 cm with a fresh weight of 0.7 g [93]. Rhizome sprouting occurs from April and continues until the canopy closes between June and July [87]. The optimal soil depth for rhizome emergence is 2 cm, but emergence can occur from 1 m depth; rhizomes become 15–20 m long and penetrate the soil at 2–3 m depth [20]. Clonal populations develop very dense stands that accumulate large amounts of biomass [94]. A mature stand can produce up to 1467 g m^{-2} dry weight [95]. Flowering occurs between August and mid-September, when fruit set begins; storage of fruit in stems during winter is another common feature of populations in invaded areas [88]. Leaves are photosynthetically active until the onset of senescence and drop in winter [96]. When shoots die in October, a few dormant buds develop on the rhizome [97]. In May and June, 80–90% of the freshly synthesized carbohydrates remain in the shoots, while in August and September, proportions of 35% and 70%, respectively, are diverted to the rhizomes [98]. The stand can persist for more than 50 years as old and young rhizome parts remain physiologically connected [96].

Seed production per stem can reach 127,000 [89]. Seeds are reported to have high germination rates (≥90%) in central Europe [99]. They can be easily dispersed by wind in all directions at a distance of 16 m [90]. Seeds can also float in streams, which facilitates the colonization of riparian corridors [20]. Typical dispersal routes for seeds include roadsides and railways [90]. Regarding the population dynamics of clones, stems with at least two rhizome nodes can lead to the production of 2.3 new shoots m^{-2}, and the stem density per unit area can range from 8.9 to 42 stems m^{-2}, leading to the formation of 86–407 rhizome nodes m^{-2} [95]. Stems with rhizome fragments can spread via fresh and salt water, while anthropogenic activities such as illegal use as an ornamental plant or transport of soil profiles contaminated with rhizome fragments facilitate introduction into new areas [100]. Shallow tillage also leads to excessive spread at infested sites [88].

As for Bohemian knotweed, hexaploid ($2n = 66$) populations of the hybrid are most common in Europe [86,90]. Correct identification of this weed is difficult given its similarities to Japanese knotweed [101]. Bailey et al. [25] summarized the main morphological differences between the hybrid and Japanese knotweed. First, the leaves of Bohemian knotweed are larger (23 × 19 cm) than those of Japanese knotweed. Second, the leaves do not have the truncate bases typical of Japanese knotweed but have a weakly to moderately cordate base and are of medium shape. Furthermore, unlike Japanese knotweed, the leaves are characterized by a medium texture and appearance. Another difference is found on the lower abaxial leaf surface; there are very distinct, stout, celled pointed hairs in Bohemian knotweed, while there are single-celled "bumps" on the leaf veins of Japanese knotweed. In addition, the outlines of epidermal cells are clearly visible in Japanese knotweed, while they are clearly intermediate in Bohemian knotweed. The stems of Bohemian knotweed also vary more in their height (2.5–4.0 m) than those of Japanese knotweed. Both sexes occur in hybrid populations, and hermaphrodites, which are self-incompatible, appear to be outnumber male-sterile plants [87]. Flowering begins in mid-August to late September and continues for several weeks or until stems are cut down due to frost; reproduction occurs both by rhizomes and seed. Seed production is an important means of dispersal [102]. Seeds can be dispersed over long distances by wind [103]. The water of rivers and streams is also very important for seed dispersal as the seeds can float in water for more than two days and remain highly germinable [104]. As for its ability to propagate vegetatively by rhizomes, Bímová et al. [93] found that Bohemian knotweed has higher rhizome regeneration rates compared to Japanese knotweed and other closely related species of the genus *Reynoutria* spp. The spread of the hybrid is also favored by anthropogenic activities [105].

3.4. Giant Hogweed

Giant hogweed is a perennial broadleaved weed classified as a hemicryptophyte; it is diploid ($2n = 22$) and belongs to the Apiaceae family [106]. Unlike other species included in this review, it reproduces only by seed. Because it is a monocarpic species, the life cycle is complete after flowering and seed formation. This forb originates from the southern part of the western Great Caucasus in Russia and northeastern Georgia [45]. The most striking feature of giant hogweed is its conspicuous appearance; plants are usually 3–4 m tall at flowering time but can reach a height of up to 5.5 m [107]. In fact, giant hogweed is among the tallest and largest herbs in Europe. Plants grow from branched taproots that develop at soil depths up to 45–60 cm; the crown develops 10 cm below the soil surface [106]. The crown surface is woody for years, and the crown diameter reaches 15 cm at flowering time [107]. The well-developed lower leaves of adult plants become huge, e.g., 3 m long and 1.7 m wide. Vegetative growth begins in winter and flowering is in early June, 3–5 years after the establishment year, with whitish flowers. The stems grow rapidly in the year of flowering. The diameter of the stem is up to 10 cm [106]. The terminal inflorescences are large, compound umbels that are 80 cm wide and contain 50–150 hairy umbel rays that are 15–40 cm long and terminate in smaller umbellets [107]. There is a compound terminal umbel surrounded by satellite umbels. The satellite umbels overtop the terminal umbel. Numerous umbels form below it, which branch. Tertiary and quaternary umbels are also formed in each branch [108].

A total of 81,519 flowers can be produced per plant [106]. They are insect-pollinated, hermaphroditic, and protandrous [108]. However, flower production is not a reliable indicator of fertility and seed production. The flowers of the main umbel are hermaphrodite, but those of the lower umbels mature earlier in the season, and many of them are male. According to Perglová et al. [109], a single mature plant had 1 terminal umbel, 4.3 satellite umbels, 3.5 branch umbels, 17.3 tertiary umbels, and 2.8 quaternary umbels. Seed production per umbel was 9216, 1288, 1157, and 32 seeds for terminal, satellite, branch, and tertiary umbels, respectively, giving an average production of 20,671 seeds per plant. Other studies determined an average production of 15,729 seeds per plant [110]. Under real field conditions, seed banks are only persistent in the short term, as most seeds ($\geq$90%) germinate after the first winter [111]. The majority of seeds ($\geq$95%) are concentrated in the top 5 cm of the soil layer. As for seed dispersal potential, although seeds are flat with winged edges, wind is not the most effective means of long-distance dispersal. Most seeds (75%) fall near the parent plants within a 1.2 m radius [112]. There is evidence that seed release from 2 m height at wind speeds of 10 and 14 m s^{-1} resulted in dispersal over 2 and 10 m only [113]. According to Moravcová et al. [111], seeds can float in water for at least 8 h, resulting in long-distance dispersal when seeds are dispersed by fast-flowing streams. Wadsworth et al. [114] suggest that dispersal by water can lead to seed dispersal over a distance of 10 km. Anthropogenic activities are also an important means of colonizing new areas. Seeds can adhere to car tires along roads, while the slip streams of trains also contribute to seed dispersal along railways [111]. In the European Union (EU), there is a ban on the sale, planting, and keeping of the plant, as any potential use as an ornamental plant is very likely to lead to further spread of giant hogweed in new areas [115].

3.5. *Giant Goldenrod and Canadian Goldenrod*

Giant goldenrod and Canadian goldenrod are perennial, herbaceous, rhizomatous species belonging to the botanical family Asteraceae. Both species are native to North America, are classified as hemicryptophytes, and have a C_3 photosynthetic pathway. Their base chromosome number is $n = 9$; there are diploid, tetraploid, and hexaploid forms in both species [116,117]. Tetraploid giant goldenrod populations are most common in Europe, while diploids occur with lower frequency [116,118]. As with Canadian goldenrod, the dominant populations in Europe are diploid [119,120]. Stems are erect and unbranched, except in inflorescence. In giant goldenrod, the stems are up to 2.80 m tall, glabrous from base to inflorescence, and purple in color [23]. The stems of Canadian goldenrod are up to 2.5 m tall, glabrous at the base, often reddish, and weakly to densely hairy at least in the upper half [121]. Leaves are simple, alternate, and oblong to lanceolate, trifoliate, with the largest leaves in the middle of the stem and decreasing in size toward the top. They are 80–180 mm long and 10–30 mm wide in giant goldenrod and 30–150 mm long and 5–22 mm wide in Canadian goldenrod [23,24].

Inflorescences form broad pyramidal panicles with recurved branches and a central axis; they are produced from June. One difference between these two species is that the inflorescence architecture is denser in giant goldenrod [23]. The main flowering period is between mid-August and late September, but flowering may continue into October. The flowers of giant goldenrod are golden yellow, and those of Canadian goldenrod are lemon yellow. Both species are self-sterile, insect-pollinated, and have fertile female flowers and fertile bisexual disk flowers; achenes are pubescent and have a pappus. Another morphological difference is the brownish-white color of the pappus in giant goldenrod and the whitish color in Canadian goldenrod [23]. New seeds, shoots, and rhizomes are produced each year, and all aboveground shoots die in the fall. Most plants of giant goldenrod reproduce in the first growing season, and a single branch produces up to 19,000 achenes. Seed germination is highest in spring but continues through summer. Flowering in the first year, on the other hand, is not common in Canadian goldenrod; up to 13,000 achenes can be produced per branch. In dry weather, wind is an important aid in colonizing new areas. Seedlings sprout in early summer. In both species, the seeds of European plants do not show dormancy [23,24]. It should be noted that both species can hybridize with European goldenrod (*Solidago vigaurea* L.), which is native to Europe [122]. In particular, hybridization between giant goldenrod and European goldenrod results in the hybrid *Solidago* × *snarskisii* Gudžinskas & Žalneravičius, while outcrosses between Canadian goldenrod and European goldenrod produce the hybrid *Solidago* × *niederederi* Khek [122].

Shoot elongation begins in spring, and shoot height increases almost linearly until the end of July [123,124]. In giant goldenrod, rhizomes can be formed within four weeks of germination. In contrast, rhizomes of Canada goldenrod are usually formed after the first growing season [24]. They are much longer compared to those of Canadian goldenrod, and this is an important distinguishing feature between these two species [23]. The root system is branched, and secondary rhizomes are formed. Rhizome buds appear in spring, and new rhizomes (from the same crown) begin to sprout in summer, forming circular stem clusters.

Jakobs et al. [125] reported a production of 16.8 rhizomes per ramet in giant goldenrod. In Canadian goldenrod, the production of rhizomes per ramet ranges from 2 to 11 [126]. Rhizomes break dormancy in mid-April of the following year after production, and each rhizome produces a shoot from its tip; after a short period as a rosette with scale leaves, the shoot expands rapidly, and the total leaf area is largest in midsummer [127]. New rhizome buds are usually visible in August, but most rhizome growth occurs in the fall after fruit set is complete [23,24]. For both species, vegetative propagation is very important for spreading over shorter distances (up to 0.8 m per year for giant goldenrod) and establishing dense populations. In the study by Jakobs et al. [125], the mean density of giant goldenrod reached up to 78.5 stems m^{-2}, while Dudek et al. [128] found a mean density of 24.7 stems m^{-2} for Canadian goldenrod. Populations of giant goldenrod spread by rhizome propagation along riverbanks, while populations of Canadian goldenrod thrive best in loose and drier soils found near urban areas, roadsides, and railways [125,129]. Giant goldenrod spreads more rapidly due to increased rhizome length (up to 90 cm), increased secondary rhizome production, and rhizome development over a larger radius [130].

3.6. Bermuda Buttercup

Bermuda buttercup is a perennial C_3 geophyte native to South Africa with an underground bulb system. Biotypes of this broadleaved weed species include diploids ($2n = 14$), tetraploids ($2n = 28$), and pentaploids ($2n = 35$) [131]. Pentaploid biotypes dominate in invaded areas worldwide, while tetraploids are also found in the Mediterranean region [132]. Shoots arise from short vertical stems attached to pale brown underground bulbs [133]. Bulbs are borne at the tip of vertical rhizomes and form contractile roots annually that pull bulbs deeper each year, and bulbils are formed in axillary buds of vertical rhizomes [134,135]. Bulbs and bulbils can be buried in the soil at a depth of 2–50 cm or even deeper [22]. Plants grow up to 40 cm tall, and their aerial parts are bright green in color; leaves are trifoliate, arise from an enlarged basal petiole apex, and are arranged in a loose basal rosette [136]. The petioles are up to 12 cm long, and the clover-like leaves bear 3–4 leaflets that are 3.5 cm long, which are often hairy on the underside [137]. Each plant bears 1–5 axillary inflorescences on a 15–30 cm long peduncle arising from the basal leaf rosette [138]. The umbel-like inflorescences bear 4–10 funnel-shaped flowers of 2.5–3.8 cm in diameter with 5 obovate petals that are 2.0–2.5 cm long and intensely yellow in color [133]. The fruits are pointed, pubescent capsules that are about 0.6 cm long [137].

This is a tristylous species with a heteromorphic self-incompatibility system [139]. The plants exhibit the phenomenon of heterostyly, with at least two flower types—some with short styles that do not exceed the length of the anthers, others with longer styles. This hinders self-fertility and makes cross-pollination necessary [140]. In the Mediterranean region, only the type with a short style occurs, so no seeds are produced, and reproduction is only by vegetative means [131]. Bermuda buttercup bulbs sprout around mid-October before the first autumn rains, when soil temperatures are low enough to overcome bulb dormancy [141]. This usually coincides with the formation of contractile roots [22]. Flowering begins in early February and lasts until April. At the end of flowering, in late spring, the aboveground part of the bulb disappears to reappear in autumn [132]. Paspatis [22] found that each plant produces 1–40 bulbs during a single growing season. Verdaguer et al. [142] recorded a total production of more than 35 bulbs per plant, of which more than 20 were hypogeous, larger bulbs, and the rest were epigeous, smaller bulbs. The same authors also noted that the larger the plant, the more bulbs are produced. More than 20 bulbils are formed from each bulb per year, usually between 80 and 114 days after the sprouting of the parent bulbs [133]. Damanakis and Markaki [137] reported that bulb density can exceed 841 bulbs m^{-2}.

The spread of the bulbs can be localized due to the action of the contractile roots, which pull the bulbs away from the mother plant at distance of 40 cm; in just a few years, the soil is completely covered with a dense carpet of Bermuda buttercup [22]. In three years, plant density will exceed 100 plants m^{-2} [141]. Once established, the stand is robust and persistent for many years. Bulbs and bulbils, located in the uppermost soil layers, are light and can be dispersed over long distances by wind and birds; water is another important dispersal medium because the vegetative organs can float [133]. Anthropogenic activities greatly facilitate the spread of Bermuda buttercup. The transportation of contaminated soil or garden waste, spread by mechanical tillage, and use of contaminated tillage equipment are some examples of human-assisted spread [143].

4. Impact on Biodiversity—Evidence from the Broader European Territory

All selected alien plant species pose a major threat to biodiversity and consequently EC in the Adriatic-Ionian region, as representative case studies from across Europe show (Table 2).

Common milkweed is a drought-tolerant, hardy species that thrives on dry, sandy soils and adapts well to loamy, fertile, and even moist soils [55]. It can also grow in poor soils with low nitrogen and phosphorus availability and form dense populations, varying the degree of mineralization of organic carbon and nitrogenous compounds [63]. The plants tolerate pH values between 4 and 7.6, while seedlings can survive in soils with pH values between 2 and 12 [144,145]. Its competitiveness in invaded habitats is attributed to its ability to form dense canopies that shade the soil and suppress the emergence and growth of native plants, as well as its effective use of resources and allelopathic potential [21]. This species is a noxious invader in Europe and is on the list of Invasive Alien Species of Union concern, provided by the EU Commission [115]. In a recent study conducted on abandoned agricultural land in Hungary, common milkweed significantly reduced cover of native grassland species [146]. In Slovakia, common milkweed caused a significant reduction in ground cover of white clover (*Trifolium repens* L.), common dandelion (*Taraxacum officinale* L.), meadow brome (*Bromus commutatus* Schrad.), and orchard grass (*Dactylis glomerata* L.) on permanent grasslands and abandoned vineyards [10]. This species also alters the soil environment by increasing humus, phosphorus, and nitrate levels and reducing pH and carbonated lime in Pannonic open sand grasslands [21]. Aside from its effects on native plant communities and soil health, the presence of common milkweed may also be harmful to some arthropods. An example of this is the study by Gallé et al. [147], who investigated the effects of common milkweed infestation on ground-dwelling arthropod fauna in a poplar forest in Hungary. These authors showed that the presence of common milkweed at a density of 59–86 stems^{-2} negatively affected the abundance of two spider species.

Jerusalem artichoke is a hardy and drought-tolerant species that grows well in a variety of soils with pH values between 5.1 and 8.2 [83]. As an invader, it affects biodiversity in wet habitats, near riverbanks and watercourses, and in wet meadows and abandoned agricultural lands by suppressing the growth of native plants [15]. In Hungary, Romania, and Ukraine, Filep et al. [85] observed that Jerusalem artichoke reduced native species richness by nearly 40% at densities of 92–100 stems m^{-2} along riverbanks. Aside from its ability to develop tall, dense canopies that stifle the growth of native plants, this species is known for its strong allelopathic potential. For example, the incorporation of Jerusalem artichoke residues from populations found in agricultural fields in northwestern Italy reduced the growth of barnyardgrass and large crabgrass seedlings by 30 and 70%, respectively [28]. Main allelochemicals of leaf tissues include 2-*OH*-cinnamic acid, salicylic acid, and 4-*OH*-benzaldehyde, while 2-*OH*-cinnamic acid is the main allelochemical released by plants into the soil as root exudate [148]. In the laboratory experiments of Filep et al. [148], leaf extracts from populations collected along a stream in southern Hungary significantly reduced the germination rate of bedstraw (*Galium mollugo* L.), while root extracts inhibited the plumule growth of couch grass (*Elymus repens* (L.) Gould). The same authors also found mortalities of up to 60% and 90%, respectively, in plants of hedge bedstraw and tansy (*Tanacetum vulgare* L.) grown in competition with Jerusalem artichoke. In addition to its competitive nature, this troublesome weed also removes nutrients from the soil, slows natural colonization by trees, and promotes erosion of riverbanks [71].

Japanese knotweed is a resilient species that tolerates pH values between 3.0 and 8.5 and survives in soils polluted with heavy metals and salt or in areas with low nitrogen availability [88]. Light is the only requirement for optimal growth, and therefore this weed cannot colonize forests, but grasslands, riparian zones, and watercourses are susceptible to invasion, as are urban environments [87]. This environmental weed causes major ecological changes in invaded communities by forming persistent and extensive monospecific stands [20]. Recent pot experiments with Japanese knotweed populations from northwestern Italy indicate that this species is extremely competitive with native plant communities in grasslands. Specifically, Mincheva et al. [35] found that competition from Japanese knotweed resulted in 13%, 20%, and 68% lower specific leaf area, growth height, and aboveground biomass in white clover, respectively, and 22% lower specific leaf area in perennial ryegrass (*Lolium perenne* L.). It should be noted that the weed density was only one plant per pot, whereas the native species had been established at a density of 10 plants per pot. The authors attributed the competitiveness of Japanese knotweed to its more efficient and aggressive use of resources compared to the native species. In Slovenia, Dorigo et al. [37] mapped 10.4 ha along the Ljubljanica river where Japanese knotweed had displaced native plant species; the actual proportion of Japanese knotweed to the total species composition ranged from 50% to 100%, with an average of 85%. At five of the six sites studied by Dassonville et al. [149], the number of native plant species was 75% lower in infested plots compared to control plots. The same authors also noted that the invader can alter soil properties, particularly concentrations of the nutrients Cu, K, Mg, Mn, P, and Zn, in ways that promote its survival and self-renewal and inhibit the growth of native plant communities. Similar observations were made in a riparian zone in Normandy, France, where the presence of Japanese knotweed affects bacterivorous nematodes and microarthropods belonging to the Collembola class by releasing nutrients and root exudates (i.e., secondary metabolites) into the soil [9].

Bohemian knotweed, like Japanese knotweed, is highly adaptable to a wide range of soil and climatic conditions; the hybrid can be extremely competitive with native plant communities [150]. The study by Gerber et al. [151], conducted in riparian areas in Switzerland, Germany, and France, found that invasion by Japanese and Bohemian knotweed significantly reduced the species richness of native plant communities. It should be noted that both species have strong allelopathic potential in addition to their ability to compete for resources (light, water, and nutrients). Allelopathic compounds can be released into the soil as root exudates or through the decomposition of aboveground plant parts and can affect the germination and growth of native plants [152]. Gerber et al. [151] also found that total invertebrate biomass (including herbivorous insects, predators, and detritivores) was 45–60% lower in *Reynoutria* spp. infested plots than in non-infested plots. In the recent study by Neupert et al. [153], invasion of Bohemian knotweed resulted in 39% lower species richness for belowground invertebrates and 64% lower species richness in native plant communities in grasslands in northwestern France. Similar to its parent species, Japanese knotweed, Bohemian knotweed is also capable of altering the soil environment in favor of its growth in invaded sites. Dassonville et al. [154] observed that the hybrid reduced soil moisture and soil pH and increased soil carbon-to-nitrogen (C:N) ratio in a willow forest in Belgium and a wasteland in east-central France; soil moisture decreased and soil C:N ratio increased in a mesic-rough grassland located in the east-central part of the country. Furthermore, in all of the above sites, invasion by Bohemian knotweed resulted in lower activity of denitrifying bacteria in soil and thus nitrogen retention in the invaded ecosystems to its own advantage—a mechanism that can be seen as a kind of niche construction [154].

Giant hogweed is another species on the list of Invasive Alien Species of Union concern in the EU [115]. It is usually found on sandy or silty soils with a pH of 6.0 to 8.5. It is a nitrophilous species that grows best in nitrogen-rich soils. It is found along riverbanks, in grasslands, and in abandoned, disturbed habitats where it is very competitive with native species. In grasslands in the northwestern Czech Republic, giant hogweed reduced species richness by 24% and native plant biomass by 40%, averaged over 24 different sites in the study by Dostál et al. [155]. In three riparian zones in Ireland, the presence of 1 plant m^{-2} resulted in 46–60% and 27–44% lower species richness in May and October, respectively [156]. In Germany, Thiele and Otte [157] showed that unmanaged grasslands are more prone to invasion by giant hogweed compared to tall-herb communities and forests, and that a 50% increase in giant hogweed ground cover reduces native species richness by 2.4. Although the main competitive characteristic of giant hogweed is to shade lower-growing species, Jandová et al. [158] also pointed out that plants of this species can also release allelochemicals into the soil through their root system. It can also be assumed that these compounds are toxic to nematodes and other soil organisms [12]. Renčo et al. [12] reported that the average number of nematode species at infested sites was up to 20% lower than the diversity of nematode species at uninfested sites. The same authors also found that giant hogweed invasion increased soil pH from 6.23 to 7.20 and decreased soil carbon and nitrogen content by 24% and 31%, respectively, on riverbanks in the eastern part of Slovakia. The adverse effects of this alien species on soil health are also evident in the study by Koutika et al. [159], in which the invasion in waste lands and valleys in the central part of Belgium resulted in a significant reduction in soil organic matter content by reducing carbon mineralization.

Giant goldenrod and Canadian goldenrod are species with a broad tolerance to biological stressors and high adaptability to different soil types and environmental conditions [23,24]. One difference between these closely related species is that giant goldenrod grows best on riverbanks and watercourses, while Canadian goldenrod prefers looser, drier soils and is most common along roadsides and railways, and near urban areas, forest edges, abandoned fields, etc. [129]. In any case, both species are strong competitors of native vegetation and pose a serious threat to biodiversity in general. The results of a field study conducted in riparian zones in Hungary showed that giant goldenrod can reduce native species diversity by 68% at densities of 340–350 stems m^{-2} [160]. In the Netherlands, invasive giant goldenrod populations resulted in species richness reductions of 46% and 50% on riverbanks and in semi-natural grasslands, respectively [161]. In the above study, the presence of the invader promoted the growth of certain soil fungi and nematode groups over others. According to Koutika et al. [159], giant goldenrod delivers large amounts of residues to the soil surface. The residues are characterized by lower lignin content compared to the residues of native species and decompose rapidly, increasing the carbon content in the soil and altering the carbon and nitrogen cycles. The negative impacts of Canadian goldenrod invasion on biodiversity are similar. In abandoned agricultural fields in southern Transylvania, Romania, Canada goldenrod invasion has reduced native plant species richness, bee abundance, and visitation by wild bees, honeybees, and hoverflies at the expense of native species [162]. Fenesi et al. [163] collected populations of Canadian goldenrod that had infested an abandoned grassland, a disturbed grassland and an abandoned agricultural field in Romania to be used in competition with native species, i.e., the couch grass and health false brome (*Brachypodium pinnatum* (L.) P.Beauv.). The results of the pot experiments showed that the invader suppressed the growth of native grasses by reducing their light availability and competing strongly for soil resources. In the study by De Groot et al. [164], conducted in ruderal areas near the city of Ljubljana, the species richness of plants decreased by more than 50% due to the invasion of Canadian goldenrod. Another finding of these researchers was that the presence of Canadian goldenrod in the field resulted in 60% and 70% lower species richness and abundance of butterfly species, respectively. In addition, a very recent study by Bielecka et al. [165] on grasslands in southern and eastern Poland found that Canada goldenrod invasion leads to an increase in soil carbon content and C:N ratio, resulting in soil degradation. In the same study, plant species diversity was 20–30% lower in the invaded plots, indicating high competitiveness of the invaders over native species. It should also be mentioned here that both species are considered allelopathic to native vegetation; secondary metabolites with potential allelopathic effects have been isolated from root extracts of both species [166,167].

Table 2. Representative case studies on the impact of invasion of selected alien plant species on biodiversity in different habitats in European countries.

Species	Invaded Habitats	Countries	Affected Organisms/Parameters	Case Studies
Common milkweed	Grasslands, abandoned agricultural fields	Slovakia	Native plant communities	[10]
	Poplar forest	Hungary	Soil arthropods	[147]
Jerusalem artichoke	Riparian areas	Hungary, Romania, Ukraine	Native plant communities	[85]
Japanese knotweed	Forest edges, abandoned agricultural fields, field edges, grasslands	Belgium	Native Plant communities, soil properties	[149]
	Riparian areas	France	Soil properties, arthropods	[9]
	Riparian areas	Switzerland, Germany, France	Native plant communities, invertebrates	[151]
	Grasslands	France	Native plant communities, invertebrates	[153]
Bohemian knotweed	Willow forest, grasslands, wastelands	Belgium, France	Soil properties	[154]
	Riparian areas	Switzerland, Germany, France	Native plant communities, invertebrates	[151]
	Grasslands	Czech Republic	Native plant communities	[155]
Giant hogweed	Riparian areas	Slovakia	Soil properties, soil nematodes	[12]
Giant goldenrod	Abandoned agricultural fields, grasslands	Belgium	Soil properties	[159]
	Riparian areas	Hungary	Native plant communities	[160]
	Grasslands	Netherlands	Native plant communities, Soil properties, Soil fungi, Soil nematodes	[161]
Canadian goldenrod	Ruderal areas	Slovenia	Native plant communities, insect-pollinators	[164]
	Grasslands	Poland	Native plant communities, soil properties	[165]
Bermuda buttercup	Orchards	Greece	Native plant communities, soil properties	[143]
	Abandoned agricultural fields	Italy	Native plant communities	[11]

Bermuda buttercup is best adapted to Mediterranean habitats, where it is often found in grasslands, abandoned fields, unmanaged orchards, and other types of disturbed habitats [11,168]. It tolerates a variety of soil types but performs best in heavy, well-drained, fertile soils that are slightly acidic, neutral, or slightly alkaline [133]. There is much evidence that the plant competes with native plant species in invaded areas. For example, in the study by Lazzaro et al. [11], native species richness in abandoned orchards was severely affected by Bermuda buttercup cover. These authors observed an initial decline in native species richness when Bermuda buttercup infestations reached up to 25% and an even more significant decline in native species richness when Bermuda buttercup infestations reached 75–100%. In Greece, heavy infestations of olive groves with Bermuda buttercup reduced plant species diversity by 75% in winter and completely displaced native plant communities in April and May [143]. Apart from its ability to compete for resources, the plant may also have allelopathic effects on native flora in Mediterranean habitats [22]. In addition, the flowers of Bermuda buttercup have been reported to attract insects and reduce their visits to the flowers of native white rocket (*Diplotaxis erucoides* (L.) DC.) by about 33% in abandoned agricultural fields in Spain [168]. Even if it does not displace native plants, it has detrimental effects on soil health by enriching the soil and altering ecosystem nutrient cycling [22]. For instance, Petsikos et al. [143] noted that Bermuda buttercup invasion can reduce the decrease of carbon stored in above-ground vegetation in olive groves by 68%, suggesting that soil organic content will decrease in infested areas.

Apart from their direct negative impact on biodiversity, these perennial invaders also pose a serious threat to EC in the Adriatic-Ionian region and the wider European area. Given their high dispersal potential, it is highly likely that once these species occur in habitats adjacent to corridors, they will also invade the corridors and develop dense, monospecific populations that displace native flora, alter ecosystem functions and properties, and disrupt EC between protected areas [7,169]. As Vicente et al. [170] have shown, invasive populations of perennial alien plants pose a major threat to the establishment and maintenance of functional ecological linkages between areas of high conservation value. Species with such high dispersal potential and a competitive nature may exploit the establishment of corridors and use them as pathways for further dispersal into areas of high conservation value [52,171,172]. In such cases, the design of ecological corridors should be redesigned [172]. It should be noted here that the relationship between invasive plants and ecological linkages is a research objective that should be further explored in species-specific case studies in the near future [173]. Further work should also determine the extent to which the perennial invasive plants examined in this study reduce habitat connectivity for species that rely on displaced native plant species [7]. In any case, a general fact highlighted by Glen et al. [169] is that the establishment of ecological corridors between areas of high conservation value is not sufficient to halt biodiversity decline unless the corridors and their surrounding larger areas are protected from biological invasions. Vilà and Ibáñez [174] also stated that seed sources of alien plants should be kept at a safe distance from areas where ecological corridors have been established. In a more recent study, Gregory et al. [7] emphasized that invasive species management is one of the best management practices to create and maintain functioning ecological corridors. Therefore, it can be assumed that invasive species management should be considered as one of the priorities to improve EC throughout the Balkan Peninsula. As for the perennial herbaceous alien plants examined in the current study, their management should include both the adoption of measures that can prevent the spread of invasive populations in the Adriatic-Ionian region and the implementation of available eradication methods, as presented in the following section.

5. Available Management Practices

5.1. *Preventative Measures*

Recent literature shows that the occurrence of the species considered in this study has significantly increased in recent years in different habitats in several countries of the Balkan Peninsula. Considering their ecological characteristics, the occurrence of small patches in new locations can lead to massive infestations in subsequent years. Moreover, the restoration of heavily infested sites is a major challenge because these perennial invaders are very persistent and tend to alter the soil environment in invaded habitats and prevent recolonization of native plant communities [15,23,24,63,88,107,133]. Therefore, early management actions should be taken to prevent the spread of these environmental weeds in ecological networks and protected areas with high conservation value. The early identification and eradication of new invasive populations in the wider Adriatic-Ionian region is the first proactive step to protect EC. Experts in ecology, forestry, agriculture, and weed science should organize transboundary networks to continuously monitor the invasion of these alien plants in all associated countries. As shown by Dorigo et al. [37] in Slovenia, the use of remote sensing methods and extensive field surveys is essential to assess the extent of current invasions and to effectively map new infestations. The development of potential Decision Support Systems (DSS) capable of conducting risk assessments and predicting the further spread of these species in the Balkan Peninsula would also help decision-makers and land managers to mitigate this threat to biodiversity and EC [175].

Researchers, land managers, and decision makers also have a special responsibility to raise public awareness of the harmful effects of invasions of these species on biodiversity and EC. Various tools such as developing websites, setting up forums, organizing workshops, creating informational films, and publishing scientific articles can help inform the public and ensure that plant invasions are not further overlooked by society [176]. Public participation plays an important role in mitigating plant invasions because they are partly due to human activities. For example, the horticultural industry was one of the driving forces in the introduction of common milkweed, giant hogweed, and Japanese knotweed in the Adriatic-Ionian region [21,37,43]. Therefore, it is important to inform the public about the environmental impact of these invasive species to prevent their further spread through illegal ornamental cultivation. Another example is the growing interest in the use of Jerusalem artichoke as an energy crop [83]. Society and industry should be aware of the risks associated with cultivating a species that has great potential to escape cultivation and become a troublesome invader and encourage the cultivation of energy crops that have lower invasive potential and are suitable for cultivation on marginal lands [177]. In addition, Bermuda buttercup should not be perceived by the public as a beneficial nectar source for honeybees in the Mediterranean region but as an invasive species that affects biodiversity and should not be allowed to spread uncontrollably in orchards and grasslands [22]. Apart from raising public awareness of the consequences of their actions, it is likely that local residents and volunteer groups, properly trained by experts, can actively contribute to invasion monitoring and implementation of eradication programs. This hypothesis was recently confirmed in the study by Cordeiro et al. [178], conducted in Portugal.

It should also be noted that abandoned agricultural fields, unrestored grasslands, riparian areas, and forests are typical examples of disturbed habitats that are vulnerable to invasion by these opportunistic species. Therefore, policy measures should be taken to prevent the further increase of such disturbed habitats. Providing farmers with better access to land, new sources of investment, and incentives to diversify implement crop diversification practices (i.e., crop rotation, intercropping, and cover cropping) are critical to prevent agricultural land degradation and abandonment [179]. Reseeding of native species to maintain the diversity of native plant communities or strategic grazing are also recommended to restore grasslands and increase their resistance to plant invasions [180,181]. Maintaining the diversity of native vegetation is also important to protect riparian areas from alien invaders [182]. Forests should also be protected from wildfires during the dry summer months because fire is a form of habitat disturbance that promotes the establishment of invasive alien plants [183].

5.2. Eradication Methods

Although the use of herbicides is not the most environmentally friendly method of weed control in natural environments, chemical weed control may be necessary to eradicate perennial invasive species from sites where they have become dominant and negatively impact biodiversity [184]. Glyphosate is the most common systemic herbicide used to eradicate perennial invasive species [11,21,96,185–189]. This herbicide is a substituted glycine that controls weeds by inhibiting the activity of enolpyruvyl shikimate phosphate synthase (EPSPS), which is involved in the synthesis of phenolic amino acids. It is taken up relatively rapidly by treated plant surfaces and translocated through the phloem into young roots and leaves, storage organs, and any other actively growing tissue or organ [190]. However, the possible phase-out of this active ingredient in Europe suggests that research should focus on alternative weed control methods to eradicate perennial invasive alien plants in the countries of the Adriatic-Ionian region [191–193].

Synthetic auxins are compounds that mimic the action of natural plant auxins. They are another group of systemic herbicides that are expected to play an important role as glyphosate-alternative weed control options to eradicate invasive populations of broadleaved herbaceous weeds. For example, the application of a pre-package mixture of clopyralid, fluroxypyr, and MCPA proved to be highly effective in controlling Jerusalem artichoke in riparian areas in the Czech Republic [194]. Svehlakova et al. [195] suggested that triclopyr is a very good option to control this weed at flowering. Triclopyr, 2,4-D, aminopyralid, and clopyralid can achieve a good level of control of Japanese knotweed [96]. The efficacy of triclopyr against giant hogweed has also been reported [189]. Herbicides that inhibit acetolactate synthase (ALS) such as metsulfuron-methyl or protoporphyrinogen oxidase-inhibiting herbicides such as oxyfluorfen provide good control of Bermuda buttercup [22,133]. Regardless of the active ingredient used, the timing of treatment is critical to achieve optimal control. Treatments should be applied near the phenological growth stage of senescence, when herbicides can be better translocated into plants by the flow of photosynthetic assimilates from leaves to the underground vegetative parts of perennial weeds [22,96]. In addition, multiple herbicide treatments repeated over time are necessary to eradicate a well-developed stand [20,21].

It should be noted at this point that foliar applications of herbicides have some strong disadvantages. Foliar applications, especially of synthetic auxins, are not suitable for all weather conditions, and off-target movement of herbicide spray-drift may cause injury to nontarget plants [196]. In addition, broadcast application of herbicides can pollute soil, groundwater, and surface water and have negative side-effects on the environment [197]. Therefore, herbicide application methods other than broadcast foliar applications should be investigated. Apart from conducting applications as site-specific spot treatments, stems with a diameter of $\geq$2.0 cm can be pierced with a hollow needle, allowing the highly concentrated herbicide to enter the interior of the stem [198]. This technique, known as stem-injection, provided excellent control of Japanese knotweed in the study by Delbart et al. [186]. Another method is to cut the stems and immediately apply phloem-mobile herbicides to the freshly exposed cambial tissue [199]. The potential of both methods to control herbaceous perennial species that become woody with age should be further investigated as they pose lower environmental risks compared to conventional methods because herbicides are applied to specific parts of the target plants.

As for the mechanical methods evaluated to control the studied species, Nagy et al. [200] reported that mowing reduced the stem density of giant goldenrod by 72–81%. Conducting mowing for five consecutive years is recommended to deplete stands of Canadian goldenrod [201]. Regular mowing twice a year reduces both aboveground and belowground biomass of Jerusalem artichoke [202]. Monthly mowing treatments have been reported to reduce the biomass of stems of Japanese knotweed [186]. One explanation could be that repeated mowing depletes the resources of the underground vegetative organs of perennial weeds. Mowing plants just before flowering may also reduce the seed production of common milkweed [21]. Tillage is a non-chemical way to mechanically control giant hogweed. Cutting the taproot of plants at 15 cm soil depth in mid-spring prevents regrowth and results in good control of this species [189]. Tillage in early spring, when vegetative growth is complete, interrupts the formation of new bulbs and bulbils in Bermuda buttercup plants [22]. According to Švehláková et al. [202], digging up the tubers was among the effective non-chemical practices for the control of Jerusalem artichoke. However, as was also emphasized for this and other perennial species in this study, uprooted vegetative parts should be removed from water on riverbanks, as they can float in water and spread over new sites [15,81].

There is also evidence that mowing can be combined with physical weed control methods such as grazing and flooding. A recent example is the study by Nagy et al. [200], in which flooding with a water level of 10–15 cm above the soil surface for one month, together with a single mowing operation, reduced the stem density of giant goldenrod by 83%. The same authors observed that cattle grazing from mid-spring to early winter combined with a single mowing reduced giant goldenrod stem density by 87%. In previous studies, long-term sheep grazing reduced giant goldenrod populations and allowed native plant species to become established in meadows [203]. Ducs et al. [204] also recommended that rabbit grazing could be an effective method to facilitate the control of common milkweed. Zhang et al. [205] used a bioherbicide isolated from the fungal pathogen *Athelia rolfsii* (Curzi) C.C. Tu & Kimbr. and reduced Canadian goldenrod density by 89% in abandoned agricultural fields and artificial forests. Hagner et al. [206] used a mulch material consisting of peat fibers, tap water, and birch (*Betula* spp.) slow pyrolysis liquids. The results showed that the above mulch material completely eliminated giant hogweed seedlings. Therefore, further research should be conducted on the development of bioherbicides as an environmentally friendly, physical method to suppress noxious invasive weed specious.

Biological control, i.e., the release of biological agents associated with the species considered invasive at a particular site, is another non-chemical option for eradicating perennial invaders in invaded habitats. For example, the psyllid *Aphalara itadori* Shinjii serves as a biological agent against Japanese knotweed, while the stem-boring weevil *Lixus iridis* Olivier, 1807 is considered a natural enemy of giant hogweed [207,208]. Tóth et al. [209] pointed out that *Fusarium sporotrichioides* Sherb. is a fungal plant pathogen that facilitates biological control of giant hogweed. Podroužková et al. [210] showed that two snail species, namely *Succinea putris* Linnaeus, 1758 and *Urticicola umbrosus* (C. Pfeiffer, 1828), climb up the hairy stems of Jerusalem artichoke and feed on the aerial parts of the plants. In the Mediterranean region, boomrapes (*Orobanche* spp.) emerge at the time of flowering of Bermuda buttercup. These phanerogamous parasites attach to the root system of Bermuda buttercup plants and utilize the nutrients necessary for bulb formation. As a result, Bermuda buttercup plants form small bulbs and show poor vegetative growth the following fall [22].

To summarize, both herbicides and non-chemical weed control options should be considered as components of integrated eradication programs (Figure 2).

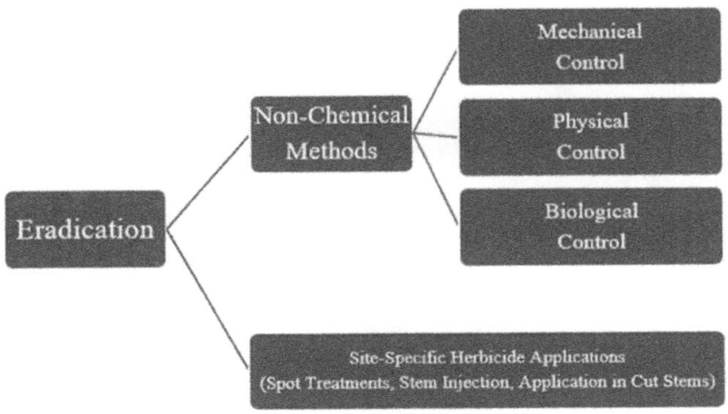

Figure 2. Weed control practices that should be combined to eradicate populations of the selected invasive perennial herbaceous alien plant species included in the current study.

6. Conclusions

The occurrence of the perennial herbaceous species studied in this review has increased in the last two decades in the Adriatic-Ionian region. All species have a high dispersal potential on grasslands, abandoned agricultural lands, forest edges, and riparian areas and pose a significant threat to native plant communities, the environment, and biodiversity. Therefore, early action should be taken to prevent the spread of these weeds in ecological networks and protected areas of high conservation value. Researchers have a special responsibility to raise public awareness of the harmful effects of invasions of these species on biodiversity and EC. Other objectives should be the early identification and mapping of invasive populations and the development of successful eradication programs integrating site-specific herbicide application practices, mechanical, physical, and biological control methods.

Author Contributions: Conceptualization, I.T., I.B. and I.G.; methodology, N.A. and I.G.; software, N.K.; validation, N.A. and P.K., formal analysis, I.B.; investigation, P.K. and I.G.; resources, N.K.; data curation, I.G.; writing—original draft preparation, I.G.; writing—review and editing, I.G.; visualization, N.A.; supervision, I.T.; project administration, I.B.; funding acquisition, I.T. All authors have read and agreed to the published version of the manuscript.

Funding: This research was funded by the Interreg Adrion DINALPCONNECT project (865).

Institutional Review Board Statement: Not applicable.

Data Availability Statement: Not applicable.

Acknowledgments: We thank all Interreg Adrion DINALPCONNECT project partners who shared their thoughts on invasive species that are problematic in their respective countries.

Conflicts of Interest: The authors declare no conflict of interest.

References

1. Haddad, N.M.; Brudvig, L.A.; Clobert, J.; Davies, K.F.; Gonzalez, A.; Holt, R.D.; Lovejoy, T.E.; Sexton, J.O.; Austin, M.P.; Collins, C.D.; et al. Habitat fragmentation and its lasting impact on earth's ecosystems. *Sci. Adv.* **2015**, *1*, e1500052. [CrossRef] [PubMed]
2. Perrin, M.; Bertrand, N.; Vanpeene, S. Ecological connectivity in spatial planning: From the EU framework to its territorial implementation in the French context. *Environ. Sci. Policy* **2022**, *129*, 118–125. [CrossRef]
3. Transboundary Ecological Connectivity of Alps and Dinaric Mountains. Available online: https://dinalpconnect.adrioninterreg.eu/ (accessed on 13 April 2022).
4. Hanski, I.; Ovaskainen, O. The metapopulation capacity of a fragmented landscape. *Nature* **2000**, *404*, 755–758. [CrossRef] [PubMed]
5. Montgelard, C.; Zenboudji, S.; Ferchaud, A.-L.; Arnal, V.; van Vuuren, B.J. Landscape genetics in mammals. *Mammalia* **2014**, *78*, 139–157. [CrossRef]
6. De la Fuente, B.; Beck, P.S. Invasive species may disrupt protected area networks: Insights from the pine wood nematode spread in Portugal. *Forests* **2018**, *9*, 282. [CrossRef]
7. Gregory, A.; Spence, E.; Beier, P.; Garding, E. Toward best management practices for ecological corridors. *Land* **2021**, *10*, 140. [CrossRef]
8. DINALPCONNECT, 2021: Report on Situation Analysis of Current Agricultural and Forestry Practices Affecting EC in Pilot Regions. EU Interreg Adrion; DINALPCONNECT Project. Available online: https://3.basecamp.com/3617449/buckets/15708478/documents/3948097125 (accessed on 13 April 2022).
9. Abgrall, C.; Forey, E.; Mignot, L.; Chauvat, M. Invasion by *Fallopia japonica* alters soil food webs through secondary metabolites. *Soil Biol. Biochem.* **2018**, *127*, 100–109. [CrossRef]
10. Jurová, J.; Renčo, M.; Gömöryová, E.; Čerevková, A. Effects of the invasive common milkweed (*Asclepias syriaca*) on nematode communities in natural grasslands. *Nematology* **2020**, *22*, 423–438. [CrossRef]
11. Lazzaro, L.; Ferretti, G.; Bianchi, E.; Benesperi, R. Treatment by glyphosate—Based herbicide allowed recovering native species after *Oxalis pes-caprae* L. invasion: Indications from a Mediterranean island. *Plant Biosyst.* **2019**, *153*, 651–659. [CrossRef]
12. Renčo, M.; Jurová, J.; Gömöryová, E.; Čerevková, A. Long-term giant hogweed invasion contributes to the structural changes of soil nematofauna. *Plants* **2021**, *10*, 2103. [CrossRef]
13. Arianoutsou, M.; Bazos, I.; Delipetrou, P.; Kokkoris, Y. The alien flora of Greece: Taxonomy, life traits and habitat preferences. *Biol. Invasions* **2010**, *12*, 3525–3549. [CrossRef]
14. Barina, Z.; Rakaj, M.; Somogyi, G.; Erős-Honti, Z.; Pifkó, D. The alien flora of Albania: History, current status and future trends. *Weed Res.* **2014**, *54*, 196–215. [CrossRef]
15. Barudanović, S.; Zečić, E.; Macanović, A.; Duraković, B.; Mašić, E. Invasive alien plant species in global perspectives with special references to Bosnia and Herzegovina. In *Invasive Alien Species: Observations and Issues from around the World*, 1st ed.; Pullaiah, T., Ielmini, M.R., Eds.; John Wiley & Sons Ltd.: Hoboken, NJ, USA, 2021; pp. 215–252.
16. Boršić, I.; Ješovnik, A.; Mihinjač, T.; Kutleša, P.; Slivar, S.; Cigrovski Mustafić, M.; Desnica, S. Invasive alien species of Union concern (Regulation 1143/2014) in Croatia. *Nat. Croat.* **2018**, *27*, 357–398. [CrossRef]
17. Galasso, G.; Conti, F.; Peruzzi, L.; Ardenghi, N.M.G.; Banfi, E.; Celesti–Grapow, L.; Albano, A.; Alessandrini, A.; Bacchetta, G.; Ballelli, S.; et al. An updated checklist of the vascular flora alien to Italy. *Plant Biosyst.* **2018**, *152*, 556–592. [CrossRef]
18. Küzmič, F.; Šilc, U. Alien species in different habitat types of Slovenia: Analysis of vegetation database. *Period. Biol.* **2017**, *119*, 199–208. [CrossRef]
19. Stešević, D.; Petrović, D. Preliminary list of plant invaders in Montenegro. *Biol. Nyssana* **2010**, *1*, 35–42.

20. Barney, J.N.; Tharayil, N.; DiTommaso, A.; Bhowmik, P.C. The biology of invasive alien plants in Canada. 5. *Polygonum cuspidatum* Sieb. & Zucc.[= *Fallopia japonica* (Houtt.) Ronse Decr.]. *Can. J. Plant Sci.* **2006**, *86*, 887–906.
21. Follak, S.; Bakacsy, L.; Essl, F.; Hochfellner, L.; Lapin, K.; Schwarz, M.; Tokarska-Guzik, B.; Wołkowycki, D. Monograph of invasive plants in Europe N°6: *Asclepias syriaca* L. *Bot. Lett.* **2021**, *168*, 422–451. [CrossRef]
22. Paspatis, E.A. Chemical, cultural and biological control of *Oxalis pes-caprae* in vineyards in Greece. In *Weed Control on Vine and Soft Fruits*, 1st ed.; Cavalloro, R., Robinson, D.W., Eds.; CRC Press: Boca Raton, FL, USA, 1987; pp. 27–29.
23. Weber, E.; Jakobs, G. Biological flora of central Europe: *Solidago gigantea* Aiton. *Flora Morphol. Distrib. Funct. Ecol. Plants* **2005**, *200*, 109–118. [CrossRef]
24. Werner, P.A.; Gross, R.S.; Bradbury, I.K. The biology of Canadian weeds.: 45. *Solidago canadensis* L. *Can. J. Plant Sci.* **1980**, *60*, 1393–1409. [CrossRef]
25. Bailey, J.P.; Bímová, K.; Mandák, B. Asexual spread versus sexual reproduction and evolution in Japanese Knotweed *s.l.* sets the stage for the "Battle of the Clones". *Biol. Invasions* **2009**, *11*, 1189–1203. [CrossRef]
26. Ministrstvo za Okolje in Proctor. Strokovne Podlage za Program Ukrepov za Obvladovanje Invazivnih Tujerodnih vrst za vrsto Sirska Svilnica (*Asclepias syriaca* L.) [Expert Basis for the Program of Measures for the Control of Invasive Alien Species for the Species Common Milkweed (*Asclepias syriaca* L.)]. Ljubljana (Slovenia): Zavod Republike Slovenije za Varstvo Narave. 2018. Available online: https://zrsvn-varstvonarave.si/wp-content/uploads/2020/08/Strokovne-podlage-za-sirsko-svilnico.pdf (accessed on 11 April 2022).
27. Maslo, S. Preliminary list of invasive alien plant species (IAS) in Bosnia and Herzegovina. *Herbologia* **2016**, *16*, 1–14. [CrossRef]
28. Tesio, F.; Weston, L.A.; Vidotto, F.; Ferrero, A. Potential allelopathic effects of Jerusalem artichoke (*Helianthus tuberosus*) leaf tissues. *Weed Technol.* **2010**, *24*, 378–385. [CrossRef]
29. Mori, E.; Mazza, G.; Galimberti, A.; Angiolini, C.; Bonari, G. The porcupine as "Little Thumbling": The role of *Hystrix cristata* in the spread of *Helianthus tuberosus*. *Biologia* **2017**, *72*, 1211–1216. [CrossRef]
30. Zelnik, I. The presence of invasive alien plant species in different habitats: Case study from Slovenia. *Acta Biol. Slov.* **2012**, *55*, 25–38.
31. Follak, S.; Eberius, M.; Essl, F.; Fürdös, A.; Sedlacek, N.; Trognitz, F. Invasive alien plants along roadsides in Europe. *EPPO Bull.* **2018**, *48*, 256–265. [CrossRef]
32. Radisek, S.; Jakse, J.; Zhao, T.T.; Cho, S.E.; Shin, H.D. First report of powdery mildew of *Helianthus tuberosus* caused by *Golovinomyces ambrosiae* in Slovenia. *J. Plant Pathol.* **2018**, *100*, 331. [CrossRef]
33. Hulina, N. "Planta hortifuga" in flora of the continental part of Croatia. *Agric. Conspec. Sci.* **2010**, *75*, 57–65.
34. Vuković, N.; Bernardić, A.; Nikolić, T.; Hršak, V.; Plazibat, M.; Jelaska, S.D. Analysis and distributional patterns of the invasive flora in a protected mountain area—A case study of Medvednica Nature Park (Croatia). *Acta Soc. Bot. Pol.* **2010**, *79*, 285–294. [CrossRef]
35. Mincheva, T.; Barni, E.; Siniscalco, C. From plant traits to invasion success: Impacts of the alien *Fallopia japonica* (Houtt.) Ronse Decraene on two native grassland species. *Plant Biosyst.* **2016**, *150*, 1348–1357. [CrossRef]
36. Siniscalco, C.; Barni, E. Are non-native plant species a threat to the Alps? Insights and perspectives. In *Climate Gradients and Biodiversity in Mountains of Italy*, 1st ed.; Pedrotti, F., Ed.; Springer: Cham, Switzerland, 2018; pp. 91–107.
37. Dorigo, W.; Lucieer, A.; Podobnikar, T.; Čarni, A. Mapping invasive *Fallopia japonica* by combined spectral, spatial, and temporal analysis of digital orthophotos. *Int. J. Appl. Earth Obs. Geoinf.* **2012**, *19*, 185–195. [CrossRef]
38. Vuković, N.; Šegota, V.; Alegro, A.; Koletić, N.; Rimac, A.; Dekanić, S. "Flying under the radar"—How misleading distributional data led to wrong appreciation of knotweeds invasion (*Reynoutria* spp.) in Croatia. *Bioinvasions Rec.* **2019**, *8*, 175–189. [CrossRef]
39. Bailey, J. The Japanese knotweed invasion of Europe: The potential for further evolution in non-native regions. In *Biological Invasion of Ecosystems by Pests and Beneficial Organisms*, 1st ed.; Yano, E., Matsuo, K., Shiyomi, M., Andow, D., Eds.; NIAES Series 3; NIAES: Tsukuba, Japan, 1999; pp. 27–37.
40. Jovanović, S.; Hlavati-Širka, V.; Lakušić, D.; Jogan, N.; Nikolić, T.; Anastasiu, P.; Vladimirov, V.; Šinžar-Sekulić, J. *Reynoutria* niche modelling and protected area prioritization for restoration and protection from invasion: A Southeastern Europe case study. *J. Nat. Conserv.* **2018**, *41*, 1–15. [CrossRef]
41. Giuliani, C.; Lastrucci, L.; Cresti, L.; Santini, G.; Fogg, B.; Lippi, M.M. The morphology and activity of the extrafloral nectaries in *Reynoutria x bohemica* (*Polygonaceae*). *Plant Biol.* **2019**, *21*, 975–985. [CrossRef] [PubMed]
42. Lazzaro, L.; Bolpagni, R.; Barni, E.; Brundu, G.; Blasi, C.; Siniscalco, C.; Celesti-Grapow, L. Towards alien plant prioritization in Italy: Methodological issues and first results. *Plant Biosyst.* **2019**, *153*, 740–746. [CrossRef]
43. Boršić, I.; Borovečki-Voska, L.; Kutleša, P.; Šemnički, P. New localities of *Heracleum mantegazzianum* Sommier et Levier (Apiaceae) in Croatia and control measures taken. *Period. Biol.* **2015**, *117*, 449–452. [CrossRef]
44. Maslo, S. Giant hogweed *Heracleum mantegazzianum* Sommier &Levier—A new non-indigenous species in the flora of Bosnia and Herzegovina. *Herbologia* **2010**, *11*, 17–24.
45. Jahodová, Š.; Fröberg, L.; Pyšek, P.; Geltman, D.; Trybush, S.; Karp, A. Taxonomy, identification, genetic relationships and distribution of large *Heracleum* species in Europe. In *Ecology and Management of Giant Hogweed (Heracleum mantegazzianum)*; Pyšek, P., Cock, M.J.W., Nentwig, W., Ravn, H.P., Eds.; CABI: Cambridge, MA, USA, 2007; pp. 1–19.
46. Marinšek, A.; Kutnar, L. Occurrence of invasive alien plant species in the floodplain forests along the Mura River in Slovenia. *Period. Biol.* **2017**, *119*, 251–260. [CrossRef]

47. Šajna, N. Alien plant species invading rare and protected habitats in Slovenia. In *Weed and Pest Control—Molecular Biology, Practices and Environmental Impact*, 1st ed.; Travlos, I.S., Bilalis, D., Chachalis, D., Eds.; Nova Publishers: New York, NY, USA, 2016; pp. 35–54.
48. Boršić, I.; Milović, M.; Dujmović, I.; Bogdanović, S.; Cigić, P.; Rešetnik, I.; Nikolić, T.; Mitić, B. Preliminary check-list of invasive alien plant species (IAS) in Croatia. *Nat. Croat.* **2008**, *17*, 55–71.
49. Novak, N.; Novak, M. The differences in the invasiveness of some alien plant species between continental and coastal part of Croatia. *Poljoprivreda* **2018**, *24*, 63–69. [CrossRef]
50. Stešević, D.; Caković, D. Contribution to the alien flora of Montenegro and Supplementum to the Preliminary list of plant invaders. *Biol. Nyssana* **2013**, *4*, 1–7.
51. Stesevic, D.; Bubanja, N. Five new alien species in the flora of Montenegro: *Coreopsis tinctoria* Nutt., *Ipomoea indica* (Burm.) Merr., *Lupinus* × *regalis* Bergmans, *Physalis angulata* L., and *Solidago canadensis* L. and new possible threats to the biodiversity. *Acta Bot. Croat.* **2017**, *76*, 98–102. [CrossRef]
52. Lazzaro, L.; Bolpagni, R.; Buffa, G.; Gentili, R.; Lonati, M.; Stinca, A.; Acosta, A.T.R.; Adorni, M.; Aleffi, M.; Allegrezza, M.; et al. Impact of invasive alien plants on native plant communities and Natura 2000 habitats: State of the art, gap analysis and perspectives in Italy. *J. Environ. Manag.* **2020**, *274*, 111140. [CrossRef] [PubMed]
53. Straub, S.C.; Parks, M.; Weitemier, K.; Fishbein, M.; Cronn, R.C.; Liston, A. Navigating the tip of the genomic iceberg: Next-generation sequencing for plant systematics. *Am. J. Bot.* **2012**, *99*, 349–364. [CrossRef] [PubMed]
54. United States Department of Agriculture (USDA). PLANTS Database. Common Milkweed—*Asclepias syriaca* L. Available online: https://plants.usda.gov/home/plantProfile?symbol=assy (accessed on 9 February 2022).
55. Bhowmik, P.C.; Bandeen, J.D. The biology of Canadian weeds: 19. *Asclepias syriaca* L. *Can. J. Plant Sci.* **1976**, *56*, 579–589. [CrossRef]
56. United States Department of Agriculture (USDA). Plant Guide. Common Milkweed—*Asclepias syriaca* L. Available online: https://plants.usda.gov/DocumentLibrary/plantguide/pdf/pg_assy.pdf (accessed on 9 February 2022).
57. Jeffery, L.S.; Robison, L.R. Growth characteristics of common milkweed. *Weed Sci.* **1971**, *19*, 193–196. [CrossRef]
58. Gudžinskas, Z.; Petrulaitis, L.; Žalneravičius, E. *Asclepias speciosa* (Apocynaceae, Asclepiadoideae): A rare or unrecognized alien species in Europe? *PhytoKeys* **2019**, *121*, 29–41. [CrossRef]
59. Wyatt, R.; Broyles, S.B. Ecology and evolution of reproduction in milkweeds. *Annu. Rev. Ecol. Syst.* **1994**, *25*, 423–441. [CrossRef]
60. Anderson, W.P. Perennial Weeds. In *Characteristics and Identification of Selected Herbaceous Species*, 1st ed.; Iowa State University Press: Ames, IA, USA, 1999; p. 228.
61. Csontos, P.; Bozsing, E.; Cseresnyes, I.; Penksza, K. Reproductive potential of the alien species *Asclepias syriaca* (Asclepiadaceae) in the rural landscape. *Pol. J. Ecol.* **2009**, *57*, 383–388.
62. Oegama, T.; Fleitcher, R.A. Factors that influence dormancy in Milkweed seeds. *Can. J. Bot.* **1972**, *50*, 713–718. [CrossRef]
63. Bagi, I. Common milkweed (*Asclepias syriaca* L.). In *The Most Important Invasive Plants in Hungary*, 1st ed.; Botta-Dukát, Z., Balogh, L., Eds.; HAS Institute of Ecology and Botany: Vácrátót, Hungary, 2008; pp. 151–158.
64. Platt, W.J.; Weis, I.M. Resource partitioning and competition within a guild of fugitive Prairie plants. *Am. Nat.* **1977**, *111*, 479–513. [CrossRef]
65. Moravcová, L.; Pyšek, P.; Jarošík, V.; Havlíčková, V.; Zákravský, P. Reproductive characteristics of neophytes in the Czech Republic: Traits of invasive and non-invasive species. *Preslia* **2010**, *82*, 365–390.
66. Novák, R.; Dancza, I.; Szentey, L.; Karamán, J.; Béres, I.; Kazinczi, G.; Gólya, G. Arable Weeds of Hungary. In *Fifth National Weed Survey (2007–2008)*; Ministry of Agriculture and Rural Development: Budapest, Hungary, 2009; p. 95.
67. Szilassi, P.; Szatmári, G.; Pásztor, L.; Árvai, M.; Szatmári, J.; Szitár, K.; Papp, L. Understanding the environmental background of an invasive plant species (*Asclepias syriaca*) for the future: An application of LUCAS field photographs and machine learning algorithm methods. *Plants* **2019**, *8*, 593. [CrossRef]
68. Tokarska-Guzik, B.; Pisarczyk, E. Risk Assessment of *Asclepias syriaca*. 2015. Available online: https://www.codeplantesenvahissantes.fr/fileadmin/PEE_Ressources/TELECHARGEMENT/Asclepias_syriaca_RA.pdf (accessed on 30 June 2020).
69. Kalloo, G. Jerusalem artichoke: *Helianthus tuberosus* L. In *Genetic Improvement of Vegetable Crops*, 1st ed.; Kalloo, G., Bergh, B.O., Eds.; Pergamon Press: Oxford, UK, 1993; pp. 747–750.
70. Kocurek, M.; Pilarski, J. Activity of C_4 enzymes in C_3-type herbaceous plants. *Photosynthetica* **2011**, *49*, 473–477. [CrossRef]
71. Pacanoski, Z.; Mehmeti, A. The first report of the invasive alien weed Jerusalem artichoke (*Helianthus tuberosus* L.) in the Republic of North Macedonia. *Agric. For.* **2020**, *66*, 115–127. [CrossRef]
72. Swanton, C.J.; Clements, D.R.; Moore, M.J.; Cavers, P.B. The biology of Canadian weeds. 101. *Helianthus tuberosus* L. *Can. J. Plant Sci.* **1992**, *72*, 1367–1382. [CrossRef]
73. Bussmann, R.W.; Batsatsashvili, K.; Kikvidze, Z.; Paniagua-Zambrana, N.Y.; Khutsishvili, M.; Maisaia, I.; Sikharulidze, S.; Tchelidze, D. Helianthus annuus L. *Helianthus tuberosus* L. Asteraceae. In *Ethnobotany of the Mountain Regions of Far Eastern Europe: Ural, Northern Caucasus, Turkey, and Iran*, 1st ed.; Batsatsashvili, K., Kikvidze, Z., Bussmann, R.W., Eds.; Springer: Cham, Switzerland, 2020; pp. 453–458.
74. Denisow, B.; Tymoszuk, K.; Dmitruk, M. Nectar and pollen production of *Helianthus tuberosus* L.—An exotic plant with invasiveness potential. *Acta Bot. Croat.* **2019**, *78*, 135–141. [CrossRef]

75. Kanatas, P.; Gazoulis, I.; Zannopoulos, S.; Tataridas, A.; Tsekoura, A.; Antonopoulos, N.; Travlos, I. Shattercane (*Sorghum bicolor* (L.) Moench Subsp. *Drummondii*) and weedy sunflower (*Helianthus annuus* L.)—Crop Wild Relatives (CWRs) as weeds in agriculture. *Diversity* **2021**, *13*, 463. [CrossRef]
76. Kays, S.J.; Nottingham, S.F. *Biology and Chemistry of Jerusalem Artichoke: Helianthus tuberosus L.*; Taylor & Francis Group: London, UK, 2008.
77. Kosaric, N.; Cosentino, G.P.; Wieczorek, A.; Duvnjak, Z. The Jerusalem artichoke as an agricultural crop. *Biomass* **1984**, *5*, 1–36. [CrossRef]
78. Kliszcz, A. Phenological Growth Stages and BBCH—Identification Keys of Jerusalem artichoke (*Helianthus Tuberosus* L.). *Ann. Univ. Paedagog. Crac. Stud. Nat.* **2021**, *6*, 203–225. [CrossRef]
79. Podlaski, S.; Pietkiewicz, S.; Choluj, D.; Horaczek, T.; Wisniewski, G.; Gozdowski, D.; Kalaji, H.M. The relationship between the soil water storage and water-use efficiency of seven energy crops. *Photosynthetica* **2017**, *55*, 210–218. [CrossRef]
80. Terzić, S.; Altagić, J.; Maksimović, I.; Zeremski, T.; Petrović, S.; Dedić, B. Influence of photoperiod on vegetation phases and tuber development in topinambour (*Helianthus tuberosus* L.). *Arch. Biol. Sci.* **2012**, *64*, 175–182. [CrossRef]
81. Žgančíková, I.; Vereš, T.; Čurná, V. Monitoring of the *Helianthus tuberosus* (L.)—As an invasive weed of natural ecosystems. *Res. J. Agric. Sci.* **2012**, *44*, 127–130.
82. Kompała-Bąba, A.; Błońska, A. Plant communities with *Helianthus tuberosus* L. in the towns of the Upper Silesian Industrial Region (southern Poland). *Biodivers. Res. Conserv.* **2008**, *11*, 57–64.
83. Liava, V.; Karkanis, A.; Danalatos, N.; Tsiropoulos, N. Cultivation practices, adaptability and phytochemical composition of Jerusalem artichoke (*Helianthus tuberosus* L.): A weed with economic value. *Agronomy* **2021**, *11*, 914. [CrossRef]
84. Rossini, F.; Provenzano, M.E.; Kuzmanović, L.; Ruggeri, R. Jerusalem artichoke (*Helianthus tuberosus* L.): A versatile and sustainable crop for renewable energy production in Europe. *Agronomy* **2019**, *9*, 528. [CrossRef]
85. Filep, R.; Balogh, L.; Balázs, V.L.; Farkas, Á.; Pal, R.W.; Czigle, S.; Czégényi, D.; Papp, N. *Helianthus tuberosus* L. agg. in the Carpathian Basin: A blessing or a curse? *Genet. Resour. Crop Evol.* **2018**, *65*, 865–879. [CrossRef]
86. Mandák, B.; Pyšek, P.; Lysák, M.; Suda, J.; Krahulcová, A.; Bímová, K. Variation in DNA—Ploidy levels of *Reynoutria* taxa in the Czech Republic. *Ann. Bot.* **2003**, *92*, 265–272. [CrossRef]
87. Shaw, D. Fallopia japonica (Japanese knotweed). In *Invasive Species Compendium*; CABI: Wallingford, UK, 2013.
88. Beerling, D.J.; Bailey, J.P.; Conolly, A.P. *Fallopia japonica* (Houtt.) Ronse Decraene. *J. Ecol.* **1994**, *82*, 959–979. [CrossRef]
89. Bram, M.R.; McNair, J.N. Seed germinability and its seasonal onset of Japanese knotweed (*Polygonum cuspidatum*). *Weed Sci.* **2004**, *52*, 759–767. [CrossRef]
90. Tiébré, M.S.; Bizoux, J.P.; Hardy, O.J.; Bailey, J.P.; Mahy, G. Hybridization and morphogenetic variation in the invasive alien *Fallopia* (Polygonaceae) complex in Belgium. *Am. J. Bot.* **2007**, *94*, 1900–1910. [CrossRef] [PubMed]
91. Forman, J.; Kesseli, R.V. Sexual reproduction in the invasive species *Fallopia japonica* (Polygonaceae). *Am. J. Bot.* **2003**, *90*, 586–592. [CrossRef] [PubMed]
92. Engler, J.; Abt, K.; Buhk, C. Seed characteristics and germination limitations in the highly invasive *Fallopia japonica* (Polygonaceae). *Ecol. Res.* **2011**, *26*, 555–562. [CrossRef]
93. Bímová, K.; Mandák, B.; Pyšek, P. Experimental study of vegetative regeneration in four invasive *Reynoutria* taxa (Polygonaceae). *Plant Ecol.* **2003**, *166*, 1–11. [CrossRef]
94. Hollingsworth, M.L.; Bailey, J.P. Evidence for massive clonal growth in the invasive weed *Fallopia japonica* (Japanese Knotweed). *Bot. J. Linn. Soc.* **2000**, *133*, 463–472. [CrossRef]
95. De Waal, L.C. A viability study of *Fallopia japonica* stem tissue. *Weed Res.* **2001**, *41*, 447–460. [CrossRef]
96. Bashtanova, U.B.; Beckett, K.P.; Flowers, T.J. Physiological approaches to the improvement of chemical control of Japanese knotweed (*Fallopia japonica*). *Weed Sci.* **2009**, *57*, 584–592. [CrossRef]
97. Adachi, N.; Terashima, I.; Takahashi, M. Central die–back of monoclonal stands of *Reynoutria japonica* in an early stage of primary succession on Mount Fuji. *Ann. Bot.* **1996**, *77*, 477–486. [CrossRef]
98. Price, E.; Gamble, R.; Williams, G.; Marshall, C. Seasonal patterns of partitioning and remobilization of ^{14}C in the invasive rhizomatous perennial Japanese knotweed. *Evol. Ecol.* **2002**, *15*, 347–362. [CrossRef]
99. Funkenberg, T.I.M.; Roderus, D.; Buhk, C. Effects of climatic factors on *Fallopia japonica* seedling establishment: Evidence from laboratory experiments. *Plant Species Biol.* **2012**, *27*, 218–225. [CrossRef]
100. Weston, L.A.; Barney, J.N.; DiTommaso, A. A review of the biology and ecology three invasive perennials in New York State: Japanese knotweed (*Polygonum cuspidatum*), mugwort (*Artemisia vulgaris*) and pale swallow—Wort (*Vincetoxicum rossicum*). *Plant Soil* **2005**, *277*, 53–69. [CrossRef]
101. Drazan, D.; Smith, A.G.; Anderson, N.O.; Becker, R.; Clark, M. History of knotweed (*Fallopia* spp.) invasiveness. *Weed Sci.* **2021**, *69*, 617–623. [CrossRef]
102. Gaskin, J.; Schwarzländer, M.; Grevstad, F.; Haverhals, M.; Bourchier, R.; Miller, T. Extreme differences in population structure and genetic diversity for three invasive congeners: Knotweeds in western North America. *Biol. Invasions* **2014**, *16*, 2127–2136. [CrossRef]
103. Gillies, S.; Clements, D.R.; Grenz, J. Knotweed (*Fallopia* spp.) invasion of North America utilizes hybridization, epigenetics, seed dispersal (unexpectedly), and an arsenal of physiological tactics. *Invasive Plant Sci. Manag.* **2016**, *9*, 71–80. [CrossRef]

104. Rouifed, S.; Puijalon, S.; Viricel, M.-R.; Piola, F. Achene buoyancy and germinability of the terrestrial invasive *Fallopia × bohemica* in aquatic environment: A new vector of dispersion? *Écoscience* **2011**, *18*, 79–84. [CrossRef]
105. Mandák, B.; Pyšek, P.; Bímová, K. History of the invasion and distribution of *Reynoutria* taxa in the Czech Republic: A hybrid spreading faster than its parents. *Preslia* **2004**, *76*, 15–64.
106. Tiley, G.E.D.; Dodd, F.S.; Wade, P.M. *Heracleum mantegazzianum* Sommier & Levier. *J. Ecol.* **1996**, *84*, 297–319.
107. Page, N.A.; Wall, R.E.; Darbyshire, S.J.; Mulligan, G.A. The biology of invasive alien plants in Canada. 4. *Heracleum mantegazzianum* Sommier & Levier. *Can. J. Plant Sci.* **2006**, *86*, 569–589.
108. Moravcová, L.; Perglová, I.; Pyšek, P.; Jarošík, V.; Pergl, J. Effects of fruit position on fruit mass and seed germination in the alien species *Heracleum mantegazzianum* (Apiaceae) and the implications for its invasion. *Acta Oecol.* **2005**, *28*, 1–10. [CrossRef]
109. Perglová, I.; Pergl, J.; Pyšek, P. Flowering phenology and reproductive effort of the invasive alien plant *Heracleum mantegazzianum*. *Preslia* **2006**, *78*, 265–285.
110. Tiley, G.E.D.; Philp, B. Effects of cutting flowering stems of Giant Hogweed *Heracleum mantegazzianum* on reproductive performance. *Asp. Appl. Biol.* **2000**, *58*, 77–80.
111. Moravcovà, L.; Pyšek, P.; Krinke, L.; Pergl, J.; Perglová, I.; Thompson, K. Seed germination, dispersal and seed bank in *Heracleum mantegazzianum*. In *Ecology and Management of Giant Hogweed (Heracleum mantegazzianum)*; CABI: Wallingford, UK, 2007; pp. 74–91.
112. Otte, A.; Franke, R. The ecology of the Caucasian herbaceous perennial *Heracleum mantegazzianum* Somm. et Lev. (Giant Hogweed) in cultural ecosystems of Central Europe. *Phytocoenologia* **1998**, *28*, 205–232. [CrossRef]
113. Neiland, M.R.M.; Proctor, J.; Sexton, R. Giant hogweed (*H. mantegazzianum* Somm. & Levier) by the River Allan and part of the River Forth. *Forth Nat. Hist.* **1987**, *9*, 51–56.
114. Wadsworth, R.A.; Collingham, Y.C.; Willis, S.G.; Huntley, B.; Hulme, P.E. Simulating the spread and management of alien riparian weeds: Are they out of control? *J. Appl. Ecol.* **2000**, *37*, 28–38. [CrossRef]
115. Invasive Alien Species of Union Concern. Available online: https://ec.europa.eu/environment/nature/pdf/IAS_brochure_species.pdf (accessed on 8 April 2022).
116. Hull-Sanders, H.M.; Johnson, R.H.; Owen, H.A.; Meyer, G.A. Effects of polyploidy on secondary chemistry, physiology, and performance of native and invasive genotypes of *Solidago gigantea* (Asteraceae). *Am. J. Bot.* **2009**, *96*, 762–770. [CrossRef]
117. Smart, A.J.; Larson, G.E.; Bauman, P.J. Grass and Canada goldenrod (*Solidago canadensis*) competition and implications for management in the northern tallgrass Prairie. *Prairie Nat.* **2012**, *45*, 4–12.
118. Schlaepfer, D.R.; Edwards, P.J.; Semple, J.C.; Billeter, R. Cytogeography of *Solidago gigantea* (Asteraceae) and its invasive ploidy level. *J. Biogeogr.* **2008**, *35*, 2119–2127. [CrossRef]
119. Malecka, J. Studies on the genus *Solidago* L. IV. Cytoembryology of *Solidago canadensis* L. var. *"scabra"*. *Acta Biol. Cracoviensa Ser. Bot.* **1989**, *31*, 85–95.
120. Musial, K. Studies on the genus *Solidago* L. III. Embryology of *Solidago canadensis* L. var. *canadensis*. *Acta Biol. Cracoviensa Ser. Bot.* **1989**, *31*, 73–84.
121. Šutovská, M.; Capek, P.; Kocmálová, M.; Fraňová, S.; Pawlaczyk, I.; Gancarz, R. Characterization and biological activity of *Solidago canadensis* complex. *Int. J. Biol. Macromol.* **2013**, *52*, 192–197. [CrossRef]
122. Musiał, K.; Pagitz, K.; Gudžinskas, Z.; Łazarski, G.; Pliszko, A. Chromosome numbers in hybrids between invasive and native Solidago (Asteraceae) species in Europe. *Phytotaxa* **2020**, *471*, 267–275. [CrossRef]
123. Huang, H.; Guo, S.; Chen, G. Reproductive biology in an invasive plant *Solidago canadensis*. *Front. Biol. China* **2007**, *2*, 196–204. [CrossRef]
124. Weber, E. Strong regeneration ability from rhizome fragments in two invasive clonal plants (*Solidago canadensis* and *S. gigantea*). *Biol. Invasions* **2011**, *13*, 2947–2955. [CrossRef]
125. Jakobs, G.; Weber, E.; Edwards, P.J. Introduced plants of the invasive *Solidago gigantea* (Asteraceae) are larger and grow denser than conspecifics in the native range. *Divers. Distrib.* **2004**, *10*, 11–19. [CrossRef]
126. Cheng, J.; Yang, X.; Xue, L.; Yao, B.; Lu, H.; Tian, Z.; Li, J.; Zhou, X.; Zhang, Y.; Zia-ul-Haq, M.; et al. Polyploidization contributes to evolution of competitive ability: A long term common garden study on the invasive *Solidago canadensis* in China. *Oikos* **2020**, *129*, 700–713. [CrossRef]
127. Botta-Dukat, Z.; Dancza, I. Morphological plasticity in the rhizome system of Solidago gigantea: Comparison of wet and dry habitats. In Proceedings of the Sixth International Conference on the Ecology and Management of Invasive Plants, Emapi, Loughborough, UK, 12–15 September 2001.
128. Dudek, K.; Michlewicz, M.; Dudek, M.; Tryjanowski, P. Invasive Canadian goldenrod (*Solidago canadensis* L.) as a preferred foraging habitat for spiders. *Arthropod Plant Interact.* **2016**, *10*, 377–381. [CrossRef]
129. Benelli, G.; Pavela, R.; Cianfaglione, K.; Nagy, D.U.; Canale, A.; Maggi, F. Evaluation of two invasive plant invaders in Europe (*Solidago canadensis* and *Solidago gigantea*) as possible sources of botanical insecticides. *J. Pest Sci.* **2019**, *92*, 805–821. [CrossRef]
130. Schmid, B.; Puttick, G.M.; Burgess, K.H.; Bazzaz, F.A. Correlations between genet architecture and some life history features in three species of *Solidago*. *Oecologia* **1988**, *75*, 459–464. [CrossRef]
131. Castro, S.; Ferrero, V.; Costa, J.; Sousa, A.J.; Navarro, L.; Loureiro, J. Reproductive strategy of the invasive *Oxalis pes-caprae*: Distribution patterns of flower morphs, ploidy levels and sexual reproduction. *Biol. Invasions* **2013**, *15*, 1863–1875. [CrossRef]

132. Castro, S.; Castro, M.; Ferrero, V.; Costa, J.; Tavares, D.; Navarro, L.; Loureiro, J. Invasion fosters change: Independent evolutionary shifts in reproductive traits after *Oxalis pes-caprae* L. introduction. *Front. Plant Sci.* **2016**, *7*, 874. [CrossRef] [PubMed]
133. Peirce, J.R. The biology of Australian weeds. 31. *Oxalis pes-caprae* L. *Plant Prot. Q.* **1997**, *12*, 110–119.
134. Oberlander, K.C.; Emshwiller, E.; Bellstedt, D.U.; Dreyer, L.L. A model of bulb evolution in the eudicot genus *Oxalis* L. *Mol. Phylogenet. Evol.* **2009**, *51*, 54–63. [CrossRef] [PubMed]
135. Pütz, N. Vegetative spreading of *Oxalis pes-caprae* (Oxalidaceae). *Plant Syst. Evol.* **1994**, *191*, 57–67. [CrossRef]
136. Brandes, D. Sociology and ecology of *Oxalis pes-caprae* L. in the Mediterranean region with special attention to Malta. *Phytocoenologia* **1991**, *19*, 285–306. [CrossRef]
137. Damanakis, M.; Markaki, M. Studies on the biology of *Oxalis pes-caprae* L. under field conditions in Crete Greece. *Zizaniology* **1990**, *2*, 145–154.
138. Roets, F.; Oberlander, K.C.; Dreyer, L.L. New relatives of *Oxalis pes-caprae* (Oxalidaceae) from South Africa. *Blumea* **2014**, *59*, 131–138. [CrossRef]
139. Ornduff, R. Reproductive systems and chromossome races of *Oxalis pes-caprae* L. and their bearing on the genesis of a noxious weed. *Ann. Mo. Bot. Gard.* **1987**, *74*, 79–84. [CrossRef]
140. Castro, S.; Loureiro, J.; Santos, C.; Ater, M.; Ayensa, G.; Navarro, L. Distribution of flower morphs, ploidy level and sexual reproduction of the invasive weed *Oxalis pes-caprae* in the Western area of the Mediterranean region. *Ann. Bot.* **2007**, *99*, 507–517. [CrossRef]
141. Vilà, M.; Bartomeus, I.; Gimeno, I.; Traveset, A.; Moragues, E.V.A. Demography of the invasive geophyte *Oxalis pes-caprae* across a Mediterranean island. *Ann. Bot.* **2006**, *97*, 1055–1062. [CrossRef]
142. Verdaguer, D.; Sala, A.; Vila, M. Effect of environmental factors and bulb mass on the invasive geophyte *Oxalis pes-caprae* development. *Acta Oecol.* **2010**, *36*, 92–99. [CrossRef]
143. Petsikos, C.; Dalias, P.; Troumbis, A.Y. Effects of *Oxalis pes-caprae* L. invasion in olive groves. *Agric. Ecosyst. Environ.* **2007**, *120*, 325–329. [CrossRef]
144. Bacieczko, W.; Borcz, A. Structure of *Asclepias syriaca* L. population against phytocenotic and habitat conditions in Widuchowa (West Pomerania). *Biodivers. Res. Conserv.* **2015**, *40*, 69–75. [CrossRef]
145. Evetts, L.L.; Burnside, O.C. Germination and seedling development of common milkweed and other species. *Weed Sci.* **1972**, *20*, 371–378. [CrossRef]
146. Kelemen, A.; Valkó, O.; Kröel-Dulay, G.; Deák, B.; Török, P.; Tóth, K.; Miglécz, T.; Tóthmérész, B. The invasion of common milkweed (*Asclepias syriaca*) in sandy old-fields—Is it a threat to the native flora? *Appl. Veg. Sci.* **2016**, *19*, 218–224. [CrossRef]
147. Gallé, R.; Erdélyi, N.; Szpisjak, N.; Tölgyesi, C.; Maák, I. The effect of the invasive *Asclepias syriaca* on the ground—Dwelling arthropod fauna. *Biologia* **2015**, *70*, 104–112. [CrossRef]
148. Filep, R.; Pal, R.W.; Balázs, V.L.; Mayer, M.; Nagy, D.U.; Cook, B.J.; Farkas, Á. Can seasonal dynamics of allelochemicals play a role in plant invasions? A case study with *Helianthus tuberosus* L. *Plant Ecol.* **2016**, *217*, 1489–1501. [CrossRef]
149. Dassonville, N.; Vanderhoeven, S.; Vanparys, V.; Hayez, M.; Gruber, W.; Meerts, P. Impacts of alien invasive plants on soil nutrients are correlated with initial site conditions in NW Europe. *Oecologia* **2008**, *157*, 131–140. [CrossRef]
150. Lavoie, C. The impact of invasive knotweed species (*Reynoutria* spp.) on the environment: Review and research perspectives. *Biol. Invasions* **2017**, *19*, 2319–2337. [CrossRef]
151. Gerber, E.; Krebs, C.; Murrell, C.; Moretti, M.; Rocklin, R.; Schaffner, U. Exotic invasive knotweeds (*Fallopia* spp.) negatively affect native plant and invertebrate assemblages in European riparian habitats. *Biol. Conserv.* **2008**, *141*, 646–654. [CrossRef]
152. Kato-Noguchi, H. Allelopathy of knotweeds as invasive plants. *Plants* **2021**, *11*, 3. [CrossRef]
153. Neupert, M.; Margerie, P.; Forey, E.; Chauvat, M.; Bureau, F.; Aubert, M.; Prével, S.; Langlois, E.; Vincenot, L. The best of both worlds? Hybridization potentiates exotic Bohemian Knotweed's (*Reynoutria* × *bohemica*) impacts on native plant and faunal communities. *Biol. Life Sci. Forum* **2021**, *2*, 20.
154. Dassonville, N.; Guillaumaud, N.; Piola, F.; Meerts, P.; Poly, F. Niche construction by the invasive Asian knotweeds (species complex *Fallopia*): Impact on activity, abundance and community structure of denitrifiers and nitrifiers. *Biol. Invasions* **2011**, *13*, 1115–1133. [CrossRef]
155. Dostál, P.; Müllerová, J.; Pyšek, P.; Pergl, J.; Klinerová, T. The impact of an invasive plant changes over time. *Ecol. Lett.* **2013**, *16*, 1277–1284. [CrossRef] [PubMed]
156. Gioria, M.; Osborne, B. Assessing the impact of plant invasions on soil seed bank communities: Use of univariate and multivariate statistical approaches. *J. Veg. Sci.* **2009**, *20*, 547–556. [CrossRef]
157. Thiele, J.; Otte, A. Impact of Heracleum mantegazzianum on invaded vegetation and human activities. In *Ecology and Management of Giant Hogweed (Heracleum mantegazzianum)*; Pyšek, P., Cock, M.J.W., Nentwig, W., Ravn, H.P., Eds.; CABI: Cambridge, MA, USA, 2007; pp. 144–156.
158. Jandová, K.; Dostál, P.; Cajthaml, T. Searching for *Heracleum mantegazzianum* allelopathy in vitro and in a garden experiment. *Biol. Invasions* **2015**, *17*, 987–1003. [CrossRef]
159. Koutika, L.S.; Vanderhoeven, S.; Chapuis-Lardy, L.; Dassonville, N.; Meerts, P. Assessment of changes in soil organic matter after invasion by exotic plant species. *Biol. Fertil. Soils* **2007**, *44*, 331–341. [CrossRef]
160. Pal, R.W.; Chen, S.; Nagy, D.U.; Callaway, R.M. Impacts of *Solidago gigantea* on other species at home and away. *Biol. Invasions* **2015**, *17*, 3317–3325. [CrossRef]

161. Quist, C.W.; Vervoort, M.T.; Van Megen, H.; Gort, G.; Bakker, J.; Van der Putten, W.H.; Helder, J. Selective alteration of soil food web components by invasive giant goldenrod *Solidago gigantea* in two distinct habitat types. *Oikos* **2014**, *123*, 837–845. [CrossRef]
162. Fenesi, A.; Vágási, C.I.; Beldean, M.; Földesi, R.; Kolcsár, L.-P.; Shapiro, J.T.; Török, E.; Kovács-Hostyánszki, A. *Solidago canadensis* impacts on native plant and pollinator communities in different—Aged old fields. *Basic Appl. Ecol.* **2015**, *16*, 335–346. [CrossRef]
163. Fenesi, A.; Geréd, J.; Meiners, S.J.; Tóthmérész, B.; Török, P.; Ruprecht, E. Does disturbance enhance the competitive effect of the invasive Solidago canadensis on the performance of two native grasses? *Biol. Invasions* **2015**, *17*, 3303–3315. [CrossRef]
164. De Groot, M.; Kleijn, D.; Jogan, N. Species groups occupying different trophic levels respond differently to the invasion of semi-natural vegetation by *Solidago canadensis*. *Biol. Conserv.* **2007**, *136*, 612–617. [CrossRef]
165. Bielecka, A.; Borkowska, L.; Królak, E. Environmental changes caused by the clonal invasive plant *Solidago canadensis*. *Ann. Bot. Fenn.* **2020**, *57*, 33–48. [CrossRef]
166. Abhilasha, D.; Quintana, N.; Vivanco, J.; Joshi, J. Do allelopathic compounds in invasive *Solidago canadensis* restrain the native European flora? *J. Ecol.* **2008**, *96*, 993–1001. [CrossRef]
167. Novak, N.; Novak, M.; Barić, K.; Šćepanović, M.; Ivić, D. Allelopathic potential of segetal and ruderal invasive alien plants. *J. Cent. Eur. Agric.* **2018**, *19*, 408–422. [CrossRef]
168. Jakobsson, A.; Padrón, B.; Traveset, A. Competition for pollinators between invasive and native plants: Effects of spatial scale of investigation (note). *Écoscience* **2009**, *16*, 138–141. [CrossRef]
169. Glen, A.S.; Pech, R.P.; Byrom, A.E. Connectivity and invasive species management: Towards an integrated landscape approach. *Biol. Invasions* **2013**, *15*, 2127–2138. [CrossRef]
170. Vicente, J.R.; Fernandes, R.F.; Randin, C.F.; Broennimann, O.; Gonçalves, J.; Marcos, B.; Pôças, I.; Alves, P.; Guisan, A.; Honrado, J.P. Will climate change drive alien invasive plants into areas of high protection value? An improved model-based regional assessment to prioritise the management of invasions. *J. Environ. Manag.* **2013**, *131*, 185–195. [CrossRef]
171. Christopoulou, A.; Christopoulou, A.; Fyllas, N.M.; Dimitrakopoulos, P.G.; Arianoutsou, M. How effective are the protected areas of the Natura 2000 network in halting biological invasions? A case study in Greece. *Plants* **2021**, *10*, 2113. [CrossRef]
172. Szilassi, P.; Soóky, A.; Bátori, Z.; Hábenczyus, A.A.; Frei, K.; Tölgyesi, C.; van Leeuwen, B.; Tobak, Z.; Csikós, N. Natura 2000 areas, road, railway, water, and ecological networks may provide pathways for biological invasion: A country scale analysis. *Plants* **2021**, *10*, 2670. [CrossRef]
173. Wilkerson, M.L. Invasive plants in conservation linkages: A conceptual model that addresses an underappreciated conservation issue. *Ecography* **2013**, *36*, 1319–1330. [CrossRef]
174. Vilà, M.; Ibáñez, I. Plant invasions in the landscape. *Landsc. Ecol.* **2011**, *26*, 461–472. [CrossRef]
175. Kanatas, P.; Travlos, I.S.; Gazoulis, I.; Tataridas, A.; Tsekoura, A.; Antonopoulos, N. Benefits and limitations of decision support systems (DSS) with a special emphasis on weeds. *Agronomy* **2020**, *10*, 548. [CrossRef]
176. Novoa, A.; Dehnen-Schmutz, K.; Fried, J.; Vimercati, G. Does public awareness increase support for invasive species management? Promising evidence across taxa and landscape types. *Biol. Invasions* **2017**, *19*, 3691–3705. [CrossRef]
177. Gazoulis, I.; Kanatas, P.; Papastylianou, P.; Tataridas, A.; Alexopoulou, E.; Travlos, I. Weed management practices to improve establishment of selected lignocellulosic crops. *Energies* **2021**, *14*, 2478. [CrossRef]
178. Cordeiro, B.; Marchante, H.; Castro, P.; Marchante, E. Does public awareness about invasive plants pays off? An analysis of knowledge and perceptions of environmentally aware citizens in Portugal. *Biol. Invasions* **2020**, *22*, 2267–2281. [CrossRef]
179. Schuh, B.; Dax, T.; Andronic, C.; Derszniak-Noirjean, M.; Gaupp-Berghausen, M.; Hsiung, C.-H.; Münch, A.; Machold, I.; Schroll, K.; Brkanovic, S. The Challenge of Land Abandonment after 2020 and Options for Mitigating Measures. In *Research for AGRI-Committee, 2020*; European Parliament, Policy Department for Structural and Cohesion Policies, Directorate—General for Internal Policies: Brussels, Belgium, 2020; Available online: https://bit.ly/39ElcFJ (accessed on 17 April 2022).
180. Barr, S.; Jonas, J.L.; Paschke, M.W. Optimizing seed mixture diversity and seeding rates for grassland restoration. *Restor. Ecol.* **2017**, *25*, 396–404. [CrossRef]
181. Firn, J.; Price, J.N.; Whalley, R.D. Using strategically applied grazing to manage invasive alien plants in novel grasslands. *Ecol. Process.* **2013**, *2*, 26. [CrossRef]
182. Pattison, Z.; Vallejo-Marín, M.; Willby, N. Riverbanks as battlegrounds: Why does the abundance of native and invasive plants vary? *Ecosystems* **2019**, *22*, 578–586. [CrossRef]
183. Úbeda, X.; Sarricolea, P. Wildfires in Chile: A review. *Glob. Planet. Change* **2016**, *146*, 152–161. [CrossRef]
184. Fogliatto, S.; Milan, M.; Vidotto, F. Control of *Ailanthus altissima* using cut stump and basal bark herbicide applications in an eighteenth-century fortress. *Weed Res.* **2020**, *60*, 425–434. [CrossRef]
185. Bakacsy, L.; Bagi, I. Survival and regeneration ability of clonal common milkweed (*Asclepias syriaca* L.) after a single herbicide treatment in natural open sand grasslands. *Sci. Rep.* **2020**, *10*, 14222. [CrossRef] [PubMed]
186. Delbart, E.; Mahy, G.; Weickmans, B.; Henriet, F.; Crémer, S.; Pieret, N.; Vanderhoeven, S.; Monty, A. Can land managers control Japanese knotweed? Lessons from control tests in Belgium. *Environ. Manag.* **2012**, *50*, 1089–1097. [CrossRef] [PubMed]
187. Guo, S.L.; Jiang, H.W.; Fang, F.; Chen, G.Q. Influences of herbicides, uprooting and use as cut flowers on sexual reproduction of *Solidago canadensis*. *Weed Res.* **2009**, *49*, 291–299. [CrossRef]
188. Labant-Hoffmann, É.; Kazinczi, G. Chemical and mechanical methods for suppression of Jerusalem artichoke (*Helianthus tuberosus* L.). *Herbologia* **2014**, *14*, 63–71. [CrossRef]

189. Nielsen, C.; Vanaga, I.; Treikale, O.; Priekule, I. Mechanical and chemical control of *Heracleum mantegazzianum* and *H. sosnowskyi*. In *Ecology and Management of Giant Hogweed (Heracleum mantegazzianum)*; Pyšek, P., Cock, M.J.W., Nentwig, W., Ravn, H.P., Eds.; CABI: Cambridge, MA, USA, 2007; pp. 226–239.
190. Duke, S.O.; Powles, S.B. Glyphosate: A once-in-a-century herbicide. *Pest Manag. Sci.* **2008**, *64*, 319–325. [CrossRef]
191. Fogliatto, S.; Ferrero, A.; Vidotto, F. Current and future scenarios of glyphosate use in Europe: Are there alternatives? In *Advances in Agronomy*; Sparks, D.L., Ed.; Academic Press: Cambridge, MA, USA, 2020; Volume 163, pp. 219–278.
192. Kanatas, P.; Antonopoulos, N.; Gazoulis, I.; Travlos, I.S. Screening glyphosate—Alternative weed control options in important perennial crops. *Weed Sci.* **2021**, *69*, 704–718. [CrossRef]
193. Travlos, I.; Kanatas, P.; Tsekoura, A.; Gazoulis, I.; Papastylianou, P.; Kakabouki, I.; Antonopoulos, N. Efficacy of different herbicides on *Echinochloa colona* (L.) Link control and the first case of its glyphosate resistance in Greece. *Agronomy* **2020**, *10*, 1056. [CrossRef]
194. Janikova, A.; Svehlakova, H.; Turcova, B.; Stalmachova, B. Influence of management on vegetative reproduction of invasive species of *Helianthus tuberosus* in Poodri PLA. *IOP Conf. Ser. Earth Environ. Sci.* **2020**, *444*, 012025. [CrossRef]
195. Svehlakova, H.; Janikova, A.; Kupka, J.; Sotkova, N.; Rajdus, T. Possibilities of the management of *Helianthus tuberosus* species in Poodri PLA (Czech Republic). *IOP Conf. Ser. Earth Environ. Sci.* **2017**, *92*, 012066. [CrossRef]
196. Brühl, C.A.; Zaller, J.G. Indirect herbicide effects on biodiversity, ecosystem functions, and interactions with global changes. In *Herbicides*, 1st ed.; Mesnage, R., Ed.; Elsevier: Amsterdam, The Netherlands, 2021; pp. 231–272.
197. Mendes, K.F. *Pesticides in Agriculture and Environment*, 1st ed.; B P International: West Bengal, India, 2022; p. 1.
198. Schiffleithner, V.; Essl, F. Is it worth the effort? Spread and management success of invasive alien plant species in a Central European National Park. *NeoBiota* **2016**, *31*, 43–61.
199. Miller, T.W. Integrated strategies for management of perennial weeds. *Invasive Plant Sci. Manag.* **2016**, *9*, 148–158. [CrossRef]
200. Nagy, D.U.; Rauschert, E.S.; Henn, T.; Cianfaglione, K.; Stranczinger, S.; Pal, R.W. The more we do, the less we gain? Balancing effort and efficacy in managing the *Solidago gigantea* invasion. *Weed Res.* **2020**, *60*, 232–240. [CrossRef]
201. Rajdus, T.; Svehlakova, H.; Plohak, P.; Stalmachova, B. Management of invasive species *Solidago canadensis* in Ostrava region (Czech Republic). *IOP Conf. Ser. Earth Environ. Sci.* **2020**, *444*, 012046. [CrossRef]
202. Švehláková, H.; Turčová, B.; Rajdus, T.; Plohák, P.; Nováková, J. Effective combination of management methods suppresses invasive Jerusalem artichoke. *IOP Conf. Ser. Earth Environ. Sci.* **2021**, *900*, 012045. [CrossRef]
203. Andersen, U.V.; Calov, B. Long-term effects of sheep grazing on giant hogweed (*Heracleum mantegazzianum*). *Hydrobiologia* **1996**, *340*, 277–284. [CrossRef]
204. Ducs, A.; Kazi, A.; Bilko, A.; Altbaecker, V. Milkweed control by food imprinted rabbits. *Behav. Processes* **2016**, *130*, 75–80. [CrossRef]
205. Zhang, Y.; Yang, X.; Zhu, Y.; Li, L.; Zhang, Y.; Li, J.; Qiang, S. Biological control of *Solidago canadensis* using a bioherbicide isolate of Sclerotium rolfsii SC64 increased the biodiversity in invaded habitats. *Biol. Control* **2019**, *139*, 104093. [CrossRef]
206. Hagner, M.; Lindqvist, B.; Vepsäläinen, J.; Samorì, C.; Keskinen, R.; Rasa, K.; Hyvönen, T. Potential of pyrolysis liquids to control the environmental weed *Heracleum mantegazzianum*. *Environ. Technol. Innov.* **2020**, *20*, 101154. [CrossRef]
207. Seier, M.K.; Wittenberg, R.; Ellison, C.A.; Djeddour, D.H.; Evans, H.C. Surveys for natural enemies of giant hogweed (*Heracleum mantegazzianum*) in the Caucasus Region and assessment for their classical biological control potential in Europe. In Proceedings of the XI International Symposium on Biological Control of Weeds, Canberra, Australia, 27 April–2 May 2003; SCIRO Entomology: Canberra, Australia, 2004; p. 149.
208. Shaw, R.H.; Tanner, R.; Djeddour, D.; Cortat, G. Classical biological control of *Fallopia japonica* in the United Kingdom—Lessons for Europe. *Weed Res.* **2011**, *51*, 552–558. [CrossRef]
209. Tóth, T.; Szilágyi, A.; Kövics, G. Preliminary estimation of the efficacy of *Fusarium sporotrichioides* Sherb. as biological control agent against common milkweed (*Asclepias syriaca* L.). *Acta Agrar. Debr.* **2018**, *74*, 201–204. [CrossRef]
210. Podroužková, Š.; Janovský, Z.; Horáčková, J.; Juřičková, L. Do snails eat exotic plant species invading river floodplains? *J. Molluscan Stud.* **2015**, *81*, 139–146. [CrossRef]

diversity

MDPI

Review

Farming Intensity Affects Soil Seedbank Composition and Spontaneous Vegetation of Arable Weeds

Philipp Köllmann * and Rainer Waldhardt

Division of Landscape Ecology and Landscape Planning, Department of Landscape Ecology and Resources Management, Justus-Liebig-University Giessen, Heinrich-Buff-Ring 26-32, 35392 Giessen, Germany; rainer.waldhardt@umwelt.uni-giessen.de

* Correspondence: philipp.koellmann@umwelt.uni-giessen.de

Abstract: Former studies carried out in the 2000s in the Lahn-Dill region located in the middle-east of the German state Hesse stated a depletion of arable weeds on the field scale and more diverse weed flora on the landscape scale. Current study, having started in 2018, aims to contribute to a better understanding of the interactions between arable weed species diversity, farming intensity, grown crops and landscape area. Moreover, the potential of organic farming methods for conservation and promotion the arable weed diversity is aimed to be assessed with the study. In total, 42 fields in two landscape regions were sampled—six seedbank samples were collected from each field; additionally, data on spontaneous arable weed flora were recorded each spring from 2019 to 2021; emerged aboveground weeds were identified in the fields and their coverage was documented. Four factors were considered in the field trial: Farming practice, landscape area, soil depth and the current crop. Effects of these factors on arable weed species diversity were calculated with a Generalized Linear Model (GLM), resulting in significant effects of the management system, the area and the current crop. Among the four organic farming systems that were sampled, the time period of organic growing had a significant effect on weed seed numbers in the soil with an increase in seed numbers. Average seedbank species numbers were around twice as high in organic farming systems (18 species) compared to conventional managed fields (nine species). Evidence of an ongoing species decline in the region on the landscape scale could be detected by comparison with a former study. Especially rare and endangered weed species are a concern due to seedbank and current vegetation depletion tendencies.

Keywords: weed seedbank; farming intensity; organic farming; segetal species; spontaneous vegetation; weed control; nature conservation

Citation: Köllmann, P.; Waldhardt, R. Farming Intensity Affects Soil Seedbank Composition and Spontaneous Vegetation of Arable Weeds. *Diversity* **2022**, *14*, 111. https://doi.org/10.3390/d14020111

Academic Editors: Ilias Travlos and Michael Wink

Received: 21 September 2021
Accepted: 28 January 2022
Published: 4 February 2022

1. Introduction

Due to the intensification in agricultural systems, there is a loss of biodiversity in the agricultural landscape on a global scale [1]. Segetal species found in the agro-ecosystems are strongly affected by this process, resulting in reductions in total species and individual numbers per species as well [2–4]. Arable weeds are strongly endangered in middle Europe [5]. In general, there is a depletion of arable plant species richness in farming systems with a high intensity level, thus conventional cropping systems are concerned by this tendency primarily [6]. Studies Albrecht [7] could state the positive effects of organic farming methods on arable weed diversity and its potential to preserve arable weed richness by the maintenance and promotion of a stable weed seedbank. Several studies have well documented the effect of the management system in agriculture on segetal species abundance [8,9]. Mostly, increasing land-use intensity of conventional farming practices resulted in a loss of arable plant species diversity [10–12].

Besides the decrease in high proportion of arable weeds, there are some species that have already adapted to more intensive farming conditions, and especially to herbicide inputs, in the case of *Alopecurus myosuroides* and some *Lolium* species [13–15]. As Storkey & Neve [16] stated there is an advantage of a diverse arable weed flora for economic and ecological reasons. They found a negative linear relationship between yield loss and the number of arable species. Yield losses were much lower in fields with higher species diversity (30% yield loss) compared to fields with a low species number (60% yield loss).

A diverse and species-rich cropping system offers several ecological benefits that are well documented [17–19]. Supply of food, shelter and habitat plays a key role of arable plant species for a diverse fauna [20,21].

Higher diversity values of organic managed land has been stated by several studies; Kolářová et al. [22] reported up to 4.5 times higher abundance of these species in organic farming compared to the conventional variants.

However, studies usually focus on the aboveground coverage of weeds. Way fewer research has been done on the soil seedbank of weeds. Studies by Rotchés-Ribalta et al. [23] stated an effect of the intensity level of the farming system on the seed abundance of the soil seed bank with higher seed abundances in organic and lower abundances in conventional farming systems. In studies of Hawes et al. [9], current arable weeds were stronger and more directly affected by the farming practice than the soil seed bank was. Studies that were conducted at the beginning of the 2000s in the region of the current project (Lahn-Dill-Bergland) on arable weed diversity found a significant depletion of arable weed richness on a small scale (field) but stated still moderate high species richness on a larger scale (landscape) [24]. Current study purposes to detect changes of the situation of arable weeds in the region between former studies and the current one. Besides that, relationships between management regime, region and the arable weed flora shall be clarified in this study. The research questions include: Which factors have the strongest impact on species distribution patterns and diversity indices? What changes compared to former studies can be detected referring to the arable weed flora? Information on that can contribute to developing strategies for recovery and conservation of arable weed diversity in the region and in regions with similar site conditions.

2. Materials and Methods

2.1. Study Sites and Experimental Setup

The study is based on a sampling of vegetation ecological data of 42 arable fields in two landscape areas in the region between Marburg and Giessen in Hessen, Germany. The region "Lahntal" is an area of the valley of the river Lahn which is characterized by fertile loamy soils and mainly alluvium materials like silt and clay. Site conditions allow intensive arable farming with a high proportion of cereals in the crop rotation. Winter wheat and winter barley are commonly grown there. Twenty-two sites were samples in this region. Average altitude of the region "Lahntal" is 175 m above sea level, with an annual average temperature of 10 °C and average annual precipitation of 750 mm. Coordinates of the center of these study sites are 50°43′1.20″ N and 8°42′21.69″ E. Study sites are located on a north-south line that stretches about 3.5 km north and 3.5 km south of the given coordinates.

The second region "Gladenbacher Bergland" is located about 10 km east of the region "Lahntal" and here, 20 fields were sampled. Center coordinates of the sites are 50°43′18.14″ N and 8°37′33.46″ E. Study sites are spread in an area of about 8 km in the east-west direction and 3 km in the north-south direction from the given center coordinates. Commonly grown crops in this region are clover, winter and spring barley and spelt. Bedrock material of soils in this region is mostly slate rock, resulting in flat soils. Due to these site conditions farming methods are less intensive than in the region of river Lahn. Ground level elevation is between 180 and 300 m above sea level. Annual average temperature is 9 °C. With an average annual precipitation of 900 mm, rainwater supply is a little higher than in the neighboring region.

In the winter of 2018/2019, six soil samples were taken from each of the 42 fields to obtain qualitative and quantitative data on the arable weed flora in the region. Figure 1 gives an overview of the sampling design of the study. Soil samples were taken with a metal sampling cylinder of 20 cm length and a diameter of 5 cm. For each field, six soil samples were taken in two different soil depths, to obtain quantitative and qualitative information on the arable weed flora in the region. Sampled soil depths were as follows:

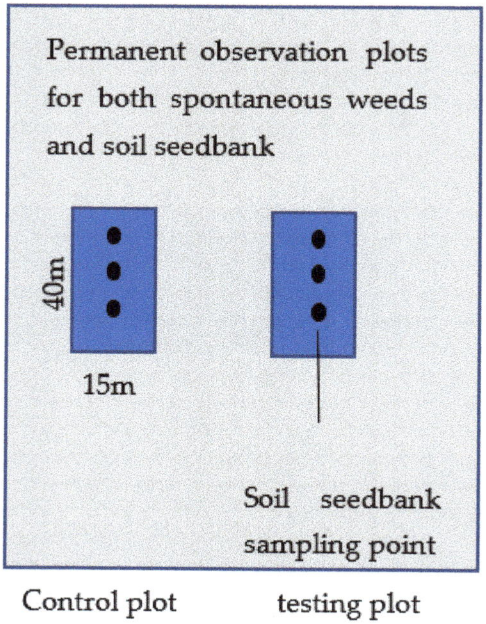

Figure 1. Sampling scheme for soil seedbank samples and relevées.

(a) 0–5 cm below soil surface and (b) 5–20 cm below soil surface. In total, there are 504 samples as data basis for seedbank analysis resulting from 42 fields multiplied with 3 repetitions per soil depth level and two soil depth levels: $42 \times 6 \times 2 = 504$.

Following the principles of the commonly used 'seedling emergence method' [25–27] a given soil volume (500 mL) was transferred into trays. During a period of four weeks identification was conducted; the first one was done between 23 April and 20 May, followed by a second period from 21 May until 24 June. In October 2019 there was a third identification of species. The trays were placed in the greenhouse before and during the period of identification in April 2019 to accelerate germination process by higher temperatures. From April onwards trays were placed outside the greenhouse, covered with a thin textile sheet to prevent inputs from foreign diaspore sources. After all seedlings had been identified in October, empty trays were left outside during winter months to overcome dormancy of remaining seeds that had not germinated until then. The last period of identification was then finished by the end of April 2020. Up to three weeks after seedlings' emergence, seedlings were identified and also individual numbers per species have been recorded. Samples were irrigated once a week before seedlings emergence and each day after germination to ensure rapid and even plant development.

Besides seedbank data, information on the spontaneous arable weed flora in the fields was collected during the spring of 2019, 2020 and 2021. Botanical surveys included a 3-week-period from the beginning of April. Spontaneous weed species were identified and also their coverage on the field was recorded on the scale of Braun–Blanquet which is commonly used for botanical surveys [28]. In the center of each field, two frames were placed with the size of 10×5 m each to identify weed species and their current coverage. Thus, in total, an area of 100 m^2 was tested for the botanical survey.

2.2. Statistical Analysis

Analyses on the data were conducted with the free software R studio, which provides several packages for analyzing vegetation ecological data; a commonly used package for data analyses on vegetation data is the package "vegan" that has been used frequently also in this study. In this package, functions for estimating ecological diversity are included such as several diversity indices (e.g., Shannon Diversity) [28]. For the assessment of the floristic diversity of the arable fields in the region, Shannon Diversity Index (H) was used which is calculated by

$$H = \sum [(pi) \times \ln(pi)] \quad (1)$$

where p_i = proportion of total sample individual number represented by species i. Evenness E was calculated by E = H/Hmax, with Hmax as the maximum diversity possible. Since the response variables 'species number' and 'individual number' of the project fields were not normally distributed, a Generalized Linear Model (GLM) was used for calculating the effects of the factors on the species and individual numbers as dependent variables. All factors were considered as fixed effects.

The full model including all factors and all factor X factor interaction effects is given in Formula (2)

$$Y = m + a + sd + m \times a + m \times sd + a \times sd \quad (2)$$

where the response variable Y = number of species, and the four factors with two factor levels each: m = management (organic, conventional), a = area (Lahntal and Gladenbacher Bergland), sd = soil depth (two factor levels 0–5; 5–20 cm). Starting from a zero model without factors, a forward selection was conducted to find the model with the best predictive power, considering the AIC values.

Studies by Hyvönen et al. [29] showed that even short periods of organic growing supports the development of a diverse arable weed flora and changes show up especially in spontaneous weed flora; in contrast, a more substantial transformation of the weed flora composition can be potentially achieved by longer periods of organic cropping. As the factor time has obviously been found to be an important factor explaining arable weed flora diversity the current study aims to assess this factor under the given site conditions. For that reason, a second Generalized Linear Model has been introduced which is targeting the factor "period of organic cropping" in case of the organic growers; for this model, only organic managed fields were included in the dataset; the samples for conventional fields were excluded though, since there has been no change of farming methods in the history of conventional farmers. The four organic farmers involved in the project started organic growing in the years 1980, 1991, 2000 and 2004; thus, organic fields have been for roughly 38, 27, 18 and 13 years in organic management at the first botanical survey in February 2019. Model Equation (1) was extended by the term t, which represents the factor 'time of organic growing':

$$Y = m + a + sd + t + m \times a \quad (3)$$

To indicate distribution patterns of arable weeds among the considered factors, a Non-metric Multi-Dimensional Scaling (NMDS) was used as ordination method due to its robustness towards input data and types of distribution of the response variable, which follows a Poisson-distribution. NMDS was conducted with two different data sets. For the NMDS ordination of the relevé data two datasets (summer and spring period) were pooled into one dataset to conduct the analyses. The analysis of the soil seedbank is based on one dataset. Similarities between sites concerning occurring weed species was calculated with the Sorensen Similarity Index [30].

As an assessment for the floristic characteristics of the sites in two different landscape regions and in two different farming systems an indicator species analysis was conducted with the R-package "indicspecies" and its function indpower that performs a correlation between target and indicator species. Indicator species analysis was conducted for the landscape level; thus, classification refers to the landscape regions or groups of landscape regions. Besides that, the farming system was included as an information for the species affinity towards the nutrient supply and their tolerance concerning farming intensity.

3. Results

3.1. Spontaneous Arable Weed Flora

Farming practice has been identified as the main factor explaining arable plant species diversity and abundance as several studies have stated [22,31,32]. As an important factor for explaining segetal species richness, the management system was accessed concerning its impact on species diversity and species abundances. The management system is an indicator of the intensity of land use. Conventional farming represents a highly intensive farming system. Organic farms decrease inputs or do not use external inputs like fertilizers and can be considered as moderate to low intensive farming systems. The factors area, management and the interaction of area x management were analyzed concerning their impact on species numbers and diversity indices in current plant coverage. It was found that management was the only factor showing a significant impact on the species numbers in the dataset of the relevés.

Current vegetation data were analyzed concerning the impact of different factors on species numbers and on Shannon Diversity Index. Following Table 1, the factors management and crop have a significant impact on species numbers with alpha = 0.001. However, the area was not found to be a significant explaining factor for species number variance. In general, species number and Shannon Diversity Index show similar patterns across the four different field trial variants (Gl-org, La-org, Gl-con, La-con). The factor crop is slightly less significant when testing its impact on Shannon Diversity Index ($\alpha = 0.05$) compared to its impact on species richness ($\alpha = 0.01$). Interaction effects were not significant in both species richness and Shannon diversity measurements indicating an independent experimental setup. Studies by Chamorro et al. [4] have shown the potential for a conversion from conventional to organic farming practices affecting the recovery of a diverse arable plant community. Species abundance could be increased from 61 to 122 within a period of five years after the transition from conventional to organic farming system Chamorro et al. [33].

Table 1. Summary of the GLM model with all considered factors and their significances (p (***) = 0, area. La = sites of the area "Lahntal", man.org = organic management system, sd.2 = soil depth 2 (5–20 cm below ground level).

	Estimate	Std.Error	z	Pr(>|z|)	Significance
(Intercept)	7.161	0.002555	2802.65	$2e \times 10^{-16}$	***
area.La	−1.252	0.004270	−293.20	$2e \times 10^{-16}$	***
man.org	0.944	0.002259	417.83	$2e \times 10^{-16}$	***
sd.2	1.031	0.002249	458.35	$2e \times 10^{-16}$	***
area.La × man.org	1.094	0.003770	290.25	$2e \times 10^{-16}$	***
area.La × sd.2	0.213	0.003399	62.56	$2e \times 10^{-16}$	***

The GLM found significant effects of all factors and all interaction effects. Species numbers in the region "Lahntal" are reduced compared to those of the other project region "Gladenbacher Bergland".

All factors except the factor area La have positive effects on the species numbers per square meter in the fields. Concerning the variance of the data, area and management are the two factors explaining the main proportion of the variance of the data. For the spontaneous weed flora, average species numbers and Shannon Diversity Indices varied significantly among the four considered variants Gl.con, Gl.org, La.con and La.org as depicted in Figure 2.

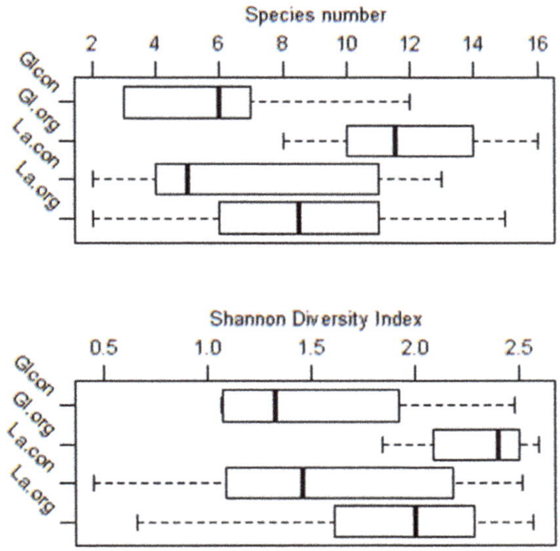

Figure 2. Diversity indices for spontaneous arable weeds. Gl and La: site indication: Gl—Gladenbacher Bergland, La—Lahntal. Org and con—management system indication: org—organic farming, con—conventional farming.

The organic variants Gl.org and La.org show significantly higher species numbers than the conventional variants. The number of species in the soil seed bank identified with the seedling-emergence method is higher than the number of species in current vegetation in three out of four variants. A reason for lower species numbers with the in-field identification might be the effect of competition between weeds and crops.

With the method of soil seed bank germinating, there is no crop, which will compete with the weeds for natural resources. This explains the higher number of wild herb species with the germinating method. In all cases of the seed bank method, the number of individuals is higher for the organic farming systems than in the conventional farming systems. The management system has a significant effect on species richness in the fields. In the investigation area of Gladenbacher Bergland there is a difference of 10 species in average between conventional and organic farming systems. The average species number in the conventional variant is eight whereas the average in the organic farming system is 18. In the other area (Marburg-Giessener Lahntal) the mean species number is lower (five species) and in the organic farming system the species number is 19. In both sites, organic farming systems show higher species richness and diversity in wild herbs. Conventional farming systems have selected a few species that do not react as sensitively as other species concerning the input of herbicides and fertilizers.

For all four groups except the conventional one in Gladenbacher Bergland, there are significant differences between the determination methods in species numbers. A reason for lower species numbers with the in-field identification could be the effect of competition between weeds and crops. With the method of soil seed bank germinating there is no crop, which will compete with the weeds for natural resources. This explains the higher number of wild herb species with the germinating method. In all cases of the seed bank method the number of individuals is higher for the organic farming systems than in the conventional farming systems.

The management system has a significant effect on species richness in the fields. In the investigation area of Gladenbacher Bergland there is a difference of 10 species in average between conventional and organic farming systems. The average species number in the conventional variant is 8 whereas the average in the organic farming system is 18. In the other area (Marburg-Giessener Lahntal) the mean species number is lower (5 species) and in the organic farming system, the species number is 19. In both sites, organic farming systems show higher species richness and diversity in wild herbs.

At the significance level of alpha = 0.05 only the factor crop was found to be a significant explanatory variable for differences in the individual and species numbers. However, area and management were not significant in the setting of the trial. For the relevé data, the effect of the factor year, area and management system were checked with a Generalized Linear Model, since the data have a Poisson distribution.

For the relevé dataset, the organic managed fields showed significantly higher average weed coverage as shown in Figure 3.

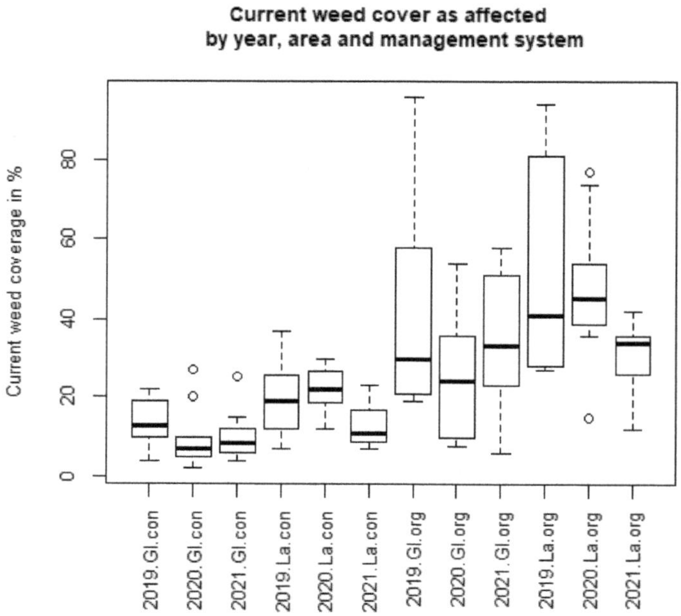

Figure 3. Weed coverage for all considered combinations of the factors: year (2019–2021), area (management system: Gl—Gladenbacher Bergland).

The coverage numbers refer to the standardized relevé plot with the size of 100 m^2. The effect of the year on the average weed cover was significant, too. For the organic managed fields, variance of the data was much wider than those for the weed coverage of the conventional fields. Results of a Generalized Linear Model on the coverage of spontaneous weeds are compiled in Table 2.

Table 2. Results of the GLM model for the coverage of spontaneous weeds as a response variable.

	Estimate	Std.Error	t-Value	Pr(>|t|)	Significance
Intercept	10	3.383.923	3.046	0.00285	*
year	−5.097	1.675	−3.043	0.00288	*
area	7.008	3.808	1.840	0.06819	n.s.
management	24.024	3.974	6.045	1.73e-08	*

Nonmetric multidimensional scaling was used to analyze current vegetation data and to identify patterns in species distribution. NMDS was conducted in R package "vegan" and resulted in Figure 4.

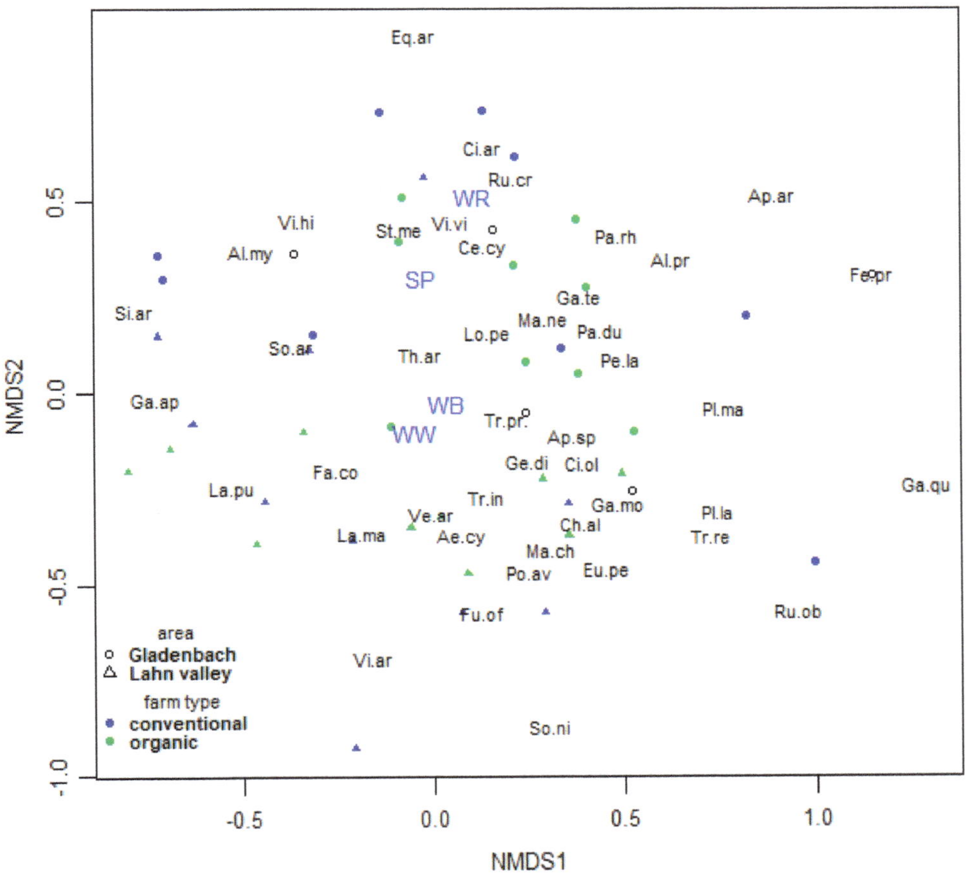

Figure 4. NMDS-Ordination of spontaneous arable weeds (relevé data); only significant factors are shown. These are winter rye (WR), spelt (SP), winter barley (WB) and winter wheat (WW). Shapes of the symbols (triangle, circle) refer to the area; colours (blue, green) refer to the farm type; e.g., blue circles represent conventional farms in the area of Gladenbach.

The best solution for the NMDS model concerning the increase in goodness of fit and the decrease in stress was found with a number of three dimensions. In this case, the correlation between the distance values and the observed dissimilarity expressed as R^2 equals 0.97 and stress is reduced to a value of 0.178. The four crops winter rye (WR), winter barley (WB), winter wheat (WW) and spelt (SP) were identified as significant factors.

Most of the arable weed species were found in both areas, in the hill-sites around Gladenbach and the valley of river Lahn and in both of the management systems.

Some species show tendencies of a more specific occurrence concerning the factors area and management—here, *Papaver dubium*, *Galeopsis tetrahit* and *Galium mollugo* were some of the species showing higher abundance patterns in organic than in conventional farming systems.

In an experiment of Hyvönen & Huusela-Veistola [1] some species with a high steadiness could be identified. *Galium aparine* was detected on 41% of the fields followed by *Chenopodium album* (68%), *Stellaria media* (76%) and *Viola arvensis* (84%) as the most frequent species in cereal cropping. This study could also identify some of these species as frequent species in cereal cropping. *Rumex crispus*, *Galium aparine* and *Alopecurus myosuroides* were more frequently found in conventionally managed fields. Moreover, *Vicia hirsuta* and *Equi-*

setum arvensis and *Cirsium arvense* and *Stellaria media* showed higher abundance patterns in more intensive cereal cropped fields.

The weed species *Cirsium arvense, Centaurea cyanus, Vicia villosa* and *Rumex crispus* were frequently associated with winter rye. In the crops winter wheat and winter barley, the weeds *Trifolium pretense, Geranium dissectum, Tripleurospermum inodorum* and *Thlaspi arvense* were frequently represented.

3.2. Soil Seedbank of the Arable Weeds

The reservoir of seeds of mostly annual arable plant species in the soil is an important parameter for analyzing the emerged arable weed flora of the past. Moreover, the soil seedbank allows forecasts regarding to the composition of futures' arable plant communities and it provides information for conservation and regulation strategies for arable plant species [34,35].

The response variable in the soil seedbank dataset (individual numbers) is not normally distributed and follows a Poisson-distribution. For that reason, a Generalized Linear Model was chosen to conduct the analyses. A model with three fixed factors and two interaction terms was found to be the model with the best performance and predictive power. The results of the model are given in Table 3.

Table 3. Results of the GLM for the seedbank data. (p (***) = 0.

	Estimate	Std.Error	z	Pr(>|z|)	Significance
(Intercept)	7.161	0.002555	2802.65	$2e \times 10^{-16}$	***
area.La	−1.252	0.004270	−293.20	$2e \times 10^{-16}$	***
man.org	0.944	0.002259	417.83	$2e \times 10^{-16}$	***
sd.2	1.031	0.002249	458.35	$2e \times 10^{-16}$	***
area.La × man.org	1.094	0.003770	290.25	$2e \times 10^{-16}$	***
area.La × sd.2	0.213	0.003399	62.56	$2e \times 10^{-16}$	***

Soil seedbank data was tested with a GLM concerning the impact of the factors area, management, soil depth and two two-way interaction effects between those mentioned factors. The Dredge-Function of the R-package "MuMIn" was used to find the model with the best performance and predictive power. Using the AIC and the AIC delta as a criterion for model selection, the model including all factors and two out three interaction effects was found to be the best one for modelling the individual numbers per square meter. For this model, AIC was 6149.2 and AIC delta was 0.17.

For the soil seed bank species numbers differed significantly from each other between the different farming methods (conventional and organic). Belowground species numbers were twice as high (18 species) as those in the conventional alternative (nine species) as depicted in the left part of Figure 5.

A Kruskal–Wallis test was conducted with the software "R studio" to assess if organic farming period has a significant effect on the seed numbers of the soil seedbank. Testing results were significant, indicating significant differences in the data. Pairwise Wilcoxon test was followed up indicating significantly higher seed numbers of 18 and 27 years of organic growing, compared to 14 years of organic growing. Other groups did not differ significantly from each other. In the study, a maximum seed density is reached between 18 and 27 years of organic farming. After this period, the tendency of a decrease in species numbers per square meter can be observed.

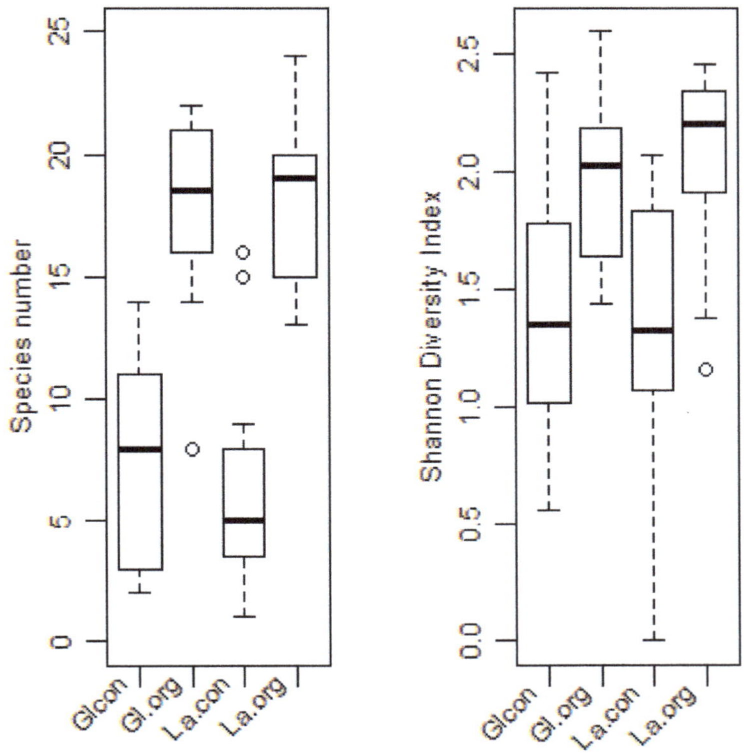

Figure 5. Weed species numbers derived from the soil seed bank data.

As shown in Figure 6, the number of weed seeds per square meter differs from organic management systems as affected by different time periods of organic growing. A Kruskal-Wallis Test was conducted to detect significant differences in the data. Results of this test are given in Table 4, indicating significant higher seed numbers after 27 years of organic growing compared to 14 years, and also, the number of weed seeds in the soil per square meter is significantly higher in systems of 18 years organic farming, compared to those of 14 years. Evidence of an increase of the soils' seed potential with an increasing time period of organic growing could be stated. This observation is applicable for the organic growing period until 27 years. Thereafter, however, the increase of seed numbers stagnates or fluctuates.

Table 4. Significance levels between the four periods of organic growing with regard to the total seed number per square meter. Significant differences on the level of $\alpha = 0.05$ are indicated with *.

	14 Years	18 Years	27 Years
14 years	-	0.035 *	-
27 years	0.025 *	0.403	-
38 years	0.403	0.402	0.207

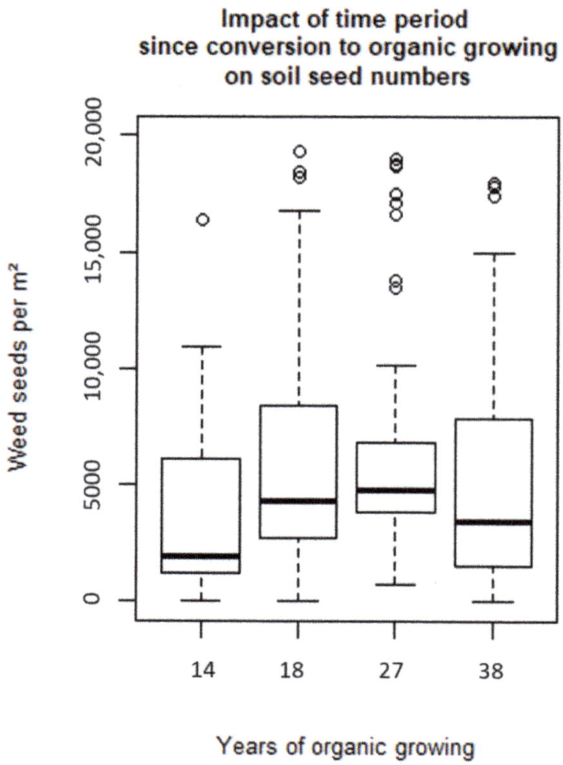

Figure 6. Comparison of time periods of organic growing with regard to the weed seed number.

3.2.1. Analysis of Spring Vegetation

The seedbank data set was divided into three subgroups to access the effect of seasonal changes in the plant species communities. The first group includes species of the seedbank that germinated during a four-week period of April 2019, the second one contains those seedlings from June 2019 and the third one contains all additional seedlings from October 2019. Main germination of seeds was in June 2019. Way fewer seedlings emerged in autumn and spring. As shown in the graphical result of an NMDS in Figure 7, Organic farming systems provide the highest species richness and species abundance. Species showing higher occurrences in conventional farming during spring are *Chenopodium polyspermum*, *Galium aparine* and *Juncus bufonius*.

In organic farming systems, *Chenopodium album*, *Plantago intermediae* and *Rumex acetosella* and *Cirsium arvense* showed higher abundances than in conventional farming systems.

Arable weeds of the soil seed bank samples were analyzed with a Nonmetric Multidimensional Scaling ordination (NMDS) as shown in Figure 8. Resulting from this analysis, the management system has the main effect on explaining the variance in species abundance of the soil seedbank.

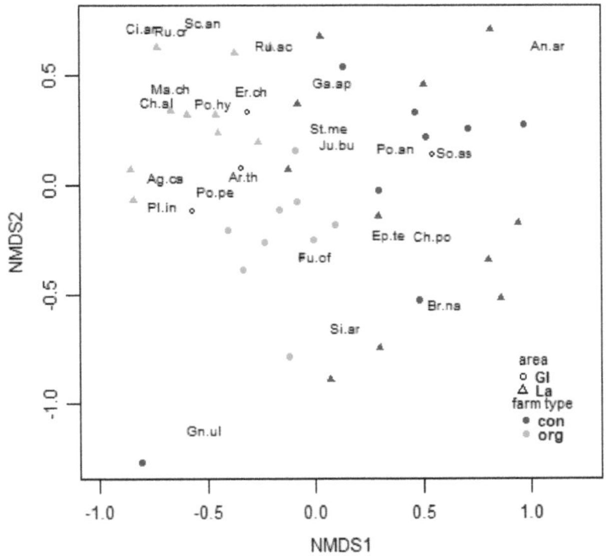

Figure 7. Species of the soil seedbank germinated in the period of May 2019. Shapes (triangels and circles) represent the two different areas; the colours green and blue indicate the two farm types. e.g., green circles mean organic farms in the region Gladenbach.

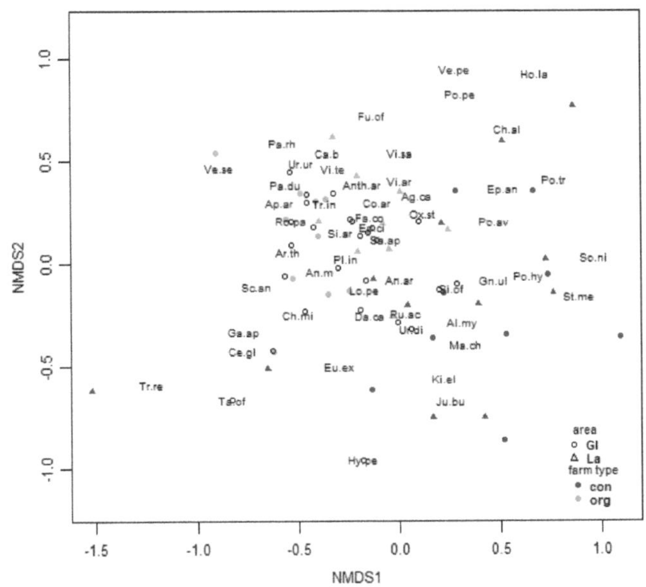

Figure 8. Species of the soil seedbank germinated in the period of May, June and October 2019.

So, the impact of the farming practice could be stated in both, current vegetation and seedbank. There is a clear grouping of species concerning their occurrence pattern. Organic farming includes the highest proportion of total species numbers. There are a few species found more frequently in conventional farming systems. Tendencies found in current vegetation data are partially reflected by the soil seedbank.

Following seedbank data, the weed species *Polygonum aviculare, Poa trivialis, Solanum nigrum, Stellaria media* and *Juncus bufonius* appear as tolerant towards intensive farming methods, whereas the highest proportion of the total species number is only found in organic farming conditions at a higher abundance level. In current vegetation on arable fields, 49 segetal species were identified, which is a share of 60%. In total, 82 species that were identified in the soil seedbank over all fields were included in the experiment. *Centaurea cyanus* and *Papaver rhoeas* were mostly found under organic farming conditions which is reflected by both seedbank data and current vegetation. However, the impact of farming practice is considerably higher in the seedbank.

The current vegetation is more directly affected by environmental conditions like climate (rainfall) which results in a higher fluctuation. In contrast, the soil seedbank is reacting way slower and delayed to a change in environmental conditions. The seedbank represents the intensity level in farming over the last couple of years. The management system is the major factor whereas the area shows a lower significance for explaining the variance.

Figure 8 shows higher species numbers in organic cropping systems, indicated by the clustering of species along the factor organic management.

Species tend to have a lower distribution and spread along the factor of environment. Some species show tendencies to occur either in more extensive or more intensive farming. Species that are more likely found in intensive farming are *Solanum nigrum, Matricharia chamomilla* and *Stellaria media.* Under less intensive conditions, *Fumaria officinalis, Centaurea cyanus* and *Holcus lanatus* can be found more likely.

3.2.2. Indicator Species Analysis

Therefore, a grouping vector of sites was built on the basis of the four field trial variants consisting of two landscape areas (Lahn valley = La, Hillsites = Gl) and two management systems (organic = org, conventional = con). The assumption was met that both management system and landscape area could have an impact on the constellation of segetal species in arable fields.

For the area of Gladenbach and Lahntal under organic farming, the indicator species *Tripleuspermum inodorum* and *Myosotis arvensis, Alopecurus myosuroides* were the most significant indicators. Following the methodology from De Cáceres (2020) an indicator species analysis was conducted using the R package "indicspecies". After running the analysis, 16 of 82 species in total were identified as indicator species. K-means algorithm was used to assign the species to four different groups. Following this procedure, four site group combinations were calculated as shown in Table 5. Most of the indicator species are found in organic farming systems (group-nr. 1), way fewer under conventional farming conditions (groups 2 + 3 +4). A principal component analysis was conducted which resulted in an optimal number of four species groups. Within the four groups the proportion of explained variance in the data is around 76.7%. An indicator species analysis was conducted with the results shown in the table below. In organic farming, there are many species that can be used as indicators for organic managed sites; in contrast there are only two species as indicators for intensive farming methods of conventional agriculture, which are in this case *Daucus carota* and *Anthemis arvensis.* In this study, *Vicia hirsuta, Cirsium arvenese* and *Rumex crispus* were detected as indicator species for organic management systems. Intensive farming methods can be indicated by the species *Daucus carota* and *Anthemis arvensis.*

Table 5. Indicator species with their indicator values and assignments to the groups. (***) significant on α = 0.99, (**) significant on α = 0.95, significant on α = 0.90.

Group nr.		A	B	Stat	*p*.Value	
1	Pap.rho	0.9732	1.0	0.987	0.001	***
	Cen.cya.	0.9784	0.8333	0.903	0.001	***
	Ver.arv.	0.6816	1.0	0.826	0.001	***
	Fal.con.	0.8128	0.5000	0.638	0.001	***
	Vio.arv.	0.4722	0.6667	0.561	0.004	**
	Dac.glo.	0.6598	0.3333	0.469	0.013	*
	Gal.apa.	0.9010	0.1667	0.388	0.011	*
	Sis.off.	0.8987	0.1667	0.387	0.039	*
	Pap.dub.	0.8812	0.1667	0.383	0.025	*
2	Sag.pr.	0.87429	0.24490	0.463	0.032	*
	Rum.obt.	100.000	0.06122	0.247	0.041	*
3	Tri.pra.	0.91054	0.43478	0.629	0.001	***
	Alop.myo.	0.75716	0.26087	0.444	0.033	*
	Laps.comm.	100.000	0.08696	0.295	0.025	*
4	Matr.cham.	0.8640	0.9636	0.912	0.001	***
	Cha.mi.	0.9445	0.1455	0.371	0.029	*

3.2.3. Impact of Crop Type on Species Richness

The effect of crop type has been stated by several studies with mainly a higher species richness and less geophytes in cereal fields than in root crops [36]. As shown in Figure 9 high variation in the weed individual numbers could be observed with significant differences between the crop types.

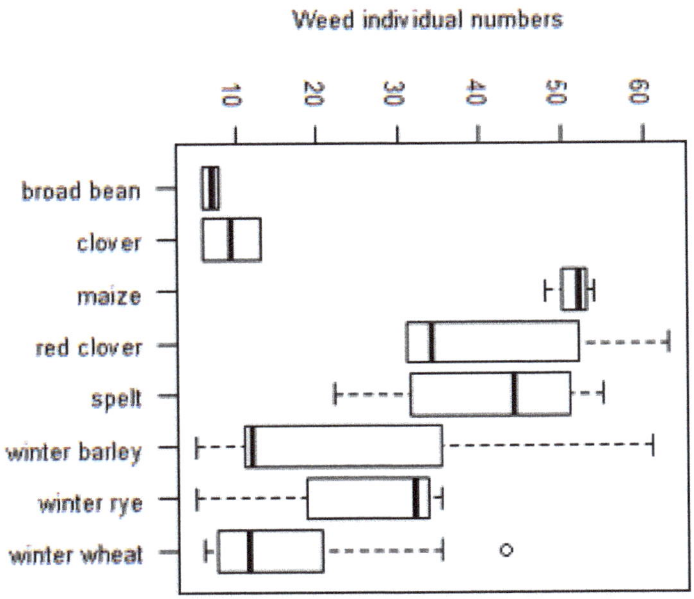

Figure 9. Comparison of average weed individual numbers in different crops on the project fields.

The highest species and individual numbers were found in the crops maize, red clover and spelt, whereas broad bean and winter wheat showed the lowest individual numbers per square meter.

To access the floristic similarity between the sites, Sorensen quantitative similarity index was calculated:

$$\text{Sorensen's qualitative index} = 2C/A + B \times 100\%$$

where:

A—the number of species in one of the two communities compared.
B—the number of species in the second community compared.
C—the number of common species in the compared communities.

In Table 6, Sorensen's Qualitative Similarity Index is given for each of the considered pairs of variants. Similarity between organic fields in two different sites was 65% and higher than those between organic and conventional sites (57.3%). Organic and conventional fields differ more in their species composition than organic fields among each other in different locations. However, the similarity between the organic fields is not considerably higher than those between conventional and organic farms, indicating that regional differences between arable weed communities might play a role although on a smaller regional scale.

Table 6. Sorensen's Qualitative Similarity Indices for the project sites.

	Conventional	Gl.Organic
organic	57.3	-
La.organic	-	65

As shown in Figure 10, Shannon Diversity varied between spring and summer season. Significant higher Diversity values were found during the summer season. As Mennan & Ngouajio [37] stated, there are seasonal cycles in the germination patterns of the weeds *Galium aparine* and wild mustard (*Brassica kaber*). Highest germination rates were stated during May with around 70% germination and a strong decrease in germination of around 30% in August.

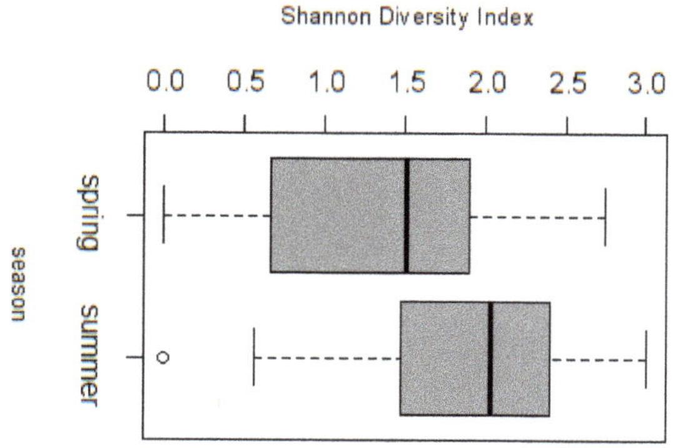

Figure 10. Comparison of average diversity indices for spring and summer germination period of arable weeds.

Diversity of arable weed communities differs between spring and summer. Higher diversity is found during the summer months, since new species have germinated and contribute to a higher index.

Studies of Lososová et al. [36] stated seasonal dynamics as a factor causing changes in arable plant community composition with higher species richness and beta diversity during the summer months. Species that were found in summer were mostly those also present in the spring germination.

There is no significant difference in species diversity between the two areas. There is no significant effect of the landscape area on the diversity of segetal species of arable fields.

Studies by Swanton et al. [38] suggest that there is an interaction effect between tillage system and soil type which may influence the distribution of soil seed bank vertically. The management showed significant effects on species diversity. Shannon Diversity Index is almost two times higher in organically managed fields than in the conventional ones. Several studies have stated the impact of an increasing farming intensity (herbicide application, plowing, fertilizing) on the diversity of arable plant communities. With higher intensity levels in farming, diversity of arable plant communities decreased significantly [10,39–42].

The hypothesis was tested that there is a different accumulation rate of weed seeds in different soil depths resulting in different diversity indices. Therefore, a Kruskal-Wallis Test was conducted which has not become significant. Shannon Diversity Indices were plotted, as shown in Figure 11. Though, species diversity has not been affected by the depth level of the soil sample. The farmers of the study fields have used a mixture of reduced tillage and ploughing. The arable fields show a homogenous mixture of seeds in the soil at a testing depth of 20 cm. A significant difference in the seed number and Shannon Diversity Index could not be found. This observation has also been stated by studies of Feledyn-Szewczyk et al. [43] which showed the effect of the tillage system on the absolute weed seed numbers in the soil and their vertical distribution in the soil of arable fields. Feledyn-Szewczyk et al. [43] showed that reduced tillage systems and ploughing systems tend to establish a mixture of seeds in the soil with higher average seed numbers in reduced tillage systems (4515 seeds m^{-2}) compared to ploughing systems (2080 m^{-2}).

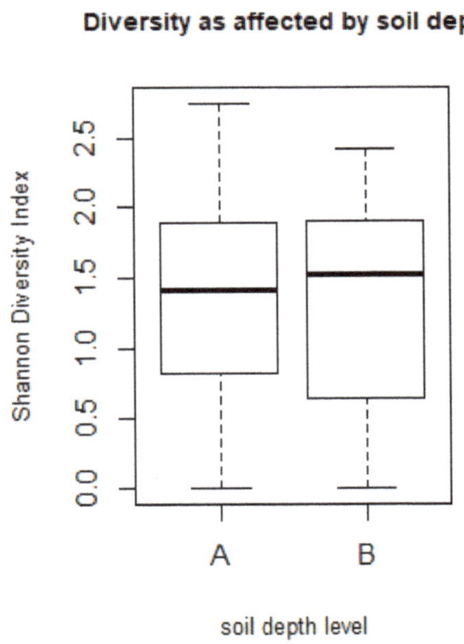

Figure 11. Comparison of two soil depths regarding their belowground species diversity. Level A: 0–5 cm below soil surface; B: 5–20 cm below soil surface.

Studies by Clements et al. [40] analyzed the impact of different soil tillage methods on the seedbank composition of different weeds. It was found that deep soil tillage, like plowing, results in a more homogeneous distribution of the seedbank over depth. In contrast, in a system with minimal soil tillage, Clements et al. [40] assumed that the upper 5 cm of the soil comprises over 60% of the seed bank. Research by Buhler et al. [41] could identify grass seed accumulation in minimal tillage systems and found higher accumulation rates at the deeper soil levels in moldboard plowing systems.

4. Discussion

Studies on the arable weed flora in the center of the German state Hesse could state a strong reaction of both spontaneous flora and soil seedbank towards the farming intensity. Among others, farming intensity has been identified as the factor with the strongest impact on species numbers, abundances and distribution patterns of arable weeds.

Species numbers derived from soil seedbank data were on average twice as high in organically managed fields (18 species) compared to conventional ones (nine species). For the relevé data, species numbers and diversity indices (Shannon Diversity) were significantly higher in organic cropping systems. Relevés included usually a proportion of the seedbank species so not all species present in the soil seedbank were also found in the relevés.

Besides management system, the crop type played a minor role in explaining differences in species richness and diversity indices. Results from NMDS analysis have shown that highest species richness can be found in organic farming systems in both landscape areas—in Valley of river Lahn and the hill site of Gladenbach. Therefore, landscape area was not detected as a factor for explaining significant differences in species distribution. The hypothesis of a difference in species and individual numbers in different soil depth levels could not be stated, however. A reason for this might be that soil tillage is applied on the study sites. The use of plow and grubber might result in a mixture of soil horizons so seeds will be mixed as well. Comparing germination numbers and diversity indices of different seasons showed a tendency of a slightly higher diversity values in summer compared to spring, significant differences could not be found.

The study could state the assumption that organic farming contributes to the regeneration of the arable seed bank. The period of organic growing affects the abundance and species richness of the arable weed species pool of the soil. Tendencies of an accumulation of seeds in the soil over time were identified. Evidence was also found in the data that, at some point in time, the seedbank increase stagnates again and stays more or less stable on a certain level of seeds per square meter.

Compared to the conventional managed fields, the organically managed fields showed significantly higher species numbers and also Shannon Diversity increased in these fields. Evidence was found that long-term intensive farming conditions result in a depletion of the seed reservoir of arable weeds in the soil, as stated by other studies as well [43–46].

Regarding the project fields, organic farming could double average seedbank species numbers in the fields from an average of 10 species per field in conventional farming systems to average of 20 species per field in organic farming systems. Crops had a significant effect on the distribution of arable weeds with higher diversity values in winter crops and lower diversity indices in spring grown crops. The season also had a significant impact on species numbers, with the highest species numbers during the summer months.

Author Contributions: Conceptualization, P.K. and R.W. All authors have read and agreed to the published version of the manuscript.

Funding: The study was carried out under the project "Impact of an innovative combine harvesting technique on biodiversity in the agrarian landscape—a pilot study in the region Marburg-Biedenkopf" funded by the German ministry of the environment (Bundesamt für Naturschutz (BfN) Konstantinstrasse 110, 53179 Bonn, Germany) with the funding number 3518892013.

Acknowledgments: Many thanks to Josef Scholz—vom Hofe for support in experimental procedures, field work and sampling and to Laura Ritz for support in data generation and digitalization.

Conflicts of Interest: The authors declare no conflict of interests.

References

1. Hyvönen, T.; Huusela-Veistola, E. Arable weeds as indicators of agricultural intensity–A case study from Finland. *Biol. Conserv.* **2008**, *141*, 2857–2864. [CrossRef]
2. Giam, X.; Bradshaw, C.J.; Tan, H.T.; Sodhi, N.S. Future habitat loss and the conservation of plant biodiversity. *Biol. Conserv.* **2010**, *143*, 1594–1602. [CrossRef]

3. Lemoine, C.; Sérusiaux, E.; Mahy, G.; Piqueray, J. Agro-environmental scheme for segetal plant conservation in Wallonia (Belgium): An assessment in conventional and organic fields. *Biotechnol. Agron. Soc. Environ.* **2018**, *22*, 35–44. [CrossRef]
4. Chamorro, L.; Masalles, R.M.; Sans, F. Arable weed decline in Northeast Spain: Does organic farming recover functional biodiversity? *Agric. Ecosyst. Environ.* **2016**, *223*, 1–9. [CrossRef]
5. Andreasen, C.; Jensen, H.A.; Jensen, S.M. Decreasing diversity in the soil seed bank after 50 years in Danish arable fields. *Agric. Ecosyst. Environ.* **2018**, *259*, 61–71. [CrossRef]
6. Hyvönen, T. Can conversion to organic farming restore the species composition of arable weed communities? *Biol. Conserv.* **2007**, *137*, 382–390. [CrossRef]
7. Albrecht, H. Development of arable weed seedbanks during the 6 years after the change from conventional to organic farming. *Weed Res.* **2005**, *45*, 339–350. [CrossRef]
8. Carmona, C.P.; Guerrero, I.; Peco, B.; Morales, M.B.; Oñate, J.J.; Pärt, T.; Tscharntke, T.; Liira, J.; Aavik, T.; Emmerson, M.; et al. Agriculture intensification reduces plant taxonomic and functional diversity across European arable systems. *Funct. Ecol.* **2020**, *34*, 1448–1460. [CrossRef]
9. Hawes, C.; Squire, G.R.; Hallett, P.D.; Watson, C.A.; Young, M. Arable plant communities as indicators of farming practice. *Agric. Ecosyst. Environ.* **2010**, *138*, 17–26. [CrossRef]
10. Fonderflick, J.; Besnard, A.; Chardès, M.-C.; Lanuzel, L.; Thill, C.; Pointereau, P. Impacts of agricultural intensification on arable plants in extensive mixed crop-livestock systems. *Agric. Ecosyst. Environ.* **2020**, *290*, 106778. [CrossRef]
11. Rotchés-Ribalta, R.; Blanco-Moreno, J.M.; Armengot, L.; Chamorro, L.; Sans, F.X. Both farming practices and landscape characteristics determine the diversity of characteristic and rare arable weeds in organically managed fields. *Appl. Veg. Sci.* **2015**, *18*, 423–431. [CrossRef]
12. Gaba, S.; Perronne, R.; Fried, G.; Gardarin, A.; Bretagnolle, F.; Biju-Duval, L.; Colbach, N.; Cordeau, S.; Fernández-Aparicio, M.; Gauvrit, C.; et al. Response and effect traits of arable weeds in agro-ecosystems: A review of current knowledge. *Weed Res.* **2017**, *57*, 123–147. [CrossRef]
13. Cavan, G.; Cussans, J.; Moss, S.R. Modelling different cultivation and herbicide strategies for their effect on herbicide resistance in Alopecurus myosuroides. *Weed Res.* **2000**, *40*, 561–568. [CrossRef]
14. Menchari, Y.; Camilleri, C.; Michel, S.; Brunel, D.; Dessaint, F.; Le Corre, V.; Délye, C. Weed response to herbicides: Regional-scale distribution of herbicide resistance alleles in the grass weed Alopecurus myosuroides. *New Phytol.* **2006**, *171*, 861–874. [CrossRef] [PubMed]
15. Yu, Q.; Abdallah, I.; Han, H.; Owen, M.; Powles, S. Distinct non-target site mechanisms endow resistance to glyphosate, ACCase and ALS-inhibiting herbicides in multiple herbicide-resistant Lolium rigidum. *Planta* **2009**, *230*, 713–723. [CrossRef]
16. Storkey, J.; Neve, P. What good is weed diversity? *Weed Res.* **2018**, *58*, 239–243. [CrossRef]
17. Seifert, C.; Leuschner, C.; Culmsee, H. Arable plant diversity on conventional cropland—The role of crop species, management and environment. *Agric. Ecosyst. Environ.* **2015**, *213*, 151–163. [CrossRef]
18. Albrecht, H.; Cambecèdes, J.; Lang, M.; Wagner, M. Management options for the conservation of rare arable plants in Europe. *Bot. Lett.* **2016**, *163*, 389–415. [CrossRef]
19. Wood, T.J.; Holland, J.M.; Goulson, D. Providing foraging resources for solitary bees on farmland: Current schemes for pollinators benefit a limited suite of species. *J. Appl. Ecol.* **2017**, *54*, 323–333. [CrossRef]
20. McGinlay, J.; Parsons, D.J.; Morris, J.; Hubatova, M.; Graves, A.; Bradbury, R.B.; Bullock, J.M. Do charismatic species groups generate more cultural ecosystem service benefits? *Ecosyst. Serv.* **2017**, *27*, 15–24. [CrossRef]
21. Smith, B.M.; Aebischer, N.J.; Ewald, J.; Moreby, S.; Potter, C.; Holland, J.M. The Potential of Arable Weeds to Reverse Invertebrate Declines and Associated Ecosystem Services in Cereal Crops. *Front. Sustain. Food Syst.* **2020**, *3*, 118. [CrossRef]
22. Kolářová, M.; Tyšer, L.; Soukup, J. Impact of site conditions and farming practices on the occurrence of rare and endangered weeds on arable land in the Czech Republic. *Weed Res.* **2013**, *53*, 489–498. [CrossRef]
23. Rotchés-Ribalta, R.; Armengot, L.; Mäder, P.; Mayer, J.; Sans, F.X. Long-Term Management Affects the Community Composition of Arable Soil Seedbanks. *Weed Sci.* **2017**, *65*, 73–82. [CrossRef]
24. Waldhardt, R.; Fuhr-Boßdorf, K.; Otte, A.; Schmidt, J.; Simmering, D. Floristisch-phytocoenotische Diversitäten einer peripheren Kulturlandschaft in Abhängigkeit von Landnutzung. *Raum und Zeit. Agrar.* **2000**, *31*, 121–147.
25. Price, J.N.; Wright, B.R.; Gross, C.L.; Whalley, W.R.D.B. Comparison of seedling emergence and seed extraction techniques for estimating the composition of soil seed banks: *Comparison of soil seed bank methods. Methods Ecol. Evol.* **2010**, *1*, 151–157. [CrossRef]
26. Rahman, A.; James, T.K.; Mellsop, J.; Grbavac, N. Effect of cultivation methods on weed seed distribution and seedling emergence. *New Zealand Plant Prot.* **2000**, *53*, 28–33. [CrossRef]
27. Ahammad, K.; Rahman, M.; Ali, M. Effect of hydropriming method on maize (Zea mays) seedling emergence. *Bangladesh J. Agric. Res.* **2014**, *39*, 143–150. [CrossRef]
28. Kolářová, M.; Tyšer, L.; Soukup, J. Weed vegetation of arable land in the Czech Republic: Environmental a management factors determining weed species composition. *Biologia* **2014**, *69*, 443–448. [CrossRef]
29. Hyvönen, T.; Ketoja, E.; Salonen, J.; Jalli, H.; Tiainen, J. Weed species diversity and community composition in organic and conventional cropping of spring cereals. *Agric. Ecosyst. Environ.* **2003**, *97*, 131–149. [CrossRef]
30. Feledyn-Szewczyk, B.; Smagacz, J.; Kwiatkowski, C.A.; Harasim, E.; Woźniak, A. Weed Flora and Soil Seed Bank Composition as Affected by Tillage System in Three-Year Crop Rotation. *Agriculture* **2020**, *10*, 186. [CrossRef]

31. Berbeć, A.K.; Staniak, M.; Feledyn-Szewczyk, B.; Kocira, A.; Stalenga, J. Organic but Also Low-Input Conventional Farming Systems Support High Biodiversity of Weed Species in Winter Cereals. *Agriculture* **2020**, *10*, 413. [CrossRef]
32. van Elsen, T. Species diversity as a task for organic agriculture in Europe. *Agric. Ecosyst. Environ.* **2000**, *77*, 101–109. [CrossRef]
33. Chamorro, L.; Armengot, L.; José-María, L.; Sans, F.X. Organic farming enhances the recovery of ancient crops and segetal weeds in Catalonia (NE of Spain). *Build. Org. Bridges* **2014**, *3*, 979–982.
34. Mahé, I.; Cordeau, S.; Bohan, D.A.; Derrouch, D.; Dessaint, F.; Millot, D.; Chauvel, B. Soil seedbank: Old methods for new challenges in agroecology? *Ann. Appl. Biol.* **2021**, *178*, 23–38. [CrossRef]
35. Schwartz-Lazaro, L.M.; Copes, J.T. A Review of the Soil Seedbank from a Weed Scientists Perspective. *Agronomy* **2019**, *9*, 369. [CrossRef]
36. Lososová, Z.; Chytrý, M.; Cimalová, S.; Kropáč, Z.; Otýpková, Z.; Pyšek, P.; Tichý, L. Weed vegetation of arable land in Central Europe: Gradients of diversity and species composition. *J. Veg. Sci.* **2004**, *15*, 415–422. [CrossRef]
37. Mennan, H.; Ngouajio, M. Seasonal cycles in germination and seedling emergence of summer and winter populations of catchweed bedstraw (*Galium aparine*) and wild mustard (*Brassica kaber*). *Weed Sci.* **2006**, *54*, 114–120. [CrossRef]
38. Swanton, C.J.; Shrestha, A.; Knezevic, S.Z.; Roy, R.C.; Ball-Coelho, B.R. Influence of tillage type on vertical weed seedbank distribution in a sandy soil. *Can. J. Plant Sci.* **2000**, *80*, 455–457. [CrossRef]
39. Schmitzberger, I.; Wrbka, T.; Steurer, B.; Aschenbrenner, G.; Peterseil, J.; Zechmeister, H.G. How farming styles influence biodiversity maintenance in Austrian agricultural landscapes. *Agric. Ecosyst. Environ.* **2005**, *108*, 274–290. [CrossRef]
40. Clements, D.R.; Benoit, D.L.; Murphy, S.D.; Swanton, C.J. Tillage Effects on Weed Seed Return and Seedbank Composition. *Weed Sci.* **1996**, *44*, 314–322. [CrossRef]
41. Buhler, D.D.; Kohler, K.A.; Thompson, R.L. Weed Seed Bank Dynamics During a Five-Year Crop Rotation. *Weed Technol.* **2001**, *15*, 170–176. [CrossRef]
42. Armengot, L.; José-María, L.; Blanco-Moreno, J.M.; Bassa, M.; Chamorro, L.; Sans, F.X. A novel index of land use intensity for organic and conventional farming of Mediterranean cereal fields. *Agron. Sustain. Dev.* **2011**, *31*, 699–707. [CrossRef]
43. Waldhardt, R.; Otte, A. Abschätzung der zur Erhaltung einer lokaltypischen Ackerflora benötigten Fläche am Beispiel der Gemarkung Erda (Lahn-Dill-Bergland, Hessen). *Peckiana* **2001**, *1*, 101–108.
44. Chytrý, M.; Schaminée, J.H.J.; Schwabe, A. Vegetation survey: A new focus for Applied Vegetation Science: Vegetation survey. *Appl. Veg. Sci.* **2011**, *14*, 435–439. [CrossRef]
45. Kuht, J.; Eremeev, V.; Talgre, L.; Madsen, H.; Toom, M.; Mäeorg, E. Soil weed seed bank and factors influencing the number of weeds at the end of conversion period to organic production. *Agron. Res.* **2016**, *14*, 1372–1379.
46. Oksanen, J. Multivariate Analysis of Ecological Communities in R: Vegan tutorial. *R Package Version* **2011**, *1*, 1–43.

diversity

MDPI

Article

Wild Plant Diversity and Soil Characteristics of Desert Roadside Vegetation in the Eastern Desert

Heba Sallam [1], Mashail Nasser Alzain [2,*], Amani Omar Abuzaid [3], Naglaa Loutfy [1], Mohamed O. Badry [1], Ahmed K. Osman [1] and Sabah A. Hammad [1]

[1] Botany and Microbiology Department, Faculty of Science, South Valley University, Qena 83523, Egypt; heba.sallam@yahoo.com (H.S.); naglaa.hasssn@sci.svu.edu.eg (N.L.); mohamedowis@svu.edu.eg (M.O.B.); ahmosman2000@yahoo.com (A.K.O.); sabah.hammad2@sci.svu.edu.eg (S.A.H.)

[2] Department of Biology, College of Sciences, Princess Nourah bint Abdulrahman University, P.O. Box 84428, Riyadh 11671, Saudi Arabia

[3] Biology Department, Faculty of Science, King Abdulaziz University, P.O. Box 80200, Jeddah 21589, Saudi Arabia; aabuzaid@kau.edu.sa

* Correspondence: mnalzain@pnu.edu.sa

Abstract: The wild vegetation of the Eastern Desert is characterized by openness and comprises perennials and ephemerals. The current study investigated the relationship between the edaphic factors of the natural vegetation along El Sheikh Fadl–Ras Gharib Road, Southwest Suez Gulf, in the northern sector of the Eastern Desert. The vegetation structure of the study area is relatively simple. The surveyed plants included 93 species from 22 families (51 perennials and 42 annuals). Asteraceae, Brassicaceae, Amaranthaceae, and Fabaceae were the richest families, constituting the majority of plant species (53.76%). Therophytes were the most frequent life forms. About 83.87% of the total flora were pluriregional elements of different affinities. Most of the recorded taxa occupied the Irano-Turanian/Mediterranean/Saharo-Sindian/Sudano-Zambezian chorotypes. The application of TWINSPAN classification resulted in grouping the vegetation into three main vegetation groups (A, B, and C), representing distinct microhabitats. The CCA ordination indicates diversity in vegetation group A. Group B was highly associated with Na, Mg, $CaCO_3$, silt, clay, and C/N. Group C showed a high correlation with sand, K, and N. The differences in wild plant life forms, richness, and diversity along the studied desert roadsides, in association with the soil differences, provide a good indication of plant biodiversity.

Keywords: CANOCO; chorological analysis; Egypt; soil analysis; TWINSPAN

Citation: Sallam, H.; Alzain, M.N.; Abuzaid, A.O.; Loutfy, N.; Badry, M.O.; Osman, A.K.; Hammad, S.A. Wild Plant Diversity and Soil Characteristics of Desert Roadside Vegetation in the Eastern Desert. *Diversity* **2023**, *15*, 874. https://doi.org/10.3390/d15070874

Academic Editor: Ilias Travlos

Received: 3 June 2023
Revised: 16 July 2023
Accepted: 18 July 2023
Published: 20 July 2023

1. Introduction

The desert ecosystem occupies approximately 95% of Egypt's total area. The Eastern Desert of Egypt occupies the area extending from the Nile Valley eastward to the Gulf of Suez and the Red Sea, which is about 223,000 km^2, i.e., 22.3% of the total area of Egypt. It is traversed by numerous canyonlike depressions (wadis) running to the Red Sea or the Nile Valley. It has a high backbone of high, rugged mountains running parallel to and a relatively short distance from the coast [1]. The inland part of the Eastern Desert of Egypt is divided into four main geomorphological and ecological regions, from north to south, and the Limestone Desert is the second region [2]. From the phytogeographical point of view, El Hadidi [3] divided the Eastern Desert of Egypt into two main sub-territories: the Galalah Desert, which includes Cairo–Suez, and the northern limestone plateau.

The desert vegetation is characterized by openness and is composed of a permanent framework of perennials, the interspaces of which may be occupied by ephemerals. Their duration depends on irregular rainfall and soil thickness [2]. The vegetation composition and plant distribution covering the northern inland part of the Egyptian Eastern Desert have been studied [4–6].

This ecosystem has unique and characteristic xerophytic vegetation that sustains the human population with essential goods and services. Despite these benefits, recent threats to its species and habitats have been recorded [7]. Anthropogenic activities are the cause of all the threats, in addition to the aridity of the climate. Desert degradation and fragmentation caused by human misuse and severe environmental changes are the major causes of the decline in global biodiversity and have become a critical environmental problem [8,9]. Therefore, restoring disturbed ecosystems in many areas is prioritized for biodiversity conservation and maintaining landscape productivity [10].

Egyptian deserts' natural vegetation faces two categories of threats: The first includes the natural processes of drought, floods, storms, diseases, natural enemies (rodents and insects), and invasion of exotic species [11,12]. The second category includes man-mediated threats such as road construction. Seeds can adhere to car tires along roads, leading to the spread of exotic species and colonizing new areas [11,13]. All of these threats result in changes in habitat conditions with the subsequent alteration of vegetation structure, loss of natural habitats, and decline in floristic composition [14,15]. The construction of roads is one type of human impact on the deserts of the Middle East [16]. Therefore, these areas are severely subjected to human disturbance. Despite the wealth of studies on aspects of vegetation along roads in temperate regions, little is known about arid regions [17,18]. Likewise, some studies have dealt, to some extent, with roadside vegetation in deserts in Egypt [19–21].

We hypothesize that the harsh environmental conditions affect plant diversity and vegetation structure in El Sheikh Fadl–Ras Gharib Desert Road. Therefore, this study aimed to (1) survey the roadside wild plant diversity along El Sheikh Fadl–Ras Gharib road as a part of the Egyptian Eastern desert, (2) assess the soil factors that control the roadside vegetation, and (3) identify the regional plant communities and the reasons for the vegetation changes along this desert road.

2. Materials and Methods

2.1. Site Description

This study was conducted on El Sheikh Fadl–Ras Gharib Road, Southwest Suez Gulf, in the northern sector of the Eastern Desert. The study area extends between 28°14′08.50″ and 28°26′08.40″ N and 32°41′24.30″ and 31°07′14.40″ E, with about a 224.45 km length (Figure 1).

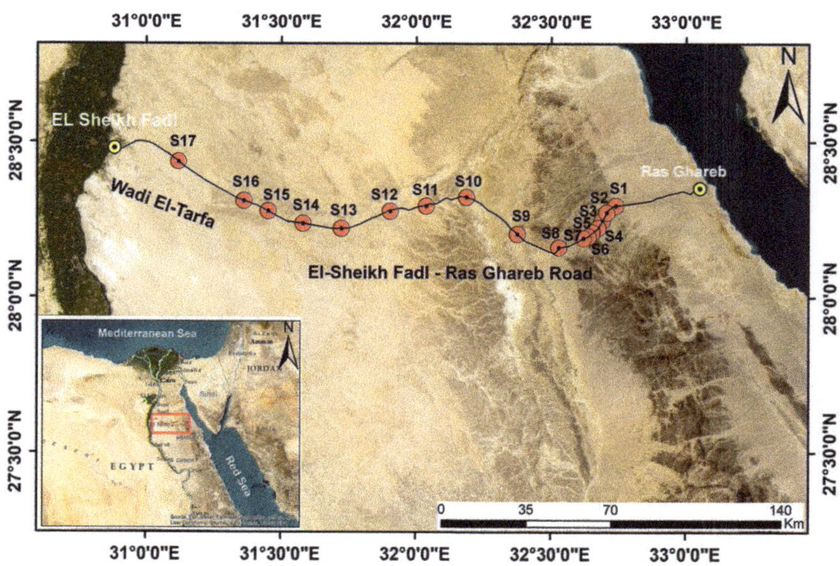

Figure 1. Location map of El Sheikh Fadl–Ras Gharib Road study area in the Eastern desert in Egypt (red rectangle) showing the sampling sites (S1–S17, in red).

The road crosses wadi El-Tarfa from the Red Sea coast in the east through the Eastern Desert in Egypt to the Nile Valley in the west. One of the most distinctive features of this road is the gravel desert, which represents a natural xeric habitat inhabited mainly by xerophytic vegetation.

Climatically, El Sheikh Fadl–Ras Gharib Road lies within a hyper-arid zone characterized by a mild winter and high daily temperature and evaporation rate [22]. The temperature is regular in seasonality, with a 26.37 °C average annual air temperature of ten years from 2013 to 2022. Winter months are cold, with a 9.90 °C minimum average air temperature in January, whereas the summer months are hot, with a 38.90 °C maximum average air temperature in August. The rainfall is scarce and sporadic and usually cascades during the winter and is due to random cloudbursts, a general feature in arid deserts [19], with an average annual precipitation of 1.69 mm/year, with a ten-year monthly mean fluctuating between 2.70 mm in December and 3.83 mm in March. However, the area may be subjected to occasional heavy rainstorms that cause dangerous flash flooding, resulting in enriching the vegetation of some dry sites on this road. The wind velocity ranges between 9.30 km·h^{-1} to 22.70 km·h^{-1} (Umm El Sas Meteorological Station; 28°30′09.00″ N, 30°44′10.0″ E) (Figure 2).

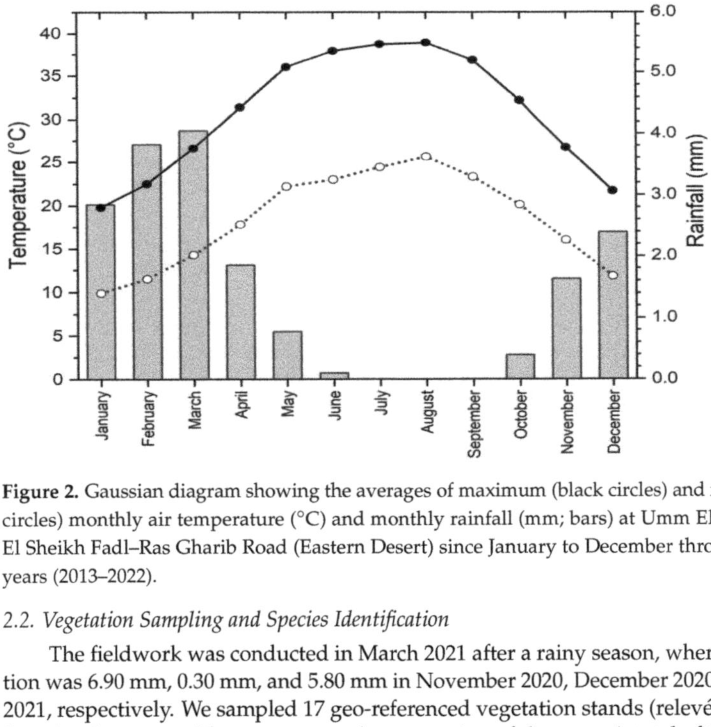

Figure 2. Gaussian diagram showing the averages of maximum (black circles) and minimum (open circles) monthly air temperature (°C) and monthly rainfall (mm; bars) at Umm El Sas city nearby El Sheikh Fadl–Ras Gharib Road (Eastern Desert) since January to December through the last ten years (2013–2022).

2.2. Vegetation Sampling and Species Identification

The fieldwork was conducted in March 2021 after a rainy season, where the precipitation was 6.90 mm, 0.30 mm, and 5.80 mm in November 2020, December 2020, and February 2021, respectively. We sampled 17 geo-referenced vegetation stands (relevés), which were found to be reasonable to represent the vegetation of the area. A total of 110 plots (each of 10 × 10 m) were randomly chosen in the selected stands (Figure 1), whenever much vegetation cover was encountered, according to the relevé method [23].

Plant taxa were collected and identified by the fifth and sixth authors (M. O. Badry and A. K. Osman) and named according to the available literature [24–28]. The taxonomic names were updated according to Plants Of The World Online [29] provided by the Royal Botanic Gardens, Kew. Growth forms of the surveyed taxa and life-form categories and their life spans were identified [30,31]. Phytogeographical affinities of the surveyed plant taxa were defined [32]. Specimens were dried, and vouchers were deposited in the herbarium of the Botany and Microbiology Department, Faculty of Science, South Valley University, Egypt.

2.3. Soil Sampling and Analyses

Two soil samples were collected at different profiles (0–25 and 25–50 cm) from randomly selected points of each relevé. The collected samples from the same site were mixed into one composite piece to analyze a given profile. Soil samples were air-dried at 105 °C in a forced-air oven for 24 to 72 h, homogenized, and passed through a 2 mm sieve to remove gravel. Soil texture was determined using the pipette method, providing quantitative data on the percentage of clay, sand, and silt [33]. The organic matter content was estimated by computing the weight loss after igniting at 600 °C for 3 h [34]. The determination of electric conductivity and pH was conducted in soil–water (1:5) extracts by the method adopted by Jackson [35]. Soil cation elements were determined. Sodium and potassium were determined by the flame emission photometry technique [36]. Calcium and magnesium were determined volumetrically by the versine titration method [37]. The analyses of soil anions included the determination of total carbonates (CO_3) and bicarbonates (HCO_3) by titration using 0.01N HCl [38], chlorides (Cl) determined volumetrically by precipitation as AgCl [35], sulfates (SO_4) estimated by turbidimetry as $BaSO_4$ using a colorimeter [39], and nitrates (NO_3) determined spectrophotometrically by sodium salicylate [40]. Total nitrogen was determined according to Bremner [41].

2.4. Data Analysis

Classification and ordination methods were applied as multivariate analysis techniques to evaluate El Sheikh Fadl–Ras Gharib Road vegetation using presence/absence data matrices. Two-Way Indicator Species Analysis (TWINSPAN) was used to classify the floristic data matrix into vegetation groups with similar species abundance patterns using the computer program CAP for Windows (community analysis package, version 1.2) [42]. The cut-off level of 'pseudo-species' followed the software's default state (one cut level set at 0). The radio box was set to Pres/Abs as the working data were presence–absence, comprising 1 s and 0 s. The maximum number of indicators was 5 per division. The maximum level of divisions was 6. The minimum group size was set to 5 per division. The relative weighting given to each pseudo-species cut level was 1 s. The indicator levels were set to all cut levels. With the minimum variance as an algorithm, a dendrogram was elaborated. The species were clustered based on the classification of the samples, following a divisive hierarchical clustering of sites. The species richness (alpha-diversity) within each TWINSPAN vegetation group was calculated as the average number of species per site, and species turnover (beta-diversity) as the ratio between the total species recorded in a certain vegetation cluster and its alpha-diversity [43]. Detrended correspondence analysis (DCA), an indirect ordination technique, was used to describe changes in the vegetation along the environmental gradients, such as altitude, water zone proximity, and soil variables [44]. Canonical correspondence analysis (CCA) was performed to depict the correlations between vegetation groups and environmental data by the CANOCO program version 4.5, using species cover, stands, and soil as variables [45]. Pearson's linear correlation coefficient (r) was used to assess the relationship of the measured edaphic variables among the vegetation assemblages using the IBM SPSS Statistics Version-20 (IBM Corp. in Armonk, NY, USA) [46].

3. Results

3.1. Floristic Composition

A total of 93 taxa of flowering plants were recorded from the study area, belonging to 63 genera in 22 families. Dicots were represented by 21 families and 87 taxa (93.55%), while monocots were represented by 1 family and by 6 taxa, representing 6.45% of the survey.

Asteraceae (17.20%), Brassicaceae (13.98), Amaranthaceae (11.83%), and Fabaceae (10.75%) were the most species-rich families (Figure 3, Appendix A).

Poaceae and Zygophyllaceae were represented by six taxa each (6.45%), while Resedaceae was represented by four species (4.30%). Three families were represented by three species each (3.23%). Meanwhile, six families were represented by two species (2.15%). On the other hand, six families were poorly represented, having one species each (1.08%). The largest families in terms of the number of genera were Asteraceae (11 genera), Brassicaceae (10 genera), and Amaranthaceae (8 genera). The most common genera with a larger number of species were *Astragalus* L. and *Zygophyllum* L., with five species each (5.38%), and *Launaea* Cass., *Diplotaxis* DC., *Lotus* L., *Reseda* Tourn. ex L., and *Tamarix* L. with three species each (3.23%). At the same time, 11 genera were represented by 2 species (2.15%). On the other hand, 45 genera were monotypic, having 1 species each (1.08%).

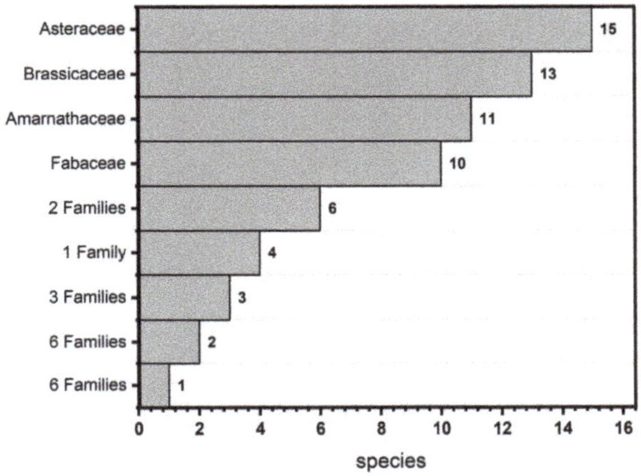

Figure 3. Histogram of the floristic composition of the 22 families surveyed in El Sheikh Fadl–Ras Gharib Road.

The surveyed plants along El Sheikh Fadl–Ras Gharib Road included 51 perennials (54.84%) and 42 annuals (45.16%), classified into 6 types of growth forms and represented by 2 trees (2.15%), 9 shrubs (9.68%), 26 subshrubs (27.96%), 1 liana (1.08%), 1 reed (1.08%) and 54 herbaceous (58.06%) species of plants (Appendix A).

3.2. *Life-Form Spectra*

Seven life-form classes were recorded in the present study. Therophytes were the most frequent life-form (42 taxa), followed by Chamaephytes (28 taxa), Hemicryptophytes (14 taxa), and Phanerophytes (6 taxa), while three life-forms were represented by single species each: Holoparasitic, (*Cistanche phelypaea* (L.) Cout.), Geophytes (*Cynodon dactylon* (L.) Pers.), and Geophytes–Helophytes (*Phragmites australis* (Cav.) Trin. ex Steud.) (Figure 4, Table S1).

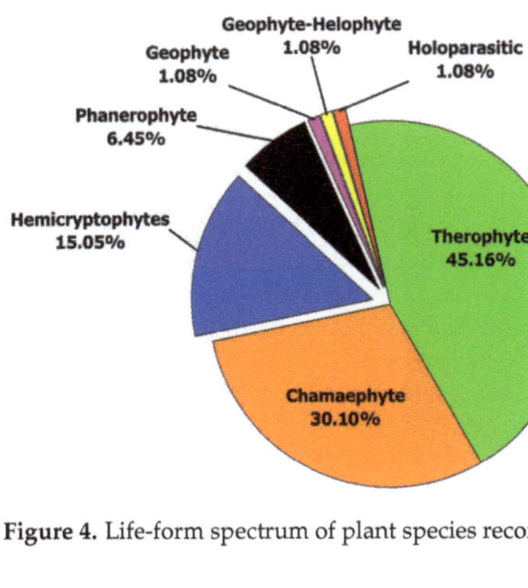

Figure 4. Life-form spectrum of plant species recorded in El Sheikh Fadl–Ras Gharib Road.

3.3. *Chorological Affinities*

The chorological analysis of the surveyed flora classified the 93 plant taxa recoded into three major phytogeographical groups: monoregional, biregional, and pluriregional. Seventy-eight taxa (83.87% of the total flora) were pluriregional elements of different affinities. These pluriregional taxa fell under 14 main chorotypes, of which 46 taxa represented the Irano-Turanian/Mediterranean/Saharo-Sindian/Sudano-Zambezian chorotypes (49.46% of the total flora), and 13 taxa represented the Irano-Turanian/Mediterranean/Saharo-Sindian chorotypes (13.97% of the total flora). The biregional elements fell under 3 main chorotypes, represented by 11 taxa (11.83% of the total flora). In contrast, the monoregional chorotype (pure Saharo-Sindian) was rarely represented in the study area, with only one species (*Matthiola arabica* Boiss.) forming 1.08% of the total number of plant species surveyed. On the other hand, the Cosmopolitan and Palaeotropical chorotypes were represented by three species (3.22%) (Figure 5, Table S2).

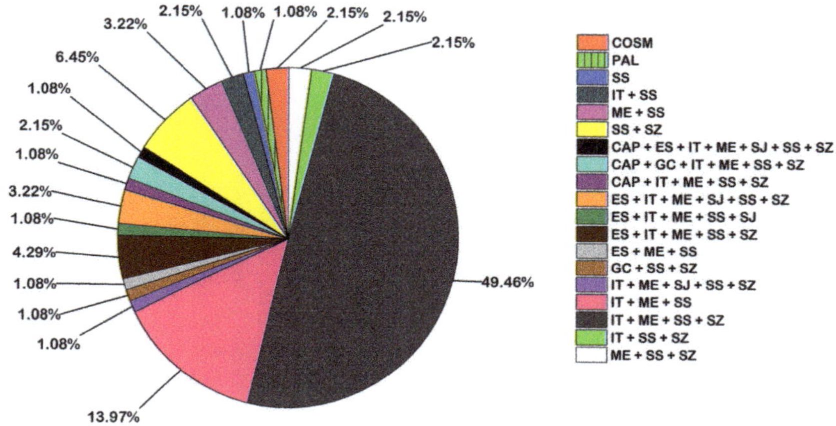

Figure 5. Phytogeographical analysis of the recorded species in El Sheikh Fadl–Ras Gharib Road. For the abbreviations, see Appendix A.

3.4. *Vegetation Analysis*

The application of the TWINSPAN classification to the 93 species recorded in the 17 stands representing the study area of El Sheikh Fadl–Ras Gharib Road resulted in grouping the vegetation into 7 clusters at level three of the hierarchical classification (Figure 6). The vegetation clusters were characterized and named after the dominant and subdominant species as follows: (I) *Halocnemum strobilaceum–Zilla spinosa;* (II) *Diplotaxis acris–Diplotaxis harra–Centaurea aegyptiaca;* (III) *Anastatica hierochuntica–Ochradenus baccatus–Rumex vesicarius;* (IV) *Caroxylon imbricatum–Ochradenus baccatus;* (V) *Artemisia judaica–Astragalus eremophilus–Atriplex turcomanica;* (VI) *Matthiola longipetala–Reseda pruinosa–Tamarix aphylla;* (VII) *Taverniera aegyptiaca–Trichodesma africanum*. These clusters were aggregated into three main groups (GA, GB, and GC).

Group A was the smallest among the separated vegetation groups. It comprised one cluster IV with three stands at the study area's beginning and end (S1, S16, and S17). It included 21 recorded species and was characterized by the lowest species richness (10 species/stands) and highest species turnover (2.10). The soils of this group were rich in Ca, K, total N, and SO_4, and recorded the highest TDS and EC (Table 1). The dominant species included *Caroxylon imbricatum, Ochradenus baccatus* (P% = 100), *Tamarix nilotica, Zygophyllum indicum, Z. coccineum, Zilla spinosa* and *Diplotaxis acris* (P% = 66.67).

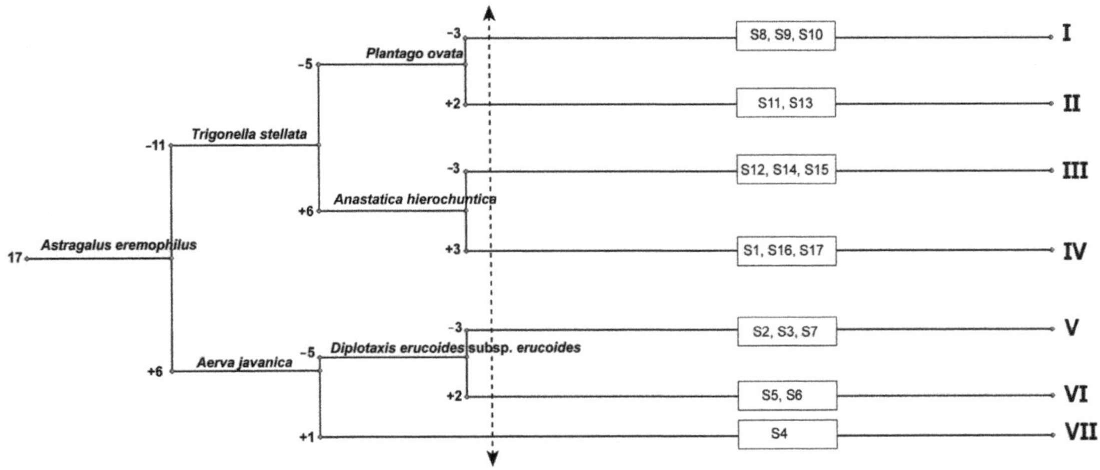

Figure 6. The dendrogram resulted from applying TWINSPAN to the 17 sampled vegetation plots (S1–S17) depicting the resulted vegetation clusters (I–VII).

Table 1. Mean values, standard deviations (SDs), and ANOVA values of the soil variables in the study area's TWINSPAN vegetation clusters (I–VII).

Edaphic Factors	I	II	III	IV	V	VI	VII	F Value	*p*
Organic matter %	2.19 ± 0.62	2.93 ± 0.82	2.36 ± 1.08	2.25 ± 1.39	2.52 ± 2.06	1.59 ± 0.35	1.38 ± 0.0	0.434	0.858
C	1.27 ± 0.36	1.70 ± 0.47	1.37 ± 0.63	1.30 ± 0.81	1.46 ± 1.20	0.92 ± 0.20	0.80 ± 0.0	0.434	0.858
N	0.027 ± 0.004	0.032 ± 0.005	0.031 ± 0.0053	0.036 ± 0.015	0.037 ± 0.01	0.028 ± 0.0	0.035 ± 0.0	0.433	0.859
$CaCO_3$ (mg/g)	30.62 ± 3.26	28.67 ± 1.29	23.00 ± 2.18	19.78 ± 6.48	17.75 ± 3.41	12.09 ± 1.53	12.50 ± 0.0	6.559 *	0.006
PH	8.44 ± 0.06	8.60 ± 0.11	8.61 ± 0.043	8.49 ± 0.19	8.47 ± 0.23	8.53 ± 0.21	8.47 ± 0.0	0.344	0.913
TDS (mg/L)	218.09 ± 32.08	563.00 ± 425.68	762.33 ± 525.65	867.74 ± 677.17	616.36 ± 633.17	137.77 ± 35.35	258.67 ± 0.0	0.647	0.711
EC (μs/cm)	363 ± 53.56	937 ± 708.52	1263 ± 861.91	1435 ± 1113.94	1025.67 ± 1053.68	228.50 ± 57.27	432 ± 0.0	0.648	0.710
Na^+ (mg/g)	1.51 ± 1.77	2.42 ± 1.17	1.70 ± 0.25	1.05 ± 1.12	2.24 ± 3.14	0.34 ± 0.20	0.36 ± 0.0	0.351	0.910
K^+ (mg/g)	1.12 ± 0.29	1.56 ± 0.42	1.62 ± 0.51	1.66 ± 0.91	1.55 ± 0.68	1.35 ± 0.47	2.27 ± 0.0	0.458	0.842
Ca^{2+} (mg/g)	0.46 ± 0.17	0.74 ± 0.53	2.87 ± 3.13	3.71 ± 3.56	1.34 ± 1.18	0.22 ± 0.0	0.62 ± 0.0	0.853	0.573
Mg^{2+} (mg/g)	0.12 ± 0.03	0.18 ± 0.14	0.85 ± 1.21	0.28 ± 0.22	0.21 ± 0.13	0.13 ± 0.56	0.10 ± 0.0	0.955	0.513
Cl^- (mg/g)	0.56 ± 0.12	1.40 ± 1.19	1.00 ± 0.07	0.91 ± 0.49	1.47 ± 1.87	0.38 ± 0.0	0.77 ± 0.0	0.341	0.915
SO_4^{2-} (mg/g)	0.49 ± 0.05	0.67 ± 0.15	0.97 ± 0.35	0.91 ± 0.62	0.60 ± 0.26	0.34 ± 0.42	0.35 ± 0.0	0.966	0.507
Clay	1.60 ± 0.46	1.20 ± 0.28	1.53 ± 0.85	1.20 ± 0.43	1.37 ± 0.15	1 ± 0.0	1 ± 0.0	2.857	0.073
Silt	10.53 ± 7.62	26.30 ± 8.34	24.40 ± 30.50	11.10 ± 14.82	14.13 ± 18.60	4.25 ± 2.75	1.90 ± 0.0	2.792	0.077
Sand	87.87 ± 7.68	72.50 ± 8.06	74.07 ± 31.35	87.70 ± 15.25	84.50 ± 18.63	94.75 ± 2.75	97.10 ± 0.0	2.859	0.072

* indicates $p < 0.01$.

Group B comprised three clusters (I, II, and III) including eight stands. Cluster I consisted of 3 stands and 44 species with a species richness of 27 and species turnover of 1.63. Cluster II consisted of 2 stands with 32 species (species richness 24.50 and turnover 1.31). Cluster III consisted of 3 stands recording 28 species; the species richness was 18.33 and species turnover was 1.53.

In this group, the soil recorded the highest Na, Mg, OM, C/N, $CaCO_3$, pH, clay, and silt (Table 1). The dominant species were *Rumex vesicarius*, *Z. spinosa* (P% = 100), *T. nilotica*, and *Z. coccineum* (P% = 87.50).

Group C was located near Ras Gharib and comprised three clusters (V, VI, VII) and six stands. Cluster V included 3 stands recording 41 species (sp. richness = 26; sp. turnover = 1.58), while cluster VI comprised 2 stands with 32 recorded species (sp. richness = 24.00 and turnover = 1.33). Cluster VII had only 1 stand with 33 species (sp. richness = 33.00 and turnover = 1.00) and high values in the sand; Cl characterized this group's soil. The dominant species were *Z. coccineum*, *Z. simplex*, *Trichodesma africanum*, *Tamarix aphylla*, *Astragalus eremophilus*, and *Artemisia judaica* (P% = 100).

The application of a DCA analysis supported the separation among the vegetation clusters (Figure 7). A distinct pattern along the gradient of DCA axes 1 and 2 (eigenvalues = 0.38 and 0.19 and cumulative percentage variance of species data = 19.30 and 29 for axes 1 and 2, respectively) indicated the relationships between the environmental gradients (proximity from water) and topographic aspects of El Sheikh Fadl–Ras Gharib Road. The TWINSPAN clusters were aggregated into three main vegetation groups, which depicted a distinct microhabitat. Group A was located at the top left part of the DCA diagram and consisted of cluster IV. Group B was segregated at the lower left part and comprised three clusters—I, II, and III—while group C included three clusters—V, VI, and VII—ordered along the gradient of DCA axis 1.

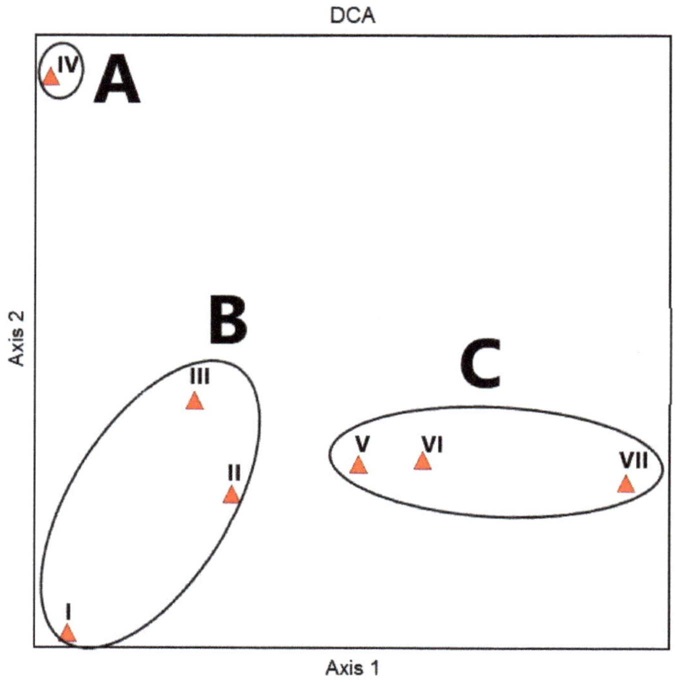

Figure 7. DCA ordination diagram on the first two axes (axes 1 and 2) of the seven vegetation clusters (I–VII, red triangles) was identified after applying TWINSPAN to the 17 sampled plots on El Sheikh Fadl–Ras Gharib Road. Black circles show DCA groups A, B, and C.

Likewise, the CCA scatter plot aggregated the seven TWINSPAN clusters into three main floristic groups (Figure 8). The length and the direction of an arrow representing a given environmental variable indicate the importance and direction of the gradient of environmental change for that variable within the set of samples measured. The cumulative percentage variance of species–environment relations for the four axes amounted to 21.5, 36.2, 46.8, and 56.2%, which suggests a strong association between vegetation and the measured parameters presented in the biplot [47]. The separation of these groups along CCA axis 1 was strongly positively affected by Na, K, Mg, PO_4, SO_3, and CO_3, and negatively by Ca, HCO_3, Cl, OM, pH, EC, and salinity. On the other hand, the separation of floristic groups along CCA axis 2 was positively correlated with Ca, Na, SO_3, PO_4, HCO_3, OM, pH, and EC, and negatively with K, Mg, Cl, CO_3, and salinity. The stands of group A were highly associated with Ca, EC, and SO_4. In comparison, the stands of group B were highly associated with Na, Mg, $CaCO_3$, silt, clay, and C/N. The stands of group C showed a high correlation with sand, K, and N.

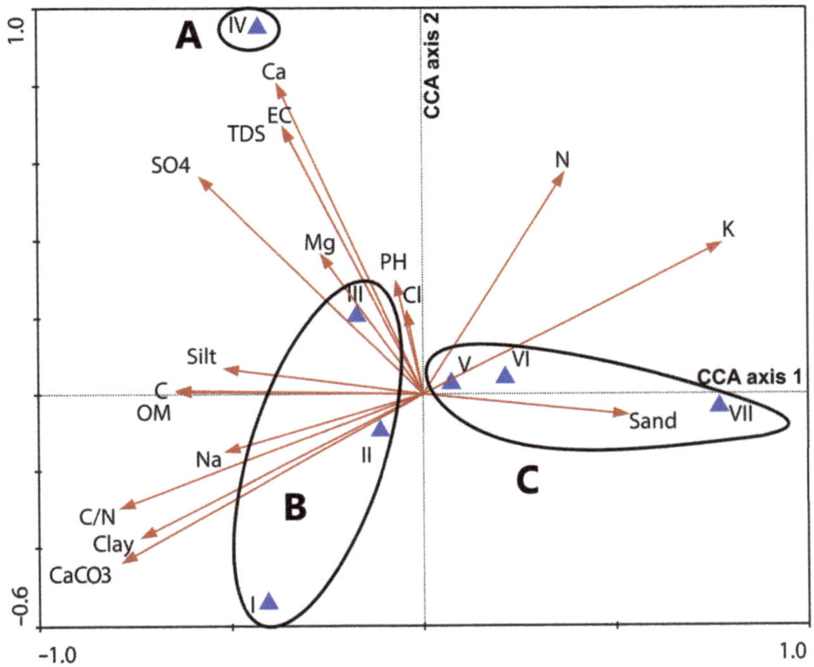

Figure 8. CCA biplot ordination of the environmental variables (arrows) and the TWINSPAN vegetation clusters (blue triangles). Floristic groups A–C. C: total carbon; N: total nitrogen; C/N: carbon–nitrogen ratio; TDS: total dissolved salts; EC: electrical conductivity; OM: organic matter; Ca: calcium; Mg: magnesium; K: potassium; Na: sodium; $CaCO_3$: calcium carbonate; Cl: chloride; SO_4: sulfate.

The correlation coefficient (*r*) between the different soil variables in the sampled stands is shown in Table 2. $CaCO_3$ showed a significant negative correlation with axis 1, while Ca ions were significantly positively correlated with axis 2. It was found that some soil variables were significantly positively correlated with other soil variables, such as OM with Na and Cl, and negatively with sand. In contrast, the TDS fraction was significantly correlated with Ca and SO_4 ions.

Table 2. Pearson correlation between soil variables and the first two CCA axes.

	AX1	AX2	OM	N	$CaCO_3$	pH	TDS	EC	Na	K	Ca	Mg	Cl	SO4	Clay	Silt	Sand
AX1		−0.300	−0.615	0.227	−0.755 *	−0.091	−0.499	−0.497	−0.451	0.644	−0.522	−0.303	−0.074	−0.666	−0.668	−0.494	0.504
AX2	−0.300		0.052	0.544	−0.263	0.100	0.716	0.713	−0.169	0.259	0.848 *	0.261	0.110	0.617	−0.246	0.011	−0.004
OM.	−0.615	0.052		0.138	0.733	0.423	0.617	0.619	0.952 **	−0.383	0.285	0.277	0.770 *	0.616	0.527	0.886 **	−0.888 **
N	0.227	0.544	0.138		−0.354	−0.154	0.590	0.592	0.160	0.652	0.461	−0.006	0.646	0.285	−0.254	0.001	0.005
$CaCO_3$	−0.755 *	−0.263	0.733	−0.354		0.188	0.218	0.219	0.697	−0.537	0.058	0.164	0.266	0.395	0.731	0.672	−0.682
pH	−0.091	0.100	0.423	−0.154	0.188		0.371	0.371	0.326	0.010	0.216	0.630	0.305	0.486	−0.007	0.742	−0.732

Table 2. *Cont.*

	AX1	AX2	OM	N	$CaCO_3$	pH	TDS	EC	Na	K	Ca	Mg	Cl	SO4	Clay	Silt	Sand
T.D.S.	−0.499	0.716	0.617	0.590	0.218	0.371		1.000 **	0.500	0.136	0.899 **	0.611	0.642	0.936 **	0.293	0.617	−0.617
EC.	−0.497	0.713	0.619	0.592	0.219	0.371	1.000 **		0.503	0.137	0.897 **	0.610	0.646	0.935 **	0.293	0.619	−0.618
Na	−0.451	−0.169	0.952 **	0.160	0.697	0.326	0.500	0.503		−0.330	0.130	0.254	0.824 *	0.479	0.602	0.844 *	−0.848 *
K	0.644	0.259	−0.383	0.652	−0.537	0.010	0.136	0.137	−0.330	1	0.148	0.021	0.189	−0.047	−0.539	−0.230	0.240
Ca	−0.522	0.848 *	0.285	0.461	0.058	0.216	0.899 **	0.897 **	0.130	0.148	1	0.647	0.256	0.898 **	0.249	0.331	−0.333
Mg	−0.303	0.261	0.277	−0.006	0.164	0.630	0.611	0.610	0.254	0.021	0.647	1	0.195	0.772 *	0.488	0.599	−0.603
Cl	−0.074	0.110	0.770 *	0.646	0.266	0.305	0.642	0.646	0.824 *	0.189	0.256	0.195	1	0.454	0.190	0.674	−0.670
SO_4	−0.666	0.617	0.616	0.285	0.395	0.486	0.936 **	0.935 **	0.479	−0.047	0.898 **	0.772 *	0.454	1	0.475	0.705	−0.708
Clay	−0.668	−0.246	0.527	−0.254	0.731	−0.007	0.293	0.293	0.602	−0.539	0.249	0.488	0.190	0.475	1	0.513	−0.531
Silt	−0.494	0.011	0.886 **	0.001	0.672	0.742	0.617	0.619	0.844 *	−0.230	0.331	0.599	0.674	0.705	0.513		−1.000 **
Sand	0.504	−0.004	−0.888 **	0.005	−0.682	−0.732	−0.617	−0.618	−0.848 *	0.240	−0.333	−0.603	−0.670	−0.708	−0.531	−1.000 **	

* indicates $p < 0.05$; ** indicates $p < 0.01$.

3.5. Plant Community and Soil Correlation

The *Taverniera aegyptiaca–Trichodesma africanum* (cluster VII) demonstrated the highest levels of species richness (33) and the lowest species turnover (1). Stands of cluster IV (*Caroxylon imbricatum–Ochradenus baccatus* community) represented the lowest species richness (10) with the highest species turnover (2.10).

In the study area, four species (*Z. indicum, Z. coccineum, Z. spinosa,* and *D. acris*) had a broad ecological amplitude and were recorded in all seven clusters. Moreover, nine species were represented in six clusters. Of them, four species did not appear in cluster IV (*Reseda pruinose, Farsetia aegyptia, D. harra,* and *Brocchia cinerea*); additionally, four species were not recorded in cluster VII (*R. vesicarius, Erodium oxyrhinchum, C. imbricatum, Atriplex turcomanica*). On the other hand, *T. africanum* appeared in all clusters except cluster III.

Seven species appeared in five clusters. They were all well-represented in cluster I but varied in the other clusters. Cluster IV was represented by only two species (*T. nilotica* and *O. baccatus,* as they both disappeared from cluster VII).

Four species were well-represented in four clusters. On the other hand, eight species were recorded in three clusters. Cluster IV was represented by only one species (*Launaea nudicaulis*).

Thirteen species appeared in only two clusters. Forty-eight species were recorded in only one cluster of this group; 14 appeared in cluster I, 11 in cluster VII, 10 in cluster V, and 1 species (*C. phelypaea*) was recorded in cluster II.

The habitats of the studied stands were composed of different soil fractions. Sand represented the greatest contributor compared to the other soil fractions. It ranged from 72.5% in cluster II to 97.1% in cluster VII. The silt ranged between 1.90% in cluster VII and 26.30% in cluster II. Clay had the lowest ratio among the other soil particles. The highest value was 1.60 in cluster I. The organic matter content in these stands was relatively constant in all clusters. The highest percentage was determined in cluster II (2.93%), while the lowest value was recorded in cluster VII (1.38%).

The total nitrogen showed very low values, not exceeding 0.037 mg/g soil (cluster V). The total dissolved salts and electric conductivity were positively correlated. The highest values were recorded in clusters III and IV, respectively, while the lowest values were attained in cluster VI. Calcium carbonate showed high percentages in the first two clusters (30.26 and 28.67% for clusters I and II, respectively).

The concentrations of soil cations varied among the different clusters. For example, Na attained high values in clusters II and V, while potassium recorded the highest concentration in cluster VII. Calcium showed the highest concentration in cluster IV (3.71 mg/g soil), but Mg was highly recorded in cluster III (0.85 mg/g soil). On the other hand, soil anions (Cl and SO_4) appeared in high concentrations in cluster V for Cl (1.47 mg/g) and cluster III for SO_4 (0.97 mg/g).

4. Discussion

The desert vegetation in arid zones is relatively homogeneous but heterogeneous on a small spatial scale. Therefore, the studies on vegetation–soil relationships in these areas were usually conducted on a small spatial scale [48,49]. In the present study area, plant life is restricted to microenvironments where runoff water collects and provides sufficient moisture for plant growth [16,50].

As in most hyper-arid desert environments, the vegetation in the study area is restricted to wadis, runnels, and depressions with deep, fine sediments that receive an adequate water supply [19]. Minimal precipitation and frequent droughts characterize the vegetation in arid regions; therefore, water availability is one of the primary factors controlling species distribution [51].

The vegetation structure in the studied area along El Sheikh Fadl–Ras Gharib Road is relatively simple, in which the species have to withstand harsh environmental conditions. The surveyed plants included 93 species from 22 different families. Asteraceae, Brassicaceae, Amaranthaceae, and Fabaceae were the richest families, constituting the most plant species (50 species; 53.76%). A study by [6] covering the Eastern Desert recorded 52 species in the same area (El Sheikh Fadl–Ras Gharib Road) and that Asteraceae was the most dominant family. The same findings were reported by [7,20,52–54].

The surveyed plants included 51 perennials and 42 annuals. The dominance of perennials (54.84% of total recorded species) may be related to the nature of the habitat types in the present study, in which reproductive capacity, ecological, morphological, and genetic plasticity are the limiting factors [55]. The high contribution of annuals (45.16% of total recorded species) can be attributed to the rain season and short life cycle that enables them to resist the instability of the agroecosystem [56].

Life-form spectra provide information that may help assess vegetation's response to variations in environmental factors. Seven life-form classes were recorded in the present study. Therophytes were the most frequent life forms (45.16%), followed by Chamaephytes (30.10%), Hemicryptophytes (15.05%), and Phanerophytes (6.45%), while three life forms were represented by a single species each (1.08%): Holoparasitic, Geophytes, and Geophytes–Helophytes. The above results agree with other studies [6,7,57]. The dominance of Therophytes over the other life forms seems to be a response to the Mediterranean climate, topography variation, and biotic influence [58]. Additionally, plant life forms showed a characteristic distribution in the different vegetation patterns. True annuals (e.g., *Astragalus vogelii, Asphodelus tenuifolius, Cotula cinerea*) occur directly after rainfall. Most other species proceed through therophytic, short-lived perennial (*Morettia philaeana, Pulicaria incisa, Z. simplex*) or even long-lived perennial life cycles (*Crotalaria aegyptiaca, Fagonia* spp., *P. incisa, Z. spinosa*) depending on the soil moisture content [59]. The highest values of Chamaephytes and Hemicryptophytes may be attributed to the ability of these species to resist drought, salinity, sand accumulation, and grazing [60].

The phytogeographical distribution of the surveyed flora classified the recorded taxa into three major phytogeographical groups: monoregional, biregional, and pluriregional. Seventy-eight taxa (83.87% of the total flora) were pluriregional elements of different affinities. Most of the recorded taxa occupied the Irano-Turanian/Mediterranean/Saharo-Sindian/Sudano-Zambezian chorotypes. These results agree with the obtained data of [6,14,20,57].

In most arid regions, multivariate analysis techniques have investigated the correlation between soil and vegetation in various habitats [51,61]. These investigations included large areas, and therefore they reported striking gradients referring to soil conditions and vegetation. Regarding vegetation and floristic composition, three main vegetation groups (plant communities) were identified and represented the plant communities that characterized the studied stands. According to [51], *Zilla spinosa–Zygophyllum coccineum* and *Haloxylon salicornicum* were the most species-rich vegetation associates occupying the Eastern Desert's northern roads. On Cairo-Desert Road, [62,63] recorded a vegetation community with a similar floristic composition, including *Acacia tortilis, Calotropis procera, Lycium shawii, Retama raetam, Tamarix nilotica, Arthrocnemum macrostachyum, Atriplex halimus, Capparis spinosa, Cornulaca monacantha, Ephedra alata, Ochradenus baccatus,* and *Pergularia tomentosa.* Most of the identified vegetation groups have very much in common with those recorded in some wadis vegetation of the Egyptian desert and the neighboring countries [64].

The first and second DCA ordination axes 1 and 2 depicted the environmental gradient expressed by the cluster analysis (Figure 7). The TWINSPAN clusters were aggregated into three main vegetation groups (A, B, and C), representing distinct microhabitats.

The CCA ordination indicated that species diversity in vegetation group A was highly associated with Ca, EC, and SO_4. At the same time, the stands of group B were highly associated with Na, Mg, $CaCO_3$, silt, clay, and C/N. The stands of group C showed a high correlation with sand, K, and N. The findings by [7] also agree with the results obtained from this study, as they reported that soil texture, salinity, and organic carbon could affect the phytodiversity of wild communities.

5. Conclusions

This study investigated the plant diversity and edaphic factors that affect the vegetation structure along the El Sheikh Fadl–Ras Gharib Road in the northeastern desert of Egypt, which has been under harsh environmental conditions, scarce and unpredictable rainfall, and anthropogenic activities that may affect their floristic and vegetation diversity. Multivariate analyses such as TWINSPAN, DCA, and CCA showed that soil variables were useful in separating three ecological species groups representing distinct microhabitats. Because of the sensitivity of desert habitats to disturbance and their slow rate of natural recovery, this study emphasized the need for incorporating plant community composition, seasonality, and different sampling methods into future studies aimed at assessing the impact of desert roads on natural desert vegetation. Moreover, our findings could be useful to detect the changes in floristic composition and improve efforts for conserving natural desert vegetation.

Supplementary Materials: The following supporting information can be downloaded at: https://www.mdpi.com/article/10.3390/d15070874/s1, Table S1: Life form spectrum of plant species recorded in El Sheikh Fadl–Ras Gharib Road; Table S2: Numbers of plant species belonging to the main floristic chorotypes and their relevant percent (%) recorded in El Sheikh Fadl–Ras Gharib Road.

Author Contributions: Conceptualization, H.S., A.K.O. and M.O.B.; methodology, H.S. and M.O.B.; formal analysis, M.N.A., A.O.A., N.L. and M.O.B.; writing—Original draft preparation, H.S., S.A.H. and M.O.B.; writing—Review and editing M.O.B.; funding acquisition, M.N.A. All authors have read and agreed to the published version of the manuscript.

Funding: This study was funded by the Princess Nourah bint Abdulrahman University Researchers Supporting Project (Grant No. PNURSP2023R103), Princess Nourah bint Abdulrahman University, Riyadh, Saudi Arabia.

Institutional Review Board Statement: Not applicable.

Data Availability Statement: Not applicable.

Acknowledgments: The authors appreciate Princess Nourah bint Abdulrahman University Researchers Supporting Project number (Grant No. PNURSP2023R103), Princess Nourah bint Abdulrahman University, Riyadh, Saudi Arabia.

Conflicts of Interest: The authors declare no conflict of interest.

Appendix A

Table A1. List of plant species recorded on El Sheikh Fadl–Ras Gharib Road along with their families, growth form, life span, life form, and chorotypes. Chorotype abbreviations: CAP: Capensis; COSM: Cosmopolitan; ES: Euro-Siberian; GC: Guino-Congo; ME: Mediterranean; PAL: Palaeotropical; PAN: Pantropical; IT: Irano-Turanian; SS: Saharo-Sindian; SJ: Sino-Japonic; SZ: Sudano-Zambezian. Life forms: Cham.: Chamaephyte; Geo.: Geophyte; Geo.–Hel.: Geophyte–Helophyte; Hemicr.: Hemicryptophyte; Holopar.: Holoparasite. Life spans: Ph.: Phanerophyte; Th.: Therophyte.

Family	Taxa	Growth Form	Life Span	Life Form	Chorotype
Acanthaceae	*Blepharis ciliaris* (L.) B.L.Burtt	Subshrub	Per.	Cham.	IT + SS + SZ
Amaranthaceae	*Aerva javanica* (Burm.f.) Juss. ex Schult.	Subshrub	Per.	Cham.	PAL
	Anabasis setifera Moq.	Subshrub	Per.	Cham.	IT + SS + SZ
	Atriplex turcomanica (Moq.) Boiss.	Subshrub	Per.	Cham.	IT + ME+ SS + SZ
	Bassia eriophora (Schrad.) AsCham.	Herb	Ann.	Ther.	IT + SS
	Caroxylon imbricatum (Forssk.) Moq.	Shrub	Per.	Cham.	IT + ME + SS + SZ
	Caroxylon volkensii (Schweinf. & AsCham.) Akhani & Roalson	Herb	Ann.	Ther.	IT + SS
	Cornulaca aucheri Moq.	Herb	Ann.	Ther.	IT + ME + SS + SZ
	Cornulaca monacantha Delile	Subshrub	Per.	Cham.	IT + ME + SS + SZ
	Halocnemum strobilaceum (Pall.) M.Bieb.	Subshrub	Per.	Cham.	ES + IT + ME + SS + SZ
	Suaeda pruinosa Lange	Shrub	Per.	Cham.	IT + ME + SS
	Suaeda vera Forssk. ex J.F.Gmel.	Shrub	Per.	Cham.	ES + ME + SS
	Aerva javanica (Burm.f.) Juss. ex Schult.	Subshrub	Per.	Cham.	PAL
Apocynaceae	*Cynanchum acutum* L.	Climbing subshrub	Per.	Cham.	ES + IT + ME + SS + SJ
Asteraceae	*Artemisia judaica* L.	Shrub	Per.	Cham.	IT + ME + SS + SZ
	Brocchia cinerea (Delile) Vis.	Herb	Ann.	Ther.	IT + ME + SS
	Centaurea aegyptiaca L.	Subshrub	Per.	Cham.	SS + SZ
	Centaurea scoparia Sieber ex Spreng.	Subshrub	Per.	Cham.	ME + SS
	Ifloga spicata (Forssk.) SCham.Bip.	Herb	Ann.	Ther.	IT + ME + SS + SZ
	Launaea capitata (Spreng.) Dandy	Herb	Ann.	Ther.	IT + ME + SS + SZ
	Launaea mucronata subsp. *cassiniana* (Jaub. & Spach) N.Kilian	Subshrub	Per.	Hemicr.	IT + ME + SS + SZ
	Launaea mucronata (Forssk.) Muschl. subsp. *mucronata*	Subshrub	Per.	Hemicr.	IT + ME + SS + SZ
	Launaea nudicaulis (L.) Hook.f.	Subshrub	Per.	Hemicr.	IT + ME + SS + SZ
	Pluchea dioscoridis (L.) DC.	Shrub	Per.	Phan.	IT + ME + SS + SZ
	Pulicaria incisa (Lam.) DC.	Herb	Per.	Hemicr.	SS + SZ
	Pulicaria undulata (L.) C.A.Mey.	Herb	Per.	Cham.	IT + ME + SS + SZ
	Reichardia tingitana (L.) Roth	Herb	Ann.	Ther.	IT + ME + SS + SZ
	Senecio glaucus subsp. *coronopifolius* (Maire) C.Alexander	Herb	Ann.	Ther.	IT + ME + SS + SZ
	Sonchus oleraceus L.	Herb	Ann.	Ther.	ES + IT + ME + SS + SZ
	Zoegea purpurea Fresen.	Herb	Ann.	Ther.	IT + ME + SS
Boraginaceae	*Gastrocotyle hispida* (Forssk.) Bunge	Herb	Ann.	Ther.	IT + ME + SS + SZ
	Trichodesma africanum (L.) Sm.	Herb	Per.	Cham.	CAP + GC + IT + ME + SS + SZ

Table A1. *Cont.*

Family	Taxa	Growth Form	Life Span	Life Form	Chorotype
Brassicaceae	*Anastatica hierochuntica* L.	Herb	Ann.	Ther.	IT + ME + SS + SZ
	Diplotaxis acris (Forssk.) Boiss.	Herb	Ann.	Ther.	IT + ME + SS + SZ
	Diplotaxis erucoides subsp. *erucoides*	Herb	Ann.	Ther.	IT + ME + SS + SZ
	Diplotaxis harra (Forssk.) Boiss.	Herb	Per.	Hemicr.	IT + ME + SS + SZ
	Eremobium aegyptiacum (Spreng.) AsCham. ex Boiss.	Herb	Ann.	Ther.	IT + ME + SS + SZ
	Farsetia aegyptia Turra	Herb	Per.	Cham.	IT + ME + SS + SZ
	Matthiola arabica Boiss.	Herb	Per.	Hemicr.	SS
	Matthiola longipetala (Vent.) DC.	Herb	Ann.	Ther.	IT + ME + SS
	Morettia philaeana (Delile) DC.	Subshrub	Per.	Hemicr.	SS + SZ
	Savignya parviflora (Delile) Webb	Herb	Ann.	Ther.	IT + ME + SS
	Schouwia purpurea (Forssk.) Schweinf.	Herb	Ann.	Ther.	IT + ME + SS + SZ
	Sisymbrium irio L.	Herb	Ann.	Ther.	ES + IT + ME + SJ + SS + SZ
	Zilla spinosa (L.) Prantl	Subshrub	Per.	Cham.	IT + ME + SS + SZ
Caryophyllaceae	*Polycarpaea repens* (Forssk.) AsCham. & Schweinf.	Subshrub	Per.	Hemicr.	IT + ME + SS + SZ
	Pteranthus dichotomus Forssk.	Herb	Ann.	Ther.	IT + ME + SS
	Spergularia diandra (Guss.) Heldr.	Herb	Ann.	Ther.	ES + IT + ME + SJ + SS + SZ
Cleomaceae	*Cleome africana* BotsCham.	Herb	Ann.	Ther.	IT + ME + SS
	Cleome amblyocarpa Barratte & Murb. Barratte & Murb.	Herb	Ann.	Ther.	IT + ME + SS + SZ
Convolvulaceae	*Convolvulus hystrix* Vahl	Shrub	Per.	Cham.	SS + SZ
	Convolvulus pilosellifolius Desr.	Subshrub	Per.	Cham.	IT + ME + SS + SZ
Euphorbiaceae	*Euphorbia retusa* Forssk.	Herb	Ann.	Ther.	IT + ME + SS
Fabaceae	*Astragalus arpilobus* subsp. *hauarensis* (Boiss.) Podlech	Herb	Ann.	Ther.	IT + ME + SS + SZ
	Astragalus eremophilus Boiss.	Herb	Ann.	Ther.	IT + ME + SS + SZ
	Astragalus spinosus (Forssk.) Muschl.	Subshrub	Per.	Cham.	IT + ME + SS
	Astragalus trigonus DC.	Subshrub	Per.	Cham.	IT + ME + SS
	Astragalus vogelii (Webb) Bornm.	Herb	Ann.	Ther.	IT + ME + SS + SZ
	Lotus arabicus Sol. ex L.	Herb	Ann.	Ther.	GC + SS + SZ
	Lotus halophilus Boiss. & Spruner	Herb	Ann.	Ther.	IT + ME + SS + SZ
	Lotus hebranicus Hochst. ex Brand	Subshrub	Per.	Hemicr.	SS + SZ
	Taverniera aegyptiaca Boiss.	Subshrub	Per.	Cham.	SS + SZ
	Trigonella stellata Forssk.	Herb	Ann.	Ther.	IT + ME + SS + SZ
Geraniaceae	*Erodium arborescens* (Desf.) Willd.	Subshrub	Per.	Hemicr.	ME + SS
	Erodium oxyrhinchum subsp. *bryoniifolium* (Boiss.) Schönb.-Tem.	Herb	Per.	Hemicr.	IT + ME + SS
	Monsonia nivea (Decne.) Webb	Herb	Per.	Hemicr.	IT + ME + SS + SZ
Malvaceae	*Malva parviflora* L.	Herb	Ann.	Ther.	ES + IT + ME + SS + SZ
Orobanchaceae	*Cistanche phelypaea* (L.) Cout.	Herb	Per.	Holopar.	IT + ME + SS + SZ
Plantaginaceae	*Plantago ciliata* Desf.	Herb	Ann.	Ther.	IT + ME + SS + SZ
	Plantago ovata Forssk.	Herb	Ann.	Ther.	IT + ME + SS + SZ

Table A1. *Cont.*

Family	Taxa	Growth Form	Life Span	Life Form	Chorotype
Poaceae	*Cynodon dactylon* (L.) Pers.	Herb	Per.	Geo.	COSM
	Polypogon monspeliensis (L.) Desf.	Herb	Ann.	Ther.	ES + IT + ME + SJ + SS + SZ
	Rostraria cristata (L.) Tzvelev	Herb	Ann.	Ther.	ES + IT + ME + SS + SZ
	Schismus barbatus (L.) Thell.	Herb	Ann.	Ther.	CAP + ES + IT + ME + SJ + SS + SZ
	Stipagrostis ciliata (Desf.) De Winter	Herb	Per.	Hemicr.	CAP + IT + ME + SS + SZ
	Phragmites australis (Cav.) Trin. ex Steud.	Reed	Per.	Geo.–Hel.	COSM
Polygonaceae	*Calligonum comosum* L'Hér.	Shrub	Per.	Phan.	IT + ME + SS + SZ
	Rumex vesicarius L.	Herb	Ann.	Ther.	IT + ME + SJ + SS + SZ
Resedaceae	*Ochradenus baccatus* Delile	Shrub	Per.	Phan.	IT + ME + SS + SZ
	Reseda muricata C.Presl	Subshrub	Per.	Cham.	ME + SS + SZ
	Reseda pruinosa Delile	Herb	Ann.	Ther.	ME + SS + SZ
	Reseda urnigera Webb	Herb	Ann.	Ther.	ME + SS
Solanaceae	*Hyoscyamus desertorum* (AsCham. ex Boiss.) Täckh.	Herb	Ann.	Ther.	IT + ME + SS
	Hyoscyamus muticus L.	Subshrub	Per.	Cham.	IT + ME + SS + SZ
Tamaricaceae	*Tamarix aphylla* (L.) H.Karst.	Tree	Per.	Phan.	IT + ME + SS + SZ
	Tamarix nilotica (Ehrenb.) Bunge	Tree	Per.	Phan.	IT + ME + SS + SZ
	Tamarix passerinoides Delile ex Decne.	Shrub	Per.	Phan.	IT + ME + SS + SZ
Urticaceae	*Forsskaolea tenacissima* L.	Herb	Per.	Hemicr.	IT + ME + SS + SZ
Zygophyllaceae	*Tribulus macropterus* Boiss.	Herb	Ann.	Ther.	IT + ME + SS + SZ
	Zygophyllum arabicum (L.) Christenh. & Byng	Subshrub	Per.	Cham.	IT + ME + SS + SZ
	Zygophyllum coccineum L.	Subshrub	Per.	Cham.	IT + ME + SS + SZ
	Zygophyllum indicum (Burm.f.) Christenh. & Byng	Subshrub	Per.	Cham.	IT + ME + SS + SZ
	Zygophyllum molle (Delile) Christenh. & Byng	Subshrub	Per.	Cham.	IT + ME + SS
	Zygophyllum simplex L.	Herb	Ann.	Ther.	CAP + GC + IT + ME + SS + SZ

References

1. Said, R. *The Geology of Egypt*; Routledge: Abingdon, UK, 2017.
2. Zahran, M.A.; Willis, A.J. *The Vegetation of Egypt*; Springer: Berlin/Heidelberg, Germany, 2009; Volume 2.
3. El Hadidi, M.N. Vegetation of the Nubian Desert (Nabta Region). In *Prehistory of the Eastern Sahara*; Wendorf, F., Schild, R., Eds.; Academic Press: London, UK, 1980; pp. 345–351.
4. Abd El-Ghani, M.M. Environmental Correlates of Species Distribution in Arid Desert Ecosystems of Eastern Egypt. *J. Arid Environ.* **1998**, *38*, 297–313. [CrossRef]
5. Fossati, J.; Pautou, G.; Peltier, J.P. Wadi Vegetation of the North-Eastern Desert of Egypt. *Feddes Repert.* **1998**, *109*, 313–327. [CrossRef]
6. El-Ghani, M.A.; Salama, F.; Salem, B.; El-Hadidy, A.; Abdel-Aleem, M. Biogeographical Relations of a Hyperarid Desert Flora in Eastern Egypt. *Afr. J. Ecol.* **2014**, *52*, 173–191. [CrossRef]
7. Abdelaal, M. Current Status of the Floristic Composition in Wadi Hagul, Northwest Suez Gulf, Egypt. *Rend. Lincei* **2017**, *28*, 81–92. [CrossRef]
8. Ward, D. *The Biology of Deserts*; Oxford University Press: Oxford, UK, 2016.
9. Abd El-Wahab, R.H. Plant Assemblage and Diversity Variation with Human Disturbances in Coastal Habitats of the Western Arabian Gulf. *J. Arid Land* **2016**, *8*, 787–798. [CrossRef]
10. Solbrig, O.T. The Origin and Function of Biodiversity. *Environment* **1991**, *33*, 16–38. [CrossRef]
11. Moustafa, A.A.; Abd El-Wahab, R.H.; Zaghloul, M.S. Conservation and Sustainable Use of Medicinal Plants in Arid and Semi-Arid Ecosystems of Egypt; Final Report; 1999. Available online: https://www.researchgate.net/publication/281438983_Conservation_and_Sustainable_use_of_Medicinal_Plants_in_Arid_and_Semi-arid_Ecosystems_of_Sinai_Egypt (accessed on 17 July 2023).

12. Adla, K.; Dejan, K.; Neira, D.; Dragana, Š. Degradation of Ecosystems and Loss of Ecosystem Services. In *One Health: Integrated Approach to 21st Century Challenges to Health*; Academic Press: Cambridge, MA, USA, 2022; pp. 281–327.
13. Gazoulis, I.; Antonopoulos, N.; Kanatas, P.; Karavas, N.; Bertoncelj, I.; Travlos, I. Invasive Alien Plant Species—Raising Awareness of a Threat to Biodiversity and Ecological Connectivity (EC) in the Adriatic-Ionian Region. *Diversity* **2022**, *14*, 387. [CrossRef]
14. Salama, F.M.; Ahmed, M.K.; El-Tayeh, N.A.; Hammad, S.A. Vegetation Analysis, Phenological Patterns and Chorological Affinities in Wadi Qena, Eastern Desert, Egypt. *Afr. J. Ecol.* **2012**, *50*, 193–204. [CrossRef]
15. Nakahama, N.; Hirasawa, Y.; Minato, T.; Hasegawa, M.; Isagi, Y.; Shiga, T. Recovery of Genetic Diversity in Threatened Plants through Use of Germinated Seeds from Herbarium Specimens. *Plant Ecol.* **2015**, *216*, 1635–1647. [CrossRef]
16. Zohary, M. Man and Vegetation in the Middle East. In *Man's Impact on Vegetation*; Holzner, W., Werger, M., Ikusima, I., Eds.; Junk: The Hague, The Netherlands, 1983; pp. 287–295.
17. Holzapfel, C.; Schmidt, W. Roadside Vegetation along Transects in the Judean Desert. *Isr. J. Bot.* **1990**, *39*, 263–270.
18. El-Ghani, M.M.A. Vegetation along a Transect in the Hijaz Mountains (Saudi Arabia). *Feddes Repert.* **1994**, *105*, 517–530. [CrossRef]
19. Abd El-Ghani, M.; Salama, F.; Salem, B.; El-Hadidy, A.; Abdel-Aleem, M. Phytogeography of the Eastern Desert Flora of Egypt. *Wulfenia* **2017**, *24*, 97–120.
20. El-Amier, Y.A.; Abdulkader, O.M. Vegetation and Species Diversity in the Northern Sector of Eastern Desert, Egypt. *West African J. Appl. Ecol.* **2015**, *23*, 75–95.
21. Shaltout, K.H.; El-Din, A.S. Habitat Types and Plant Communities along a Transect in the Nile Delta Region. *Feddes Repert.* **1988**, *99*, 153–162. [CrossRef]
22. Ayyad, M.A.; Ghabbour, S.I.; Goodall, D.W. Hot Deserts of Egypt and the Sudan. *Ecosyst. World* **1986**, *12*, 149–202.
23. Mueller-Dombois, D.; Ellenberg, H. *Aims and Methods of Vegetation Ecology*; John Wiley and Sons: New York, NY, USA, 1974.
24. Boulos, L. *Flora of Egypt, Volume 1: Azollaceae—Oxalidaceae*; Al Hadara Publishing: Cairo, Egypt, 1999.
25. Boulos, L. *Flora of Egypt, Volume 2: Geraniaceae–Boraginaceae*; Al Hadara Publishing: Cairo, Egypt, 2000.
26. Boulos, L. *Flora of Egypt, Volume 3: Verbenaceae-Compositae*; Al Hadara Publishing: Cairo, Egypt, 2002.
27. Boulos, L. *Flora of Egypt. Monocotyledons (Alimataceae-Orchidaceae)*; Al Hadara: Cairo, Egypt, 2005; Volume 4.
28. Täckholm, V. *Students' Flora of Egypt*, 2nd ed.; Cairo University: Cairo, Egypt, 1974.
29. POWO. Plants of the World Online. Facilitated by the Royal Botanic Gardens, Kew. Published on the Internet. Available online: http://powo.science.kew.org/ (accessed on 30 April 2023).
30. Hassib, M. Distribution of Plant Communities in Egypt. *Bull. Fac. Sci. Univ. Fouad 1* **1951**, *29*, 59–261.
31. Raunkiaer, C. *The Life Forms of Plants and Statistical Plant Geography*; Clarendon Press: Oxford, UK, 1934.
32. White, F.; Leonard, J.; Léonard, J. Phytogeographical Links between Africa and Southwest Asia. *Flora Veg. Mundi* **1991**, *IX*, 229–246.
33. Kilmer, V.J.; Alexander, L.T. Methods of Making Mechanical Analyses of Soils. *Soil Sci.* **1949**, *68*, 15–24. [CrossRef]
34. Sparks, D.L.; Helmke, P.A.; Page, A.L. Methods of Soil Analysis: Chemical Methods; SSSA. 1996. Available online: https://www.waterboards.ca.gov/waterrights/water_issues/programs/bay_delta/california_waterfix/exhibits/docs/Islands/II_41.pdf (accessed on 17 July 2023).
35. *Jackson ML Soil Chemical Analysis-Advanced Course*, 2nd ed.; Department of Soil Science, University of Madison: Madison, WI, USA, 1969.
36. Williams, V.; Twine, S. Flame Photometric Method for Sodium, Potassium and Calcium. *Mod. Methods Plant Anal.* **1960**, *5*, 3–5.
37. Johnson, C.M.; Ulrich, A. 2. Analytical Methods for Use in Plant Analysis. *Bull. Calif. Agric. Exp. Stn.* **1959**, *766*, 27–78.
38. Allen, S.E. Chemical Analysis. In *Methods in Plant Ecology*; Moore, P.D., Chapman, S.B., Eds.; Blackwell: Oxford, UK, 1986; pp. 285–344.
39. Black, C.A.; Evans, D.D.; Dinauer, R.C. *Methods of Soil Analysis*; American Society of Agronomy: Madison, WI, USA, 1965; Volume 9.
40. APHA, A.-W. *Standard Methods for the Examination of Water*; Ignatius Press: San Francisco, CA, USA, 1998; Volume 21, pp. 3–37.
41. Bremner, J.M. Nitrogen-Total. In *Methods of Soil Analysis, Part 3: Chemical Methods*; Sparks, D.L., Page, A.L., Helmke, P.A., Loeppert, R.H., Eds.; John Wiley & Sons, Ltd.: Hoboken, NJ, USA, 2018; pp. 1085–1121.
42. Henderson, P.A.; Seaby, R.M.H. *Community Analysis Package (CAP)*, Version 1.2; Pisces Conservation Ltd. IRC House: Lymington, UK, 1999.
43. Frosini, B.V. *Descriptive Measures of Ecological Diversity*; Università Cattolica del Sacro Cuore, Istituto di Statistica: Milan, Italy, 2003.
44. Eilertsen, O.; Okland, R.H.; Okland, T.; Pedersen, O. Data Manipulation and Gradient Length Estimation in DCA Ordination. *J. Veg. Sci.* **1990**, *1*, 261–270. [CrossRef]
45. ter Braak, C.J.F.; Smilauer, P. *CANOCO Reference Manual and CanoDraw for Windows User's Guide: Software for Canonical Community Ordination (Version 4.5)*; Microcomputer Power: Ithaca, NY, USA, 2002; Available online: www.canoco.com (accessed on 11 April 2023).
46. Coura, R.D.S.; Nardi, N.B. The State of the Art of Adeno-Associated Virus-Based Vectors in Gene Therapy. *Virol. J.* **2007**, *4*, 1–299. [CrossRef]
47. Jongman, R.H.G.; Ter Braak, C.J.F.; Van Tongeren, O.F.R. *Data Analysis in Community and Landscape Ecology*; Cambridge University Press: Cambridge, UK, 1995.

48. Xu, L.; Liu, H.; Chu, X.; Su, K. Desert Vegetation Patterns at the Northern Foot of Tianshan Mountains: The Role of Soil Conditions. *Flora Morphol. Distrib. Funct. Ecol. Plants* **2006**, *201*, 44–50. [CrossRef]
49. Rietkerk, M.; Ouedraogo, T.; Kumar, L.; Sanou, S.; Van Langevelde, F.; Kiema, A.; Van De Koppel, J.; Van Andel, J.; Hearne, J.; Skidmore, A.K.; et al. Fine-Scale Spatial Distribution of Plants and Resources on a Sandy Soil in the Sahel. *Plant Soil* **2002**, *239*, 69–77. [CrossRef]
50. Walter, H. The Water Supply of Desert Plants. In *The Water Relations of Plants*; Rutter, A.J., Whitehead, F.H., Eds.; John Wiley & Sons: New York, NY, USA, 1963; pp. 199–205.
51. Salama, F.; El-Ghani, M.A.; Gadallah, M.; EL NAGGAR, S.; Amro, A. Characteristics of Desert Vegetation along Four Transects in the Aridenvironment of Southern Egypt. *Turk. J. Bot.* **2016**, *40*, 59–73. [CrossRef]
52. Salama, F.; El-Ghani, M.A.; Gadallah, M.; El-Naggar, S.; Amro, A. Variations in Vegetation Structure, Species Dominance and Plant Communities in South of the Eastern Desert-Egypt. *Not. Sci. Biol.* **2014**, *6*, 41–58. [CrossRef]
53. Abd El-Ghani, M.M. Floristics and Environmental Relations in Two Extreme Desertzones of Western Egypt. *Glob. Ecol. Biogeogr.* **2000**, *9*, 499–516. [CrossRef]
54. El-Amier, Y.A.; El-Halawany, E.F.; Abdullah, T.J. Composition and Diversity of Plant Communities in Sand Formations along the Northern Coast of the Nile Delta in Egypt. *Res. J. Pharm. Biol. Chem. Sci.* **2014**, *5*, 826–847.
55. Grime, J. *Plant Strategies, Vegetation Processes, and Ecosystem Properties*, 2nd ed.; John Wiley & Sons: New York, NY, USA, 2006.
56. Alatar, A.; El-Sheikh, M.A.; Thomas, J. Vegetation Analysis of Wadi Al-Jufair, a Hyper-Arid Region in Najd, Saudi Arabia. *Saudi J. Biol. Sci.* **2012**, *19*, 357–368. [CrossRef]
57. Salama, F.; Abd El-Ghani, M.; El-Tayeh, N. Vegetation and Soil Relationships in the Inland Wadi Ecosystem of Central Eastern Desert, Egypt. *Turk. J. Bot.* **2013**, *37*, 489–498. [CrossRef]
58. Shaltout, K.H.; Al-Sodany, Y.M. Vegetation Analysis of Burullus Wetland: A RAMSAR Site in Egypt. *Wetl. Ecol. Manag.* **2008**, *16*, 421–439. [CrossRef]
59. Springuel, I.; Sheded, M.; Darius, F.; Bornkamm, R. Vegetation Dynamics in an Extreme Desert Wadi under the Influence of Episodic Rainfall. *Polish Bot. Stud.* **2006**, *22*, 459–472.
60. Danin, A. *Plants of Desert Dunes*; Springer Science & Business Media: Berlin/Heidelberg, Germany, 1996.
61. van Etten, E.J.B.; Fox, J.E.D. Vegetation—Environment Relationships of the Hamersley Ranges, a Mountainous Desert of North-West Australia. *Folia Geobot.* **2017**, *52*, 161–173. [CrossRef]
62. Kassas, M.; El-Abyad, M.S. On the Phytosociology of the Desert Vegetation of Egypt. *Ann. Arid Zone* **1962**, *1*, 54–83.
63. Kassas, M.; Imam, M. Habitat and Plant Communities in the Egyptian Desert: IV. The Gravel Desert. *J. Ecol.* **1959**, *47*, 289. [CrossRef]
64. Sheded, M.G.; Hamed, S.T.; Badry, M.O. Vegetation Analysis of Six Riverian Islands in Hyper-Arid Environments at Qena Governorate (Upper Egypt). *Acta Bot. Hung.* **2014**, *56*, 409–431. [CrossRef]

diversity

MDPI

Article

Exploring the Genetic Diversity among Weedy Rice Accessions Differing in Herbicide Tolerance and Allelopathic Potential

Swati Shrestha [1], Gourav Sharma [1], Shandrea Stallworth [1], Edilberto D. Redona [2] and Te Ming Tseng [1,*]

1 Department of Plant and Soil Sciences, Mississippi State University, Starkville, MS 39762, USA; s.shrestha@ufl.edu (S.S.); gourav1@vt.edu (G.S.); shandrea.stallworth@corteva.com (S.S.)
2 Agronomy-Plant Breeding, Delta Research and Extension Center, Mississippi State University, Stoneville, MS 38776, USA; edredona@drec.msstate.edu
* Correspondence: tt1024@msstate.edu

Abstract: Increasing agricultural productivity is indispensable to meet future food demand. Crop improvement programs rely heavily on genetic diversity. The success of weeds in the ecosystem can be attributed to genetic diversity and plasticity. Weedy rice, a major weed of rice, has diverse morphology and phenology, implying wide genetic diversity. Study was conducted to genotype weedy rice accessions (n = 54) previously phenotyped for herbicide tolerance and allelopathic potential using 30 SSR markers. Cultivated rice (CL163, REX) and allelopathic rice (RONDO, PI312777, PI338047) were also included in the study. Nei's genetic diversity among weedy rice (0.45) was found to be higher than cultivated rice (0.24) but less than allelopathic rice (0.56). The genetic relationship and population structure based on herbicide tolerance and allelopathic potential were evaluated. Herbicide-tolerant and susceptible accessions formed distinct clusters in the dendrogram, indicating their genetic variation, whereas no distinction was observed between allelopathic and non-allelopathic weedy rice accessions. Weedy rice accession B2, which was previously reported to have high allelopathy and herbicide tolerance, was genetically distinct from other weedy rice. Results from the study will help leverage weedy rice for rice improvement programs as both rice and weedy rice are closely related, thus having a low breeding barrier.

Keywords: crop-improvement; population genetics; weed suppression; sustainable weed management; palmer amaranth; glyphosate

Citation: Shrestha, S.; Sharma, G.; Stallworth, S.; Redona, E.D.; Tseng, T.M. Exploring the Genetic Diversity among Weedy Rice Accessions Differing in Herbicide Tolerance and Allelopathic Potential. *Diversity* **2022**, *14*, 44. https://doi.org/10.3390/d14010044

Academic Editor: Ilias Travlos

Received: 8 December 2021
Accepted: 8 January 2022
Published: 11 January 2022

1. Introduction

Commercial rice production in the US started in 1650′s and extended towards South America in the eighteenth century [1]. According to Allston (1846), weedy rice was introduced as a contaminant from Asia in 1846 and since then has been affecting US rice production [2]. Weedy rice belongs to the same genus and species as the cultivated rice [3], limiting the use of chemical control, as both the rice plants and weedy rice are susceptible to herbicides with the same mode of action. Weedy rice may be controlled using crop rotation with soybean, sorghum, maize, and other cultural practices like winter flooding and fallow tillage [4,5]. However, the popularity of rice monoculture among farmers in major rice growing areas makes the infestation of weedy rice more severe year after year. Further, with the widespread adoption of Clearfield™ rice, gene flow among the weedy rice and herbicide-tolerant Clearfield rice has been reported, complicating weedy rice management [6,7]. The average economic loss in rice due to weedy rice is 274 $/ha in Arkansas, USA [4].

Weedy rice infesting rice fields are genetically and morphologically diverse [8–10]. Weedy rice of diverse hull color (straw, black, brown, gold, gray), awn length, variable flag leaf length, and different maturity period are present [11]. Weedy rice also has higher shattering and variable dormancy [12]. These morphological variations in weedy rice can be attributed to their wide genetic variation. For example, Shivrain et al. (2010a) reported that black hull, brown hull, and straw hull weedy rice accessions are genetically diverse, with Nei's genetic diversity of 0.7 and weedy have higher genetic variation than cultivated rice [8]. Cao et al. (2006) reported that weedy rice from China have very high genetic diversity with heterozygosity of 0.313 and Shannon's diversity index of 0.572 [13]. High genetic diversity in the weedy rice, which corresponds to greater morphological variation, has pros and cons. Although the robust nature of weedy rice makes them one of the most difficult weeds to control, understanding the genetic basis of their robustness would provide valuable resources for crop improvement programs.

Weedy rice accessions have been evaluated morphologically and genetically for numerous biotic and abiotic stress tolerances. For example, 28 blast-resistant QTL were identified in two weedy rice accessions by Liu et al. (2015), which can be used in rice breeding programs to develop blast tolerant rice cultivars [14]. Ziska and Mc. Clung (2008) reported higher biomass (55%) and leaf area (62%) of weedy rice than the cultivated rice under elevated carbon dioxide conditions, indicating the ability of weedy rice to survive efficiently in the face of climate change and global warming [15].

Natural potential of plants to influence growth of neighboring plants, a phenomenon known as allelopathy, has been explored in cultivated rice since 1980s but limited research has been conducted on allelopathic potential of weedy rice [16]. Studies have shown that some rice cultivars have innate allelopathic potential which differs depending on its source, plant size, developmental stage, plant part used, and hull color [17,18]. Rice root exudates, rice straw (left in the field after harvest), and rice leaves, contain allelochemicals, such as momilactone, that suppress some weed species in rice fields [17,18]. Morphological screening of fifty-four weedy rice accessions from Arkansas, USA, showed that two weedy rice genotypes (B2 and B81) had higher allelopathic potential against barnyardgrass and Amazon sprangletop, and six weedy rice genotypes (B20, B2, S11, B49, B51, S59) had reduced sensitivity to glyphosate and flumioxazin [16,19]. These 54 weedy rice accessions were selected from larger pool of 208 weedy rice accessions collected from all the major rice growing regions or Arkansas based on their competitive traits including early flowering, high shattering, high plant biomass and high grain yield [20]. However, to the best of our knowledge, the genetic diversity in these weedy rice accession has not been explored. To determine an association between the phenotypic and genetic diversity among these weedy rice accessions from Arkansas, a study was conducted to evaluate genetic diversity in the fifty-four accessions using 30 SSR markers.

2. Materials and Methods

2.1. Plant Material and Genomic DNA Extraction

Seeds of 54 different weedy rice accessions previously characterized for herbicide response (glyphosate and flumioxazin) and allelopathic potential against weed species (barnyardgrass and Amazon sprangletop) were obtained from Dr. Burgos lab in Arkansas [16,17] (Supplementary Table S1). Seeds of commercial rice cultivars CL163 and REX were obtained from Dr. Redona's lab in Delta Research and Extension Centre, Stoneville, MS, USA, and seeds of allelopathic rice were provided by Dr. Gealy from Dale Bumpers National Rice Research Centre, Stuttgart, AR, USA.

For DNA extraction, fresh tissues were collected from young leaves of 5-week old plants grown in the greenhouse. For each biotype, three plants were grown and leaf tissue was collected from each plant to serve as biological replicate in the diversity study. The pots were placed randomly in the greenhouse to minimize the environmental effect during the study. The experimental setup in the greenhouse was completely randomized design. The tissues were stored at −80 °C overnight after collection, and DNA was extracted from each sample (three replicate for each biotype) using the CTAB method with slight modifications [21]. The quality and quantity of DNA were calculated using Nanodrop 2000 spectrophotometer. Extracted DNA was stored at −20 °C until PCR amplification.

2.2. Polymerase Chain Reaction (PCR) Using Simple Sequence Repeat (SSR) Markers

The DNA samples were diluted to 100 ng/μL before PCR. A total of 30 SSR primers (Supplementary Table S2) from the standard panel of 50 developed by McCouch et al. (2002) available publicly in the rice Gramene marker database http://archive.gramene.org/markers/microsat/50_ssr.html (accessed on 16 January 2018) was used for accessing the genetic diversity of the accessions [22]. These markers can evaluate the genetic similarity and differences among the *oryza* species with AA genome [23]. PCR reactions were carried out in 96 well plates with 25 μL reaction volume. The reaction mixture consisted of 12.5 μL of PCR master mix (Taq polymerase, dATP, dGTP, dCTP, dTTP, $MgCl_2$), forward primer 1μL, reverse primer 2 μL, DNA 1 μL and 8.5 μL of nuclease-free water. PCR profile used for DNA amplification was (i) Denaturation at 94 °C for 5 min (ii) 35 cycles of 94 °C for 1 min followed by annealing temperature from 55 °C to 67 °C, which was marker dependent (iii) final extension at 72 °C for 5 min. PCR products were separated in 6% polyacrylamide gels for 45 min at 180 volts and stained with 0.05% ethidium bromide. Stained gels were visualized under a UV trans-illuminator and photographed with a camera.

2.3. Data Analysis

Individual bands were considered as co-dominant markers and scored using Cross Checker 2.91 developed by Dr. J.B Buntjier 1999 [24]. The bands were scored as binary characters, with 1 for the presence of bands and 0 for the absence of bands to retain the allele information. POPGENE version 1.32 was used to obtain the number of alleles per locus (A), percentage of polymorphic loci (P), genetic distance (D), Nei's gene diversity (h), and Shannon's index (I) using a data matrix from Cross Checker. Nei's genetic distance was used to develop a dendrogram with the UPGMA algorithm to evaluate the genetic relationship among the accessions. STRUCTURE 2.3.4 was used for analyzing the population structure of the accessions with the genetic data generated by microsatellite SSR markers [25]. The data was run in STRUCTURE from K = 1 to K = 8 with three iterations for each K value and burn-in period of 100,000 and 500,000 replications. The best fit value of K was obtained using Structure Harvester, and Distruct was used for the diagrammatic representation of genetic data produced by STRUCTURE.

To determine the association between genotypes and herbicide tolerance, weedy rice accessions from Shrestha et al. (2019) with injury of less than 30% and more than 90% five weeks after treatment with glyphosate (1120 g a.i/ha) and/or flumioxazin (72 g a.i/ha) were selected for marker-trait association analysis. Likewise, to determine the association between weedy rice genotypes and their allelopathic potential, the most and least allelopathic weedy rice accessions (average growth inhibition of neighboring barnyardgrass and Amazon sprangletop by more than 45% and less than 20%, respectively) identified in the study by Shrestha et al. (2020) were selected for analysis [16,17].

3. Results

3.1. Genetic Diversity among Weedy Rice, Cultivated Rice, and Allelopathic Rice

Analysis of the markers showed that the alleles per locus ranged from 2–3 with an average of 2.9. The overall Nei's gene diversity (h) among the weedy rice, cultivated rice, and allelopathic rice used in the study was 0.45; lowest gene diversity of 0.14 detected by markers RM162 and RM118 and highest gene diversity of 0.65 detected by RM338 (Table 1). Weedy rice cultivated rice and allelopathic rice had h-value of 0.4, 0.24, and 0.56, respectively.

The mean Shannon Information Index (I) for the entire population was 0.74, and it ranged from 1.06 to 0.02. Higher the Shannon's Index, the greater is the genetic diversity in the population. Among the three groups, allelopathic rice had the highest I of 0.85, and cultivated rice had the lowest I, with a value of 0.38. Shannon's Information index for weedy rice was 0.66, indicating these had higher genetic diversity than southern rice cultivars CL163 and REX. Dendrogram based on Nei's genetic distance clustered weedy rice and rice cultivars (CL163 and REX) in one group, and the allelopathic rice were clustered separately (Figure 1). The genetic distance between weedy rice and rice cultivar was 0.13, and the genetic distance between weedy rice and allelopathic rice was 0.26.

Table 1. Genetic variation among the population (weedy rice, cultivated rice, and allelopathic rice) indicated through allele's number, Nei's gene diversity, and Shannon's Index.

Sl. No.	Marker	Observed Alleles	Nei's Gene Diversity	Shannon's Index
1	RM495	3	0.3098	0.5837
2	RM283	3	0.6431	1.0647
3	RM237	3	0.4464	0.7578
4	RM431	3	0.6464	1.0660
5	RM154	2	0.4989	0.6921
6	RM452	2	0.1896	0.3382
7	OSR13	2	0.1896	0.3382
8	RM338	3	0.6561	1.0820
9	RM514	3	0.2690	0.4828
10	RM124	3	0.5059	0.7461
11	RM507	3	0.6358	1.0537
12	RM413	3	0.1714	0.3666
13	RM161	3	0.4464	0.7578
14	RM133	3	0.6358	1.0537
15	RM162	3	0.1470	0.3267
16	RM125	3	0.4464	0.7578
17	RM455	3	0.6445	1.0627
18	RM118	3	0.1470	0.3267
19	RM408	3	0.6259	1.0299
20	RM152	3	0.4579	0.7810
21	RM44	3	0.3098	0.5837
21	RM284	3	0.3527	0.6383
23	RM433	3	0.6358	1.0537
24	RM447	3	0.3250	0.6035
25	RM316	3	0.5240	0.8829
26	RM215	3	0.6445	1.0627
27	RM271	3	0.3415	0.6331
28	RM484	3	0.5240	0.8829
29	RM536	3	0.6445	1.0627
30	RM277	3	0.3415	0.6331
	Mean	2.9	0.44	0.75

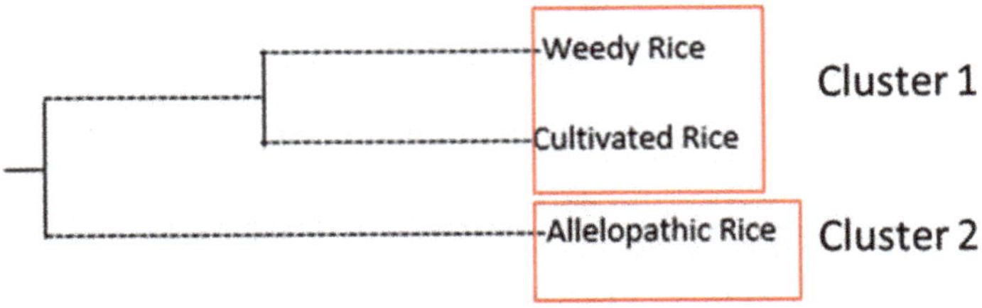

Figure 1. Dendrogram based on Nei's genetic distance indicating a genetic relationship between weedy rice, cultivated rice, and allelopathic rice. Based on Nei's genetic distance, weedy rice and cultivated rice clustered together (red box: cluster 1) and allelopathic rice (red box: cluster 2) clustered separately, indicating weedy rice and cultivated rice are more genetically similar as compared to the allelopathic rice used in the current study.

3.2. Genetic Relationship and Differentiation Based on Herbicide Tolerance

In this study, the herbicide (flumioxazin and glyphosate) tolerant and susceptible weedy rice accessions, previously described by Shrestha et al. 2019 were selected [17]. Commercial rice (CL163 and REX) which are highly susceptible to both glyphosate and flumioxazin with almost 100% injury 5 weeks after treatment with the herbicide were also included for comparison and considered as separate population (susceptible rice cultivars). The overall Nei's gene diversity among all the three populations was 0.47, varying from 0.12 to 0.66, and the Shannon's information index was 0.78 ranging from 0.2 to 1.09. Dendrogram, based on Nei's genetic distance, divided the herbicide-tolerant and susceptible accessions into four different clusters (Figure 2). Weedy rice accession ALR4 whichwas reported to be most susceptible to glyphosate and flumioxazin by Shrestha et al. (2019), clustered together with rice cultivars CL163 and REX, which are also highly susceptible to both the herbicides indicating a higher level of genetic similarity among these genotypes [17]. Accession B2, which has been reported to be highly tolerant to glyphosate by Shrestha et al. (2019), was not grouped with other herbicide-tolerant/susceptible accessions and formed a distinct cluster by itself [17]. Cluster 3 consisted of all the glyphosate and flumioxazin susceptible accessions. In contrast, cluster 2 consisted of all the accessions tolerant to herbicides indicating, association of the markers with herbicide tolerance in the current population. Individuals belonging to the same clusters had lesser genetic distance than those belonging to different clusters. The grouping of the accessions was found to be associated with herbicide tolerance, and all the herbicide-tolerant and herbicide susceptible accessions were grouped separately, implying diverse genetic backgrounds of tolerant and susceptible accessions. Results from the STRUCTURE showed correlation with the PopGene data and divided the herbicide-tolerant and susceptible accessions into K = 4 clusters, again inferring distinct clustering of accessions based on herbicide tolerance. Both black hull and straw hull herbicide-tolerant accessions showed similar coloring patterns in the figure obtained from Distruct, indicating a close genetic relationship among the weedy rice accessions in terms of herbicide tolerance irrespective of hull color (Figure 3). Likewise, both the black and straw hulled accessions susceptible to herbicides showed similarity in genetic makeup, indicating differential tolerance to herbicide is not associated with the morphological trait, hull color.

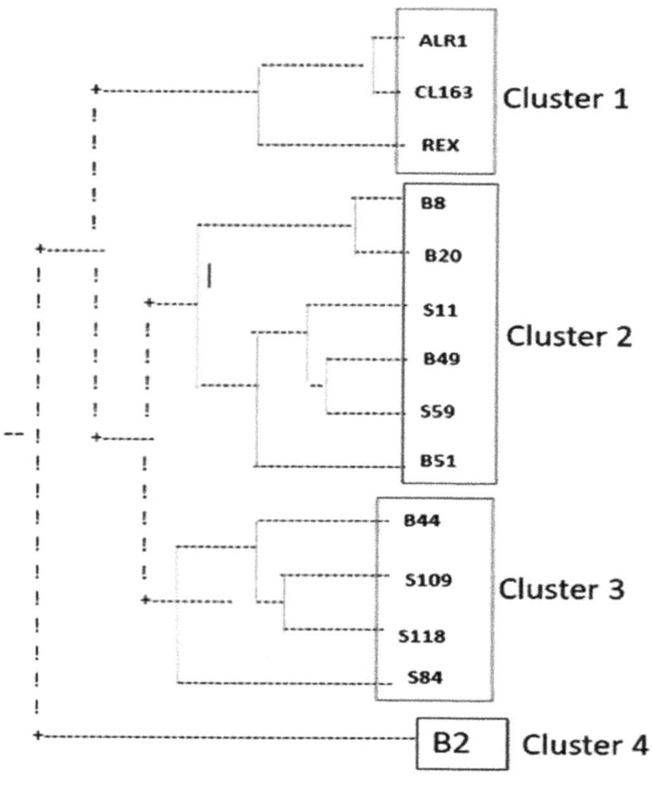

Figure 2. Dendrogram representing the relationship among the accessions with respect to herbicide tolerance. Accessions within the red box are more genetically similar as compared to accessions belonging to different red boxes, as indicated by Nei's genetic distance.

Figure 3. Population structure of accessions based on herbicide tolerance.

3.3. *Genetic Diversity among the Accessions with Respect to Allelopathic Potential*

The most and least allelopathic weedy rice accessions identified by Shrestha et al. (2020), allelopathic rice (PI338046, PI312777, Rondo), and commercial rice cultivars (CL163 and Rex) were included in this analysis [16]. The markers were not able to distinguish the accessions based on their allelopathic potential as distinct clustering pattern of allelopathic and non-allelopathic weedy rice accessions was not identified. Overall, the observed number of alleles (na) and effective number of alleles (ne) were 2.9 and 2.2, respectively. Nei's gene diversity (h) and Shannon's information index (I) for the entire population was 0.51 and 0.86, respectively. The high value of h and I indicates a high level of genetic diversity among the population. Based on Nei's genetic distance, clustering divided the accessions into three different clusters (Figure 4). Cluster one consisted of cultivated rice (CL163 and REX) and two non-allelopathic weedy rice accessions ALR-1 and ALR-4. Both CL163 and REX are reported to have no allelopathic potential; thus, cluster one had the least allelopathic accessions. Cluster two consisted of both the allelopathic and non-allelopathic accessions suggesting that the current markers were not linked strongly with the allelopathic potential of weedy rice. Cluster three consisted of allelopathic rice cultivars and one of the most allelopathic weedy rice accession B2, thus indicating genetic proximity among these accessions. Weedy rice accession B2 is reported to have high weed suppressive potential and interestingly it clustered with allelopathic rice; therefore, these might share similar genetic backgrounds. Although B2 allelopathic weedy rice was grouped with allelopathic rice cultivars, other allelopathic weedy rice did not belong to this cluster, indicating a lack of strong association of these markers with the genes controlling allelopathy in weedy rice. Population structure of the allelopathic and non-allelopathic weedy and cultivated rice showed that allelopathic rice and weedy rice had close genetic backgrounds; however, some of the allelopathic weedy rice also shared genetic similarities with non-allelopathic weedy rice (Figure 5).

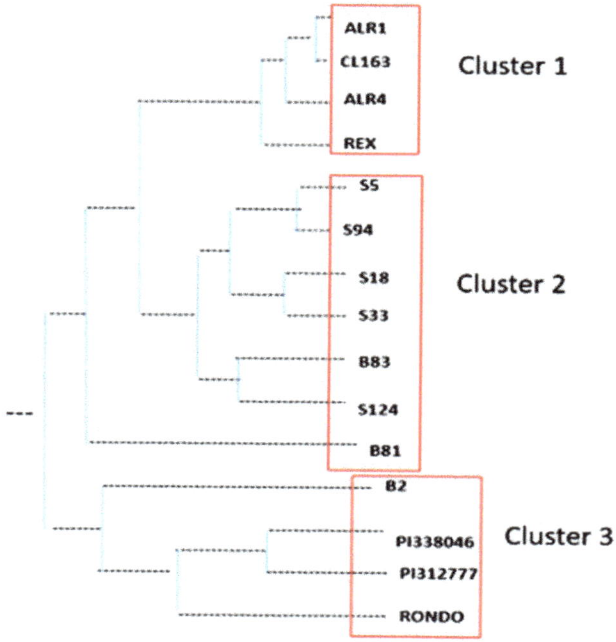

Figure 4. Dendrogram exhibiting genetic relationship among the accessions in terms of allelopathic potential. Accessions within same cluster (red box) are genetically similar as compared to the accessions in different clusters as indicated by Nei's genetic distance.

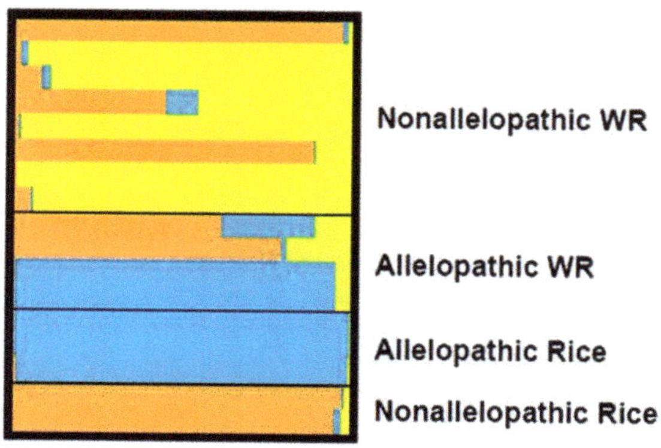

Figure 5. Population structure of accessions based on allelopathic potential.

4. Discussion

Fifty-four weedy rice accessions collected from different rice-growing regions of Arkansas, USA, were previously investigated for herbicide tolerance and allelopathic potential [16,17]. The study had identified weedy rice accessions with higher herbicide tolerance and higher weed suppressive potential. In the current study, we tried to analyze the genetic diversity in the same fifty-four weedy rice accessions along with cultivated and allelopathic rice. Nei's gene diversity measures the heterozygosity within and between the populations/individuals, and its value ranges from 0 to 1. In our study, the Nei's genetic diversity was highest in allelopathic rice and lowest in cultivated rice. Weedy rice accessions showed Nei's genetic diversity of 0.41, which is comparable to the genetic diversity of 0.31 observed in the weedy rice population from China [13]. High diversity among the weedy rice accessions might be responsible for their extensive morphological variation and adaptation in a wide range of environments [14]. The ability of weedy rice to hybridize among themselves and with the cultivated rice may have resulted in diverse genetic characteristics among the weedy rice [26]. Cultivated rice (CL 163 and REX) showed low genetic diversity of 0.24, similar to the genetic diversity of 0.26 reported for 37 rice cultivars commonly grown in Arkansas, USA, by Shivrain et al. (2010a) [8]. However, low genetic diversity of cultivated rice might also be due to only two rice cultivars used in the study. Further, both REX and CL 163 are semi-dwarf rice cultivars released in 2014/15 and 2010, respectively, primarily for cultivating in the Southern USA [27]. As both the cultivars are developed for cultivation in similar climatic conditions and are morphologically alike, lower genetic diversity among them was expected. However, for better comparison of genetic diversity between weedy rice, allelopathic rice and cultivated rice, balanced number for each of them would be ideal.

In the current study, we hypothesized that weedy rice and allelopathic rice might have higher genetic diversity than cultivated rice. The reason behind this hypothesis was that breeding for yield and other favorable characters for decades have led to narrowing of genetic diversity in cultivated rice [28]. The genetic diversity among the allelopathic rice was relatively high (0.56). Allelopathic rice cultivars used in the study, PI312777 and PI338046, are originally from the Philippines, and Rondo is an *indica* rice cultivar. Rice cultivars from Asia have high genetic diversity, which might be the reason for increased genetic diversity among the allelopathic rice used in the study [29].

Results from the dendrogram based on Nei's genetic distance indicate that weedy rice and rice cultivars CL163 and REX are closely related. Weedy rice is well adapted to flourish in the cultivated rice fields under human disturbances. As the weedy rice and cultivated rice are conspecific, gene flow from cultivated rice to weedy rice is possible [30]. The frequency of gene flow from rice cultivar (Minghui-63) to weedy rice accessions ranged from 0.011 to 0.046% [31]. As weedy rice has highly similar morphological characters with cultivated rice, the chances of gene flow between cultivated rice and weedy rice flowering simultaneously are considerably high. In such a case, genes from rice cultivars can be incorporated in the weedy rice gene pool, which might be the reason for the shorter genetic distance between the weedy rice and cultivated rice. This information suggests that weedy rice can serve as a valuable source of genetic diversity in rice improvement programs as it offers low/no genetic barrier between the two. However, the fact that weedy rice and rice are closely related to each other calls for greater attention in managing weedy rice in rice fields as high similarity among the two might create hindrance in controlling weedy rice in rice fields.

The genetic comparison of the herbicide-tolerant and susceptible weedy rice and commercial rice lines reveal that the most herbicide sensitive weedy rice accession ALR4, clustered with CL163 and REX (commercial rice lines). The single cluster of cultivated rice and herbicide-sensitive weedy rice indicates close similarity. ALR4 is a brown hull accession, and studies by Shivrain et al. (2010a) showed that brown hull accessions share a closer genetic background with cultivated rice, which may be the reason behind them clustering together in the dendrogram [8]. Further, the grouping of accession B2, which has a high tolerance to glyphosate, into a distinct cluster signifies high genetic diversity in weedy rice accessions. Accession B2 is black-hulled accession collected from Grand Prairie, AR. Weedy rice accessions from Grand Prairie are more dormant than from other locations like the White River, AR [32]. Accession B2, because of its higher dormancy period, may have been able to escape the herbicide treatments and hence survive with the rice crops in the field. This might have allowed gene flow between the cultivated rice and B2, leading to genetic changes in B2 with time. Thus, the late-emerging weedy rice seedling that grows simultaneously with cultivated rice could have a high potential of cross-pollination and introgression, leading to genetic changes in the accession isolating it from other weedy rice populations. The clustering of the accessions was associated with herbicide tolerance as all the herbicide-tolerant and herbicide-susceptible accessions grouped separately, implying diverse genetic backgrounds of tolerant and susceptible accessions. In the current study, allelopathic and non-allelopathic weedy rice accessions did not cluster separately. Allelopathy is a highly complex trait and may be associated with a minor gene effect that would not have been captured by a limited number of markers used in the current study.Additionally, in the current study only a limited number of weedy rice accessions (n = 54) were sampled. It is possible that this number of accession were not able to represt the wide genetic diversity present in weedy rice [8,9]. In the future, it is necessary to use a larger sample size, greater number of markers or employ high-throughput sequencing techniques to capture the genetic variance associated with allelopathy to get insights into the genetic basis of this trait in weedy rice.

5. Conclusions

Weedy rice belongs to the same genus and species as the cultivated rice but is more competitive than cultivated rice and can flourish in extreme environmental conditions where the cultivated rice does not perform well. Weedy rice with high herbicide tolerance and allelopathic potential have been identified but not genetically explored. The present study used 30 SSR primers to access the genetic diversity among these accessions and the molecular mechanism behind these competitive traits. Herbicide tolerance was associated with the markers irrespective of their hull color, while allelopathic potential did not show a strong association with the molecular markers used in the study. In the future, the application of whole-genome sequencing tools and genotyping by sequencing (GBS) to identify variants associated with allelopathy will help understand the genetic mechanisms of allelopathy in weedy rice.

Supplementary Materials: The following are available online at https://www.mdpi.com/article/10.3390/d14010044/s1, Table S1: List of 54 weedy rice accessions used for genotyping study, along with the herbicide tolerant and allelopathic rice identified in the study by Shrestha et al. 2019, Shrestha et al. 2020 [16,17]. Table S2: Information of 30 SSR markers used for genetic analysis in the current study.

Author Contributions: Conceptualization, T.M.T. and S.S. (Swati Shrestha); methodology, S.S. (Swati Shrestha) and S.S. (Shandrea Stallworth); formal analysis, G.S.; writing—original draft preparation, S.S. (Swati Shrestha); writing—review and editing, T.M.T., S.S. (Swati Shrestha) and E.D.R.; funding acquisition, T.M.T. All authors have read and agreed to the published version of the manuscript.

Funding: This research was funded by the Special Research Initiative Grant sponsored by the Mississippi Agricultural and Forestry Experiment Station and is based upon work that is supported by the National Institute of Food and Agriculture, U.S. Department of Agriculture, Hatch project under accession number 230100.

Institutional Review Board Statement: Not applicable.

Informed Consent Statement: Not applicable.

Data Availability Statement: Not applicable.

Conflicts of Interest: The authors declare no conflict of interest.

References

1. Singh, V.; Zhou, S.; Ganie, Z.; Valverde, B.; Avila, L.; Marchesan, E.; Merotto, A.; Zorrilla, G.; Burgos, N.; Norsworthy, J.; et al. Rice Production in the Americas. In *Rice Production Worldwide*; Springer International Publishing: Berlin/Heidelberg, Germany, 2017; pp. 137–168.
2. Allston, R.F. *The Rice Plant*; University of Michigan, Humanities Text Initiative: Ann Arbor, MI, USA, 1846.
3. Hoagland, R.E.; Paul, R.N. A Comparative SEM Study of Red Rice and Several Commerical Rice (*Oryza sativa*) Varieties. *Weed Sci.* **1978**, *26*, 619–625. [CrossRef]
4. Burgos, N.R.; Norsworthy, J.; Scott, R.C.; Smith, K.L. Red Rice (*Oryza sativa*) Status after 5 Years of Imidazolinone-Resistant Rice Technology in Arkansas. *Weed Technol.* **2008**, *22*, 200–208. [CrossRef]

5. Burgos, N.R.; Singh, V.; Tseng, T.-M.P.; Black, H.; Young, N.D.; Huang, Z.; Hyma, K.E.; Gealy, D.R.; Caicedo, A.L. The Impact of Herbicide-Resistant Rice Technology on Phenotypic Diversity and Population Structure of United States Weedy Rice. *Plant Physiol.* **2014**, *166*, 1208–1220. [CrossRef]
6. Shivrain, V.K.; Burgos, N.R.; Gealy, D.R.; A Sales, M.; Smith, K.L. Gene flow from weedy red rice (*Oryza sativa* L.) to cultivated rice and fitness of hybrids. *Pest Manag. Sci.* **2009**, *65*, 1124–1129. [CrossRef]
7. Gealy, D. Gene Movement between Rice (*Oryza sativa*) and Weedy Rice (*Oryza sativa*)—A U.S. Temperate Rice Perspective. *Crop Feral. Volunt.* **2005**, *2*, 323–354.
8. Shivrain, V.K.; Burgos, N.R.; Agrama, H.A.; Lawton-Rauh, A.; Lu, B.; Sales, M.A.; Boyett, V.; Gealy, D.R.; Moldenhauer, K.A.K. Genetic diversity of weedy red rice (*Oryza sativa*) in Arkansas, USA. *Weed Res.* **2010**, *50*, 289–302. [CrossRef]
9. Shivrain, V.K.; Burgos, N.R.; Scott, R.C.; Gbur, E.E., Jr.; Estorninos, L.E., Jr.; McClelland, M.R. Diversity of weedy red rice (*Oryza sativa* L.) in Arkansas, USA in relation to weed management. *Crop Prot.* **2010**, *29*, 721–730. [CrossRef]
10. Tseng, T.-M.; Shivrain, V.K.; Lawton-Rauh, A.; Burgos, N.R. Dormancy-linked Population Structure of Weedy Rice (*Oryza* sp.). *Weed Sci.* **2018**, *66*, 331–339. [CrossRef]
11. Noldin, J.A.; Chandler, J.M.; McCauley, G.N. Red Rice (*Oryza sativa*) Biology. I. Characterization of Red Rice Ecotypes. *Weed Technol.* **1999**, *13*, 12–18. [CrossRef]
12. Cohn, M.A.; Hughes, J.A. Seed Dormancy in Red Rice (*Oryza sativa*) I. Effect of Temperature on Dry-Afterripening. *Weed Sci.* **1981**, *29*, 402–404. [CrossRef]
13. Cao, Q.; Lu, B.-R.; Xia, H.; Rong, J.; Sala, F.; Spada, A.; Grassi, F. Genetic Diversity and Origin of Weedy Rice (*Oryza sativa* f. *spontanea*) Populations Found in North-eastern China Revealed by Simple Sequence Repeat (SSR) Markers. *Ann. Bot.* **2006**, *98*, 1241–1252. [CrossRef] [PubMed]
14. Liu, Y.; Qi, X.; Gealy, D.R.; Olsen, K.M.; Caicedo, A.L.; Jia, Y. QTL Analysis for Resistance to Blast Disease in U.S. Weedy Rice. *Mol. Plant-Microbe Interact.* **2015**, *28*, 834–844. [CrossRef] [PubMed]
15. Ziska, L.H.; McClung, A. Differential Response of Cultivated and Weedy (Red) Rice to Recent and Projected Increases in Atmospheric Carbon Dioxide. *Agron. J.* **2008**, *100*, 1259–1263. [CrossRef]
16. Shrestha, S.; Sharma, G.; Burgos, N.R.; Tseng, T.-M. Competitive ability of weedy rice: Toward breeding weed-suppressive rice cultivars. *J. Crop. Improv.* **2020**, *34*, 455–469. [CrossRef]
17. Khanh, T.D.; Chung, M.I.; Xuan, T.D.; Tawata, S. The Exploitation of Crop Allelopathy in Sustainable Agricultural Production. *J. Agron. Crop. Sci.* **2005**, *191*, 172–184. [CrossRef]
18. Ma, Y.; Zhang, M.; Li, Y.; Shui, J.; Zhou, Y. Allelopathy of rice (*Oryza sativa* L.) root exudates and its relations with *Orobanche cumana* Wallr. and Orobanche minor Sm. germination. *J. Plant Interact.* **2014**, *9*, 722–730. [CrossRef]
19. Shrestha, S.; Sharma, G.; Burgos, N.R.; Tseng, T.-M. Response of weedy rice (*Oryza* spp.) germplasm from Arkansas to glyphosate, glufosinate, and flumioxazin. *Weed Sci.* **2019**, *67*, 303–310. [CrossRef]
20. Tseng, T.M. Genetic Diversity of Seed Dormancy and Molecular Evolution of Weedy Red Rice. Ph.D. Dissertation, University of Arkansas, Fayetteville, AR, USA, 2013.
21. Doyle, J.J. A rapid DNA isolation procedure for small quantities of fresh leaf tissue. *Phytochem. Bull. Bot. Soc. Am.* **1987**, *19*, 11–15.
22. McCouch, S.R.; Teytelman, L.; Xu, Y.; Lobos, K.B.; Clare, K.; Walton, M.; Fu, B.; Maghirang, R.; Li, Z.; Xing, Y.; et al. Development and Mapping of 2240 New SSR Markers for Rice (*Oryza sativa* L.). *DNA Res.* **2002**, *9*, 199–207. [CrossRef]
23. Yang, G.P.; Maroof, M.S.; Xu, C.G.; Zhang, Q.; Biyashev, R.M. Comparative analysis of microsatellite DNA polymorphism in landraces and cultivars of rice. *Mol. Genet. Genom.* **1994**, *245*, 187–194. [CrossRef]
24. Buntjier, J.B. *Cross Checker*; Wageningen University: Wageningen, The Netherlands, 1999.
25. Pritchard, J.K.; Stephens, M.; Donnelly, P. Inference of population structure using multilocus genotype data. *Genetics* **2000**, *155*, 945–959. [CrossRef]
26. Londo, J.P.; Schaal, B.A. Origins and population genetics of weedy red rice in the USA. *Mol. Ecol.* **2007**, *16*, 4523–4535. [CrossRef] [PubMed]
27. Redona ED CL163: A New High-Amylose Content Clearfield Rice Variety. Available online: http://drec.msstate.edu/sites/default/files/cl163%20a%20new%20high-amylose%20content%20clearfield-rice%20variety.pdf (accessed on 8 September 2018).
28. Sun, C.Q.; Wang, X.K.; Li, Z.C.; Yoshimura, A.; Iwata, N. Comparison of the genetic diversity of common wild rice (Oryza rufipogon Griff.) and cultivated rice (*O. sativa* L.) using RFLP markers. *Theor. Appl. Genet.* **2001**, *102*, 157–162. [CrossRef]
29. Lapitan, V.C.; Brar, D.S.; Abe, T.; Redoña, E.D. Assessment of Genetic Diversity of Philippine Rice Cultivars Carrying Good Quality Traits using SSR Markers. *Breed. Sci.* **2007**, *57*, 263–270. [CrossRef]
30. Xia, H.-B.; Wang, W.; Xia, H.; Zhao, W.; Lu, B.-R. Conspecific Crop-Weed Introgression Influences Evolution of Weedy Rice (*Oryza sativa* f. *spontanea*) across a Geographical Range. *PLoS ONE* **2011**, *6*, e16189. [CrossRef]
31. Chen, L.J.; Lee, D.S.; Song, Z.P.; Suh, H.S.; LU, B.R. Gene flow from cultivated rice (*Oryza sativa*) to its weedy and wild relatives. *Ann. Bot.* **2004**, *93*, 67–73. [CrossRef]
32. Tseng, T.M.; Burgos, N.R.; Shivrain, V.K.; Alcober, E.A.; Mauromoustakos, A. Inter-and intrapopulation variation in dormancy of *Oryza sativa* (weedy red rice) and allelic variation in dormancy-linked loci. *Weed Res.* **2013**, *53*, 440–451. [CrossRef]

diversity MDPI

Article

Cultural Practices and Mechanical Weed Control for the Management of a Low-Diversity Weed Community in Spinach

Ioannis Gazoulis [1,*], Panagiotis Kanatas [2] and Nikolaos Antonopoulos [1]

1 Laboratory of Agronomy, Agricultural University of Athens, 11855 Athens, Greece; nikolasantwno@gmail.com
2 Department of Crop Science, University of Patras, 30200 Messolonghi, Greece; pakanatas@gmail.com
* Correspondence: giangazoulis@gmail.com

Abstract: Low-diversity weed communities are dominated by few species that are highly competitive to crops. The management of such weed communities should rely upon sustainable cultural and non-chemical practices, especially in crops such as spinach (*Spinacia oleracea* L.), where very few herbicides are available. A two-year field trial (2020 and 2021) was conducted to evaluate different fertilization practices (broadcast and banded), intra-row spacings (15 cm, 11 cm, 7 cm), and mechanical weed control treatments (untreated, one treatment, two treatments) for the management of a low-diversity weed community in spinach. Weed competition severely affected spinach commercial biomass (R^2 = 0.845). Compared to broadcast fertilization, banded fertilization reduced weed biomass and improved spinach yield and nitrogen use efficiency. Narrow intra-row spacing (7-cm) reduced weed biomass by 28 and 45% compared to intra-row spacings of 11-cm and 15-cm, respectively. Two mechanical weed control treatments resulted in 49% lower weed biomass compared to a single treatment. Commercial biomass increased with decreasing intra-row spacing (R^2 = 0.881) and increasing the number of mechanical treatments (R^2 = 0.911). More cultural and non-chemical practices should be evaluated for weed management in spinach, especially at sites infested with low-diversity weed communities.

Keywords: *Setaria viridis* (L.) P.Beauv.; *Chenopodium album* L.; banded fertilization; intra-row spacing; cage weeder; commercial biomass; nitrogen use efficiency

Citation: Gazoulis, I.; Kanatas, P.; Antonopoulos, N. Cultural Practices and Mechanical Weed Control for the Management of a Low-Diversity Weed Community in Spinach. *Diversity* **2021**, *13*, 616. https://doi.org/10.3390/d13120616

Academic Editor: Panayiotis Dimitrakopoulos

Received: 19 October 2021
Accepted: 23 November 2021
Published: 25 November 2021

1. Introduction

Spinach (*Spinacia oleracea* L.; $2n = 2x = 12$) belongs to the botanical family *Amaranthaceae* and is considered one of the most important leafy vegetables consumed in all continents of the world [1]. In Europe, spinach production reached about 700 thousand tonnes, harvested on about 41 thousand hectares in 2019 [2]. It is a valuable food source for human nutrition as its leaves are rich in minerals, vitamins, and other molecules with antioxidant properties and phenolic compounds [3]. Mature leaves can be consumed fresh or stored frozen after cooking in boiling water; cultivation can also be directed towards the production of processed spinach or fresh-cut baby leaves [4–6]. The crop can be grown both in the greenhouse and under field conditions as a cool-season leafy vegetable [7]. However, there are hybrids and cultivars that are resistant to bolting, higher temperatures, and longer photoperiods and are a viable option for summer cultivation [8].

Weeds are an important obstacle to spinach productivity, as they reduce its commercial biomass and affect the quality of the harvested product [9–11]. The impact of weed competition on spinach yield is likely to be higher in fields infested with low-diversity weed communities. Such weed communities are dominated by a small number of weed species that are highly competitive and well adapted to the soil and climatic conditions of a given agricultural area [12–14]. Storkey and Neve [13] recently emphasized that as the number of species in a given weed community decreases, crop yield losses increase. Furthermore, herbicides are not a sustainable weed control option in agricultural areas with low weed diversity because their use increases the risk of selection of herbicide-resistant weed populations [15]. Therefore, weed management in such areas should first rely upon sustainable cultural practices and non-chemical weed control methods. However, all alternative methods should aim to contribute to the restoration of weed flora composition [13]. Promoting weed species diversity should be regarded as a central issue in the development of sustainable weed management systems in agriculture [16]. This should be the target, rather than completely eliminating certain weed species from a given agricultural area and creating conditions for the invasions of other weed species that may also become dominant and highly competitive.

In any case, weed control is challenging in spinach fields since very few herbicides are available for use in this crop [17]. During past years, lenacil was the dominant herbicide for selective weed control in several spinach-producing countries in Europe [18–20]. However, it is very likely to be withdrawn from the market in European Union (EU) after the end of 2021 due to concerns for its environmental fate and impact on human health and non-target organisms [21,22]. In light of this situation, there are serious concerns among spinach growers in the EU regarding the available weed control options in the near future. Phenmedipham, a photosystem II (PS II) inhibitor, is the only option for the selective control of broadleaf weeds [23,24]. For the selective control of grass weeds, available herbicides are limited to acetyl-CoA-carboxylase (ACCase) inhibitors (i.e., cycloxydim, quizalofop-p-ethyl, and propaquizafop) [18,20]. There is much evidence that overreliance on these herbicides will lead to the rapid evolution of herbicide-resistant weed populations [25]. As a result, there is an urgent need to evaluate alternative non-chemical weed management practices to develop sustainable weed management strategies in spinach fields [26].

Increasing the populations of crop plants in the row is an effective strategy for suppressing weeds and achieving higher yields [27]. De Cauwer et al. [9] found that narrow intra-row spacings reduced weed biomass and increased total and commercial spinach biomass. Similar reports indicated that the use of increased plant populations within the row can improve the competitiveness of even low-competitive crops with slow early growth rates [28,29]. In addition, such practices can be combined with mechanical weed control treatments between crop rows to further suppress weeds [9,28–30]. The study by De Cauwer et al. [9] provided evidence on the potential of such practices for weed suppression in spinach in Central Europe. However, there are no recent relevant studies conducted on this crop in the soil and climatic conditions of the Mediterranean region. As for mechanical weed control, spring-tine harrows, split-hoes, and finger-weeders provide a sufficient level of weed control between crop rows and are safe for spinach plants grown at 20–30 cm row spacing [9,31,32]. Cage weeders are another option since they are very effective and completely safe for leafy vegetables even in early growth stages. When this kind of machinery is used, the first cage loosens the soil and the second one pulverizes it, uprooting young weed seedlings [33]. Although they are very promising for weed control in leafy vegetables, no studies have recently evaluated their performance in spinach along with crop establishment at narrow intra-row spacings.

Except for the above-mentioned practices, fertilization management is another factor that can be used for weed management purposes. Although the conventional fertilization practice in spinach is to broadcast the fertilizer at seedbed preparation, it should be remembered that weeds have a more aggressive nutrient uptake compared to crops and tend to be more competitive at higher nutrient levels [34]. Thus, broadcast fertilization should be avoided in fields with heavy weed infestation as it favors weed growth over crop growth [35]. In contrast to broadcast fertilization, banding fertilizers under the crop rows provide crops a competitive advantage over weeds by giving cultivated plants optimal access to nutrients [26]. In leafy vegetables, there is evidence that banded fertilization promotes crop growth over weed growth [36,37]. Moreover, this practice is beneficial from an agro-ecological point of view as it reduces fertilizer inputs in agriculture [38]. Considering all these reasons, banded fertilization should be considered as an alternative for fertilizing spinach and other leafy vegetables [26]. However, no studies have assessed the potential of banded fertilization for weed management purposes in spinach. There are also no studies available in this crop evaluating banded fertilization along with the factors of intra-row spacing and mechanical weed control in a single field trial.

The objective of the present study was to evaluate different fertilization practices (broadcast and banded), spinach intra-row spacings, and the number of mechanical weed control treatments for weed management in spinach. The experiment was conducted at a site infested with a low-diversity weed community dominated by two specific species. The effects of all experimental factors on spinach yield performance and nitrogen use efficiency are also presented.

2. Materials and Methods

2.1. Site Description

A two-year field trial was conducted in the prefecture of Agrinion in western Greece during the 2019–2020 and 2020–2021 growing seasons. The experimental site was located at 21°25′18″ east latitude and the north longitude was 38°32′10″. To collect precise location data, the World Geodetic System 1984 (WGS84) was the geographic coordinate system used. Spinach was the crop studied, and in particular, the hybrid *'Strongher F1'* was selected as plant material (Agrogen S.A., Athens, Koropi, Greece). This hybrid is suitable for spring cultivation in these areas and was also selected for its rapid and upright growth and resistance to 11 races of downy mildew (*Peronospora farinosa* f. sp. *spinaciae* Byford). The crop was grown in the field from April to June in both 2020 and 2021. Regarding climatic conditions during the experimental period, similar air temperature data were collected in the two growing seasons, while monthly precipitation in April and May was 29.6 and 46 mm higher, respectively, in 2020 than in 2021 (Table 1).

Table 1. Mean, mean maximum, and mean minimum monthly temperature (°C) and total monthly precipitation (mm) values observed in the experimental area.

Month	Weather Parameter							
	Mean Temperature (°C)		Mean Maximum Temperature (°C)		Mean Minimum Temperature (°C)		Total Precipitation (mm)	
	2020	2021	2020	2021	2020	2021	2020	2021
March	12.7	11.2	18.1	16.1	7.3	6.3	53.0	69.8
April	14.8	15.1	20.7	20.7	8.9	9.5	39.8	10.2
May	21.0	21.2	27.3	27.4	14.7	15.0	48.8	2.8
June	23.2	25.0	29.1	31.6	17.3	18.4	25.6	20.6

The soil type was clay loam (CL) with the following characteristics (0 to 30 cm): 28.1% clay, 25.9% silt, and 46.7% sand with a pH of 7.3 and an organic matter content of 1.0%. Spinach and swiss chard (*Beta vulgaris* L. subsp. *cicla*) were the crops grown on the soil of the experimental field in the previous growing seasons. At this site, a low-diversity weed community was established in which *Setaria viridis* (L.) P.Beauv. and *Chenopodium album* L. were the dominant weed species. These species were very persistent throughout the experimental period and accounted for more than 95% of the total weed biomass in both 2020 and 2021 (data not shown). Other weed species such as *Portulaca oleracea* L., *Tribulus terrestris* L., and *Polygonum aviculare* L. were also present but at very low densities (data not shown).

2.2. Experimental Setup

For seedbed preparation, the soil was initially plowed to a depth of 35 cm on 9 April 2020 and 10 April 2021. Soil plowing was followed by disc harrowing and spring-tooth harrowing to break up soil clods. A *'Krosker'* cultipacker (Agricultural Machinery—S. Milonas 1983 O.E., Thessaloniki, Adendro, Greece) was also run on the field as a final operation to prepare a firm seedbed. Spinach was sown on 18 April 2020 and 20 April

2021 using a manual precision seed drill *'SJ Expert'* suitable for sowing leafy vegetables (Sepeba Ibra, Les Grès, Saint Mar-tin du Fouilloux, France). The crop was sown in rows that were spaced 30 cm apart, at a sowing rate of 25 kg seed ha^{-1} and a sowing depth of 2.5 cm. After sowing, the soil was cultipacked again to achieve good seed-soil contact, and an initial irrigation was carried out with sprinklers [39].

The experiment was conducted in a three-factorial (split-split-plot arrangement) randomized complete block design (RCBD) with four replications (blocks). Two fertilization practices were assigned to the main plots, three spinach intra-row spacings were assigned to the subplots, and three different mechanical weed control treatments were assigned to the sub-subplots. Each sub-subplot included ten rows of spinach and was 3 m wide and 5 m long, giving a total sub-subplot size of 15 m^2. The subplots were 12 m wide and 15 m long giving a total subplot size of 180 m^2, while the main plots were 36 m wide and 15 m long giving a total main plot size of 540 m^2. The experimental layout included 72 experimental units (sub-subplots), and the total acreage of spinach in the experimental area was 1.080 m^2. Borders of 0.60, 1.2, and 2.4 m were also kept between adjacent sub-subplots, subplots, and main plots, respectively, in both 2020 and 2021. In 2021, new plots were established in an adjacent but different area of the same site to avoid any residual effects of the fertilizer on the studied parameters and to actually repeat the experiment in time.

The different fertilization regimes included the broadcast and banded applications of a synthetic complete fertilizer (N-P-K: 12-12-12 + 35 SO_3 + 10% organic matter; Organofert®, Hellagrolip S.A., Athens, Paleo Faliro, Greece). For broadcast application, the fertilizer was applied after the initial tillage operation to supply the soil with 120 kg N ha^{-1}, 120 kg P_2O_5, and 120 kg K_2O ha^{-1}. For banded application, the fertilizer was placed in bands at 10 cm depth below crop rows to supply the soil with 66 kg N ha^{-1}, 66 kg P_2O_5 ha^{-1}, and 66 kg K_2O ha^{-1}. Spinach emerged on 25 April 2020 and 26 April 2021, and the stand was hand-thinned (on 3 May in both years) when the plants had reached the 3-leaf growth stage (BBCH: 13). Thinning was done in such a way that we obtained three different intra-row spacings, namely 15 cm, 11 cm, and 7 cm.

As for mechanical weed control treatments, weeds were mechanically controlled between crop rows with a cage weeder (K.U.L.T. Kress Umweltschonende Landtechnik GmbH, Vaihingen an der Enz, Germany/E. Sanidas—Agricultural—Gardening Equipment & Tools, Goumenissa, Kilkis, Greece). All mechanical operations were conducted at a working speed of 3 km h^{-1} and a working depth of 3 cm. An untreated control was maintained where weeds remained uncontrolled, while the first weed control treatment included a single inter-row mechanical weeding at the 4-leaf growth stage of spinach (BBCH: 14). The second weed control treatment included two inter-row mechanical weedings at the 4- and 6-leaf growth stages of spinach (BBCH: 14–16). In both years, the dates for the first and second mechanical treatment were 13 and 20 May, when most weeds were between the 'white' or 'thread' (emergence) and 2-leaf growth stages (BBCH: 09–12).

As for other crop management practices applied during both growing seasons, irrigations were carried out weekly with sprinklers to supply the crop with 25 mm of water per week, depending on the rainfall events that occurred. To adequately meet the nitrogen requirements of spinach, a foliar application of completely water-soluble urea (N-P-K: 46-0-0; AGRI.FE.M. Ltd., Athens, Aspropirgos, Greece) was carried out 15 days after crop emergence. Urea was applied at a concentration of 1.5% (*v*/*w*) with a pressurized Gloria® 410 T sprayer (Gloria Haus & Gartengeraete GMBH, Witten, Germany) calibrated to deliver 210 L ha^{-1} of spray solution at a constant pressure of 400 kPa through a brass hollow-cone nozzle (2 mm diameter; 80° spray angle) to provide 69 kg N ha^{-1} to the crop. Spinach anthranose [*Colletotrichum dematium* f. sp. *spinaciae* (Ellis & Halsted) Arx] was controlled by two foliar pyraclostrobin (Signum® 26,7/6,7 WG, Basf Hellas S.A., Athens Greece) applications at 67 g ai ha^{-1} (400 L ha^{-1} spray solution; 275 kPa pressure) at the 4- and 6-leaf growth stages of spinach (BBCH: 14–16). Sulfoxaflor (Closer™ 120 SC, Corteva Agriscience Hellas S.A., Athens, Greece) was also applied at 24 g ai ha^{-1} (200 L ha^{-1} spray solution; 275 kPa pressure) to control green peach aphid infections [*Myzus persicae* (Sulzer, 1776)] at the 9-leaf growth stage of spinach (BBCH: 19).

2.3. Data Collection

Four 0.25 m^2 (0.5 × 0.5 m) metallic quadrats were placed in each sub-subplot the day after the last mechanical weeding, in areas with uniform weed flora and away from the margins. Each quadrat included two crop rows resulting in a total of 8, 10, and 16 spinach plants per quadrat at intra-row distances of 15 cm, 11 cm, and 7 cm, respectively. Weeds were harvested by clipping plants to a height of 2 cm with scissors, separating them by species, and placing them in numbered plastic bags. The first weed harvest was carried out in two quadrats, at approximately three weeks after the last mechanical weed control treatment and when the spinach stand was successfully established (29 May 2020 and 30 May 2021). The weed samples were then weighed to determine the fresh weed biomass per unit area using a '*KF–H2*' digital balance (Zenith S.A., Athens, Greece). The biomass of the two dominant weed species (i.e., *S. viridis* and *C. album*) was measured separately, while the total weed biomass was also measured. To do this, the biomass of other weeds that occurred at minor densities were included in the measurements of total weed biomass. Weed biomass was reassessed the day before spinach harvest (15 June in both 2020 and 2021) in the other two quadrats in each sub-subplot.

Spinach was harvested on 16 June in both years, when the plants had reached their typical leaf mass before the onset of shoot elongation (BBCH: 49). In each sub-subplot, the entire three middle rows of spinach were harvested to measure the total (fresh) spinach biomass per unit area. Harvest was done with scissors at a height of 5 cm to reduce the proportion of petioles in the harvested vegetation. Subsequently, the harvested vegetation was manually sorted into commercial (green leaves) and non-commercial spinach biomass (cotyledons, petiole parts separated from leaves, yellow leaves, etc.). Spinach waste, the proportion (%) of non-marketable spinach biomass to total spinach biomass, and the proportion (%) of total weed biomass to total spinach biomass were also assessed. In addition, nitrogen use efficiency (NUE) was estimated as the ratio of total spinach biomass to the total amount of nitrogen supplied to the crop.

2.4. Statistical Analysis

Data from each measurement were subjected to an initial 4-factor Analysis of Variance (ANOVA) conducted at $a = 0.05$ significance level. Years, fertilization practices, intra-row spinach spacings, and mechanical weed control practices were considered as fixed effects while replications (blocks) were considered as random effects. The initial ANOVAs revealed that the effects of years on the studied parameters were not significant (p-Value ≥ 0.05). Therefore, combined data analyses were conducted over the two growing seasons (2020 and 2021) for each parameter. To run the model for each parameter, fertilization, intra-row spacing, and mechanical weed control were considered as fixed effects, while replications were considered random effects. For each factor, means were separated according to Fischer's LSD (Least Significance Difference) test at an $a = 0.05$ significance level.

To express the relationships between (a) total weed biomass and the different intra-row spacings (b) total weed biomass and the number of mechanical weed control treatments, data fitted to the second order polynomial model (Equation (1)):

$$y = a + bx + cx^2 \quad (1)$$

where a is the intercept, b and c are constants, y is the dependent variable representing total weed biomass, and x is the dependent variable representing (a) intra-row spacing or (b) mechanical weed control treatments. The same model was used to express the relationships between (a) total/commercial spinach biomass (considered as the dependent variables y) and the different intra-row spacings (considered as the dependent variable x) and also between (b) total/commercial spinach biomass (y) and the number of mechanical weed control treatments (x). To express these relationships, data were pooled over all other experimental factors when intra-row spacing, or mechanical weed control was the dependent variable in each case. In addition, the relationship between total/commercial spinach biomass and total weed biomass was expressed according to the reciprocal linear model (Equation (2)):

$$y = \frac{1}{a + bx} \tag{2}$$

where a is the intercept, b is the slope of the regression line, y is the dependent variable representing total/commercial spinach biomass, and x is the independent variable representing total weed biomass. In all regression analyses where total weed biomass was used as the dependent or independent variable, data from the second evaluation of weed biomass were used. Statgraphics Centurion XVI (Statgraphics Technologies, Inc., P.O. Box 134, The Plains, VA 20198, USA) was the statistical package used for all data analyses.

3. Results

3.1. Weed Biomass

S. viridis biomass was affected by the different fertilization practices in the first (p-Value ≤ 0.01) and the second evaluation (p-Value ≤ 0.05). *C. album* biomass was not affected by the factor of fertilization (p-Value ≥ 0.05). In addition, significant were the effects of intra-row spacing and mechanical weed control on the biomass of both *S. viridis* and *C. album* (p-Value ≤ 0.001) and also on total weed biomass in both evaluations (Table 2).

Table 2. The effects of fertilization, intra-row spacing, and mechanical weed control on the biomass of *Setaria viridis* (L.) P.Beauv. and *Chenopodium album* L., and total weed biomass. For each parameter, p-Values are shown as derived from three-factor Analysis of Variance (ANOVA) (a = 0.05).

Source	Df [1]	*p*-Value					
		S. viridis Biomass		*C. album* Biomass		Total Weed Biomass	
		Eval [2] 1	Eval 2	Eval 1	Eval 2	Eval 1	Eval 2
Fertilization (F)	1	* [4]	**	0.7824	0.7506	*	*
Error [3] (a)	3						
Intra-Row Spacing (IRS)	2	***	***	***	***	***	***
F × IRS	2	0.9945	0.9369	0.6569	0.5715	0.8732	0.8519
Error (b)	12						
Mechanical Weed Control (MWC)	2	***	***	***	***	***	***
F × WC	2	0.5870	0.2438	0.9723	0.9776	0.8794	0.8717
IRS × WC	4	0.6166	0.5279	0.5493	0.4948	0.7835	0.8216
F × IRS × MWC	4	0.8795	0.7829	0.9946	0.9955	0.9996	0.9989
Error (c)	36						
Total	71						

[1] Df; Degrees of freedom. [2] Eval; Evaluation. [3] Error (a); Block × F, Error (b); Block × IRS (F), Error (c); Block × MWC (F × IRS). [4] *, **, ***; p-Value $\leq$ 0.05, 0.01, and 0.001, respectively.

Banded fertilizer application reduced *S. viridis* biomass by 15–19% compared to broadcast fertilizer application. The intra-row spacing of 11 cm resulted in 17–23% lower fresh weight for *S. viridis* compared to the intra-row spacing of 15 cm. Selecting the intra-row spacing of 7 cm caused up to 25% reductions to the biomass of this species in comparison to 11 cm. As for mechanical weed control, one treatment reduced *S. viridis* fresh weight by more than 60% in both evaluations compared to the untreated control. In sub-subplots treated mechanically two times, *S. viridis* biomass was 47% lower than in plots treated mechanically one time (Table 3).

Table 3. *Setaria viridis* (L.) P.Beauv. and *Chenopodium album* L. biomass (g m^{-2}) data obtained for each experimental factor. Total weed biomass data (g m^{-2}) are also included. Means were separated according to Fischer's Least Significance Difference (LSD) test (a = 0.05).

Factors	Weed Biomass					
	S. viridis Biomass (g m^{-2})		*C. album* Biomass (g m^{-2})		Total Weed Biomass (g m^{-2})	
	Eval [1] 1	Eval 2	Eval 1	Eval 2	Eval 1	Eval 2
Fertilization (F)						
Broadcast	97.9 a [2]	164.9 a	62.0 a	109.3 a	164.2 a	283.4 a
Banded	82.9 b	133.6 b	61.1 a	105.2 a	146.9 b	246.7 b
LSD_F	8.58	13.63	10.24	18.45	16.23	29.40
Intra-Row Spacing (IRS)						
15 cm	114.3 a	182.2 a	85.5 a	150.7 a	204.1 a	343.5 a
11 cm	87.8 b	151.1 b	60.9 b	103.9 b	151.9 b	262.7 b
7 cm	69.2 c	114.5 c	38.3 c	67.1 c	110.6 c	188.9 c
LSD_{IRS}	15.01	24.87	12.35	21.78	26.71	47.94
Mechanical Weed Control (MWC)						
Untreated (Control)	171.9 a	279.7 a	105.2 a	180.6 a	282.2 a	472.8 a
One Treatment (1×)	65.4 b	110.4 b	53.6 b	95.3 b	122.1 b	213.0 b
Two Treatments (2×)	34.6 c	57.7 c	25.9 c	45.9 c	62.4 c	109.3 c
LSD_{MWC}	12.94	21.54	24.22	42.32	33.62	60.89

[1] Eval; Evaluation. [2] Different lowercase letters in the same column indicate significant differences between means of each factor.

The intra-row spacing of 11 cm reduced *C. album* fresh weight by 29 and 31%, in the first and the second evaluation, respectively, compared to the intra-row spacing of 15 cm. Spinach establishment with 7 cm intra-row spacing resulted in the lowest values of *C. album* biomass in both evaluations. Regarding mechanical weed control, *C. album* fresh weight decreased by 47–49% in sub-subplots receiving one treatment in comparison to untreated control sub-subplots. Two mechanical treatments resulted in the lowest biomass production in this species.

In the first and the second evaluation, total weed biomass was 11 and 13% lower, respectively, in the main plots where fertilizer was applied in bands than in the main plots receiving broadcast fertilization. The results of the first evaluation revealed that narrow intra-row spacing resulted in the lowest values of total weed biomass. Total weed biomass was highest under 15 cm intra-row spacing while intermediate values corresponded to 11 cm. Moreover, weed fresh weight per unit area was lowest in sub-subplots receiving two mechanical weed control treatments and highest in untreated sub-subplots. Intermediate values corresponded to sub-subplots receiving a single treatment. The results of the second evaluation are consistent with those of the first evaluation. There were strong relationships between total weed biomass and either intra-row spacing or mechanical weed control (Figure 1).

The expression of total weed biomass as a function of intra-row spacing was performed according to the second order polynomial model: $TWB = 76.837 + (14.4706 \times IRS) + (0.220427 \times IRS^2)$ where *TWB*; total weed biomass, *IRS*; intra-row, n = 12, $p\text{-Value}_{\text{Model}} \leq 0.001$, $R^2 = 0.843$, root mean square error (RMSE) = 31.425, and mean absolute error (MAE) = 19.286. This relationship indicated that 84.3% of the variation observed in total weed biomass was due to the different intra-row spacings of spinach (Figure 1a). Pooled over all other experimental factors, weed biomass decreased by 24% in subplots where spinach was thinned to 11 cm intra-row spacing instead of 15 cm. In addition, the narrowest intra-row spacing (7 cm) resulted in 28 and 45% lower weed fresh weight per unit area compared to the wider intra-row spacings of 11 and 15 cm, respectively.

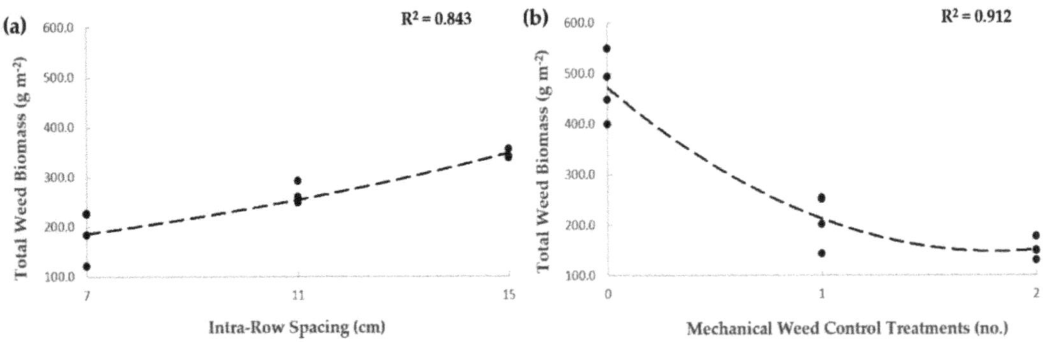

Figure 1. (**a**) The relationship between total weed biomass (g m^{-2}) and intra-row spacing (cm). (**b**) The relationship between total weed biomass (g m^{-2}) and the number of mechanical weed control treatments.

Total weed biomass was also expressed as a function of the number of mechanical weed control treatments according to the second order polynomial model: $TWB = 472.851 - (359.11 \times MWC) + (99.234 \times MWC^2)$ where *TWB*; total weed biomass, *MWC*; mechanical weed control, n = 12, $p\text{-Value}_{Model} \leq 0.001$, $R^2 = 0.912$, root mean square error (RMSE) = 48.748, and mean absolute error (MAE) = 33.745. This relationship indicated that 91.2% of the variation observed in total weed biomass was due to the number of weed control treatments (Figure 1b). Pooled over all other experimental factors, total weed biomass was 55% lower in sub-subplots treated mechanically once than in untreated sub-subplots. Moreover, two mechanical treatments resulted in 49 and 77% lower weed biomass compared to one treatment and the untreated control, respectively.

3.2. Spinach Biomass

Fertilization influenced both total and commercial spinach biomass (p-Value ≤ 0.01). The proportions of spinach waste and weed biomass to total spinach biomass were also significantly affected by the different fertilization practices (p-Value ≤ 0.05). Similar results were obtained for nitrogen use efficiency (p-Value ≤ 0.001). Significant were the effects of intra-row spacing on total and commercial spinach biomass (p-Value ≤ 0.001). The same was observed for the proportion of weed biomass in total spinach biomass and also for nitrogen use efficiency (p-Value ≤ 0.001). The proportion of weed biomass in total spinach biomass was also affected by the interaction between intra-row spacing mechanical weed control. Spinach waste proportion in total spinach biomass was another factor affected by the factor of intra-row spacing (p-Value ≤ 0.05). The factor of mechanical weed control exerted a great influence on all studied parameters (Table 4).

Before explaining the results of mean separation for each factor, it should be noted that there was a strong relationship between total spinach and total weed biomass as well as between commercial spinach and total weed biomass (Figure 2).

There was a strong negative correlation between total spinach biomass and total weed biomass described by the reciprocal linear model: $TSB = 1/[0.0284963 + (0.0000474821 \times TWB)]$ where *TSB*; total spinach biomass, *TWB*; total weed biomass, n = 72, $p\text{-Value}_{Model} \leq 0.001$, $R^2 = 0.836$, root mean square error (RMSE) = 0.00397871, and mean absolute error (MAE) = 0.00305125. Commercial spinach biomass was also negatively correlated with total weed biomass as well: $CSB = 1/[(0.0337985 + 0.0000658688 \times TWB)]$ where *CSB*; commercial spinach biomass, *TWB*; total weed biomass, n = 72, $p\text{-Value}_{Model} \leq 0.001$, $R^2 = 0.845$, root mean square error (RMSE) = 0.00533835, and mean absolute error (MAE) = 0.00413492.

Table 4. The effects of fertilization, intra-row spacing, and mechanical weed control on total spinach biomass, commercial spinach biomass, spinach waste proportion, weed biomass proportion, and nitrogen use efficiency. For each parameter, *p*-Values of are presented as derived from three-factor Analysis of Variance (ANOVA) performed (*a* = 0.05).

Source	Df [1]	Spinach Biomass				
		p-Value				
		Total Spinach Biomass	Commercial Spinach Biomass	Spinach Waste Proportion	Weed Biomass Proportion	Nitrogen Use Efficiency
Fertilization (F)	1	** [3]	**	*	*	***
Error [2] (a)	3					
Intra-Row Spacing (IRS)	2	***	***	*	***	***
F × IRS	2	0.8985	0.9598	0.8510	0.7064	0.1362
Error (b)	12					
Mechanical Weed Control (MWC)	2	***	***	***	***	***
F × MWC	2	0.8704	0.7157	0.7317	0.4892	0.1495
IRS × MWC	4	0.9944	0.9402	0.4851	*	0.8715
F × IRS × MWC	4	0.9996	0.9998	0.7078	0.9732	0.9967
Error (c)	36					
Total	71					

[1] Df; Degrees of freedom. [2] Error (a); Block × F, Error (b); Block × IRS (F), Error (c); Block × MWC (F × IRS). [3] *, **, ***; *p*-Value ≤ 0.05, 0.01, and 0.001, respectively.

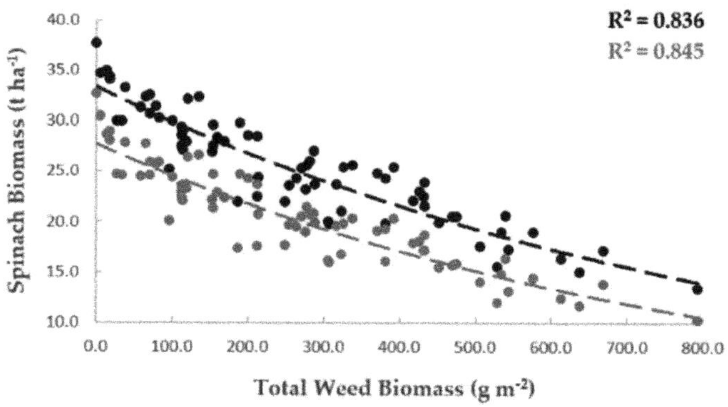

Figure 2. Black: the relationship between total spinach biomass (t ha^{-1}) and total weed biomass (g m^{-2}). Grey: the relationship between commercial spinach biomass (t ha^{-1}) and total weed biomass (g m^{-2}).

As for the effects of fertilization on crop productivity, banded fertilization increased total spinach biomass and commercial spinach biomass by 11 and 12%, respectively, compared to broadcast fertilization. Moreover, the proportions of non-commercial spinach biomass and weed biomass in total spinach biomass were significantly lower in the main plots where fertilizer was applied in bands than the values recorded in the main plots where broadcast fertilization was carried out. This agronomic practice was also beneficial for the crop since it improved its nitrogen use efficiency by 38% when applied instead of the conventional fertilization practice (Table 5).

Table 5. Total spinach biomass (t ha^{-1}), commercial spinach biomass (t ha^{-1}), spinach waste proportion (%), weed biomass proportion (%), and nitrogen use efficiency (kg kg^{-1}) data obtained for each experimental factor. Means were separated according to Fischer's Least Significance Difference (LSD) test (a = 0.05).

Factors	Spinach Biomass				
	Total Spinach Biomass (t ha^{-1})	**Commercial Spinach Biomass (t ha^{-1})**	**Spinach Waste Proportion (%)**	**Weed Biomass Proportion (%)**	**Nitrogen Use Efficiency (kg kg^{-1})**
Fertilization (F)					
Broadcast	24.1 b [1]	19.4 b	20.1 a	17.6 a	12.2 b
Banded	27.0 a	22.1 a	18.6 b	13.8 b	19.7 a
LSD_F	1.19	1.22	0.71	2.48	0.77
Intra-Row Spacing (IRS)					
15 cm	21.8 c	17.4 b	20.6 a	22.9 a	13.7 c
11 cm	26.2 b	21.5 a	18.5 b	14.2 b	16.3 b
7 cm	28.6 a	23.2 a	19.0 b	10.1 c	17.9 a
LSD_{IRS}	1.50	1.34	1.02	3.56	0.86
Mechanical Weed Control (MWC)					
Untreated (Control)	21.1 c	16.5 a	22.0 a	30.1 a	13.1 c
One Treatment (1×)	26.3 b	21.4 b	18.8 b	11.5 b	16.4 b
Two Treatments (2×)	29.2 a	24.2 a	17.3 c	5.6 c	18.3 a
LSD_{MWC}	1.96	1.57	1.00	3.79	1.22

[1] Different lowercase letters in the same column indicate significant differences between means of each factor.

Total spinach biomass was influenced by both the factors of intra-row spacing and the number of mechanical weed control treatments as well. The same was observed regarding the relationship between commercial spinach biomass and intra-row spacing. Commercial spinach biomass was also strongly related to the number of mechanical weed control treatments that were applied. The second order polynomial regressions used to express the above-mentioned relationships are presented (Figure 3).

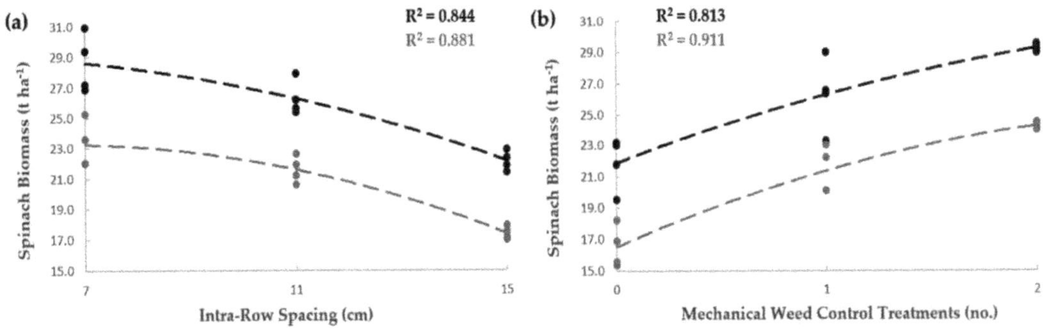

Figure 3. (**a**) Black: the relationship between total spinach biomass (t ha^{-1}) and intra-row spacing (cm). Grey: the relationship between commercial spinach biomass (t ha^{-1}) and intra-row spacing (cm). (**b**) Black: the relationship between total spinach biomass (t ha^{-1}) and the number of mechanical weed control treatments. Grey: the relationship between commercial spinach biomass (t ha^{-1}) and the number of mechanical weed control treatments.

Total spinach biomass values were related to the different intra-row spacings according to the model: $TSB = 28.3566 + (0.432474 \times IRS) - (0.0563672 \times IRS^2)$ where TSB; total spinach biomass, IRS; intra-row spacing, n = 12, $p\text{-Value}_{Model} \leq 0.001$, $R^2 = 0.844$, root mean square error (RMSE) = 1.32479, and mean absolute error (MAE) = 0.944132. The following equation was used to establish cause-effect relationships between commercial spinach biomass and intra-row spacing: $CSB = 19.9071 + (1.0405 \times IRS) - (0.0805667 \times IRS^2)$ where *CSB*; commercial spinach biomass, *IRS*; intra-row spacing, n = 12, $p\text{-Value}_{\text{Model}} \leq 0.001$, $R^2 = 0.881$, root mean square error (RMSE) = 1.04555, and mean absolute error (MAE) = 0.729403 (Figure 3a). Crop biomass was by 17% in the subplots thinned to 11 cm than in the subplots thinned to 15 cm intra-row spacing. The corresponding increase in commercial biomass reached 21%. Thinning the crop to obtain 7 cm distances between plants in the rows increased the fresh biomass of the crop by 9 and 24% in comparison to 11 and 15 cm, respectively. Commercial biomass was by 8% higher under 7 cm than under 11 cm intra-row spacing. In addition, a threefold increase was recorded under the narrowest intra-row spacing (7 cm) compared to the widest intra-row spacing (15 cm).

The equation expressing total spinach biomass as a function of mechanical weed control was: $TSB = 21.8782 + (5.1345 \times MWC) - (0.729542 \times MWC^2)$ where *TSB*; total spinach biomass, *MWC*; mechanical weed control, n = 12, $p\text{-Value}_{\text{Model}} \leq 0.001$, $R^2 = 0.813$, root mean square error (RMSE) = 1.67268, and mean absolute error (MAE) = 0.97166. It was revealed that 81.3% of the variance in total spinach biomass was due to the different number of weed control treatments. The strong relationship between commercial spinach biomass and the number of mechanical weedings is expressed as: $CSB = 16.4977 + (5.89707 \times MWC) - (1.01323 \times MWC^2)$ where *CSB*; total spinach biomass, *MWC*; mechanical weed control, n = 12, $p\text{-Value}_{\text{Model}} \leq 0.001$, $R^2 = 0.911$, root mean square error (RMSE) = 1.15514, and mean absolute error (MAE) = 0.821088 (Figure 3b). Performing one weeding increased total and commercial spinach biomass by 20 and 23%, respectively, in comparison to the untreated control. Doubling the number of mechanical treatments resulted in 9% higher crop biomass and 12% higher commercial biomass. In addition, increases of 27 and 32% were recorded for total and commercial spinach biomass, respectively, in sub-subplots receiving two mechanical operations than the values recorded in untreated plots.

As for other parameters, spinach waste proportion decreased in subplots with intra-row spacings of 7 and 11 cm compared to subplots with intra-row spacing of 15 cm. The lowest values were observed in sub-subplots treated mechanically twice while the highest values corresponded to untreated sub-subplots. Intermediate values corresponded to one weeding. Similar observations were made regarding the proportion of fresh weed biomass in total spinach biomass. In particular, the lowest values of this parameter corresponded to 7 cm intra-row spacing and two mechanical weedings. Intermediate values corresponded to 11 cm intra-row spacing and one mechanical treatment. The lowest values corresponded to subplots with 15 cm intra-row spacing and sub-subplots where weeds were left uncontrolled. Following the differences observed in spinach biomass production, nitrogen use efficiency increased by 16% under 11 cm than under 15 cm intra-row spacing. Furthermore, the highest values of this index occurred under the narrowest intra-row spacing (7 cm). Concerning the effects of mechanical weed control on nitrogen use efficiency, performing a single operation increased its value by 20%. Doubling the number of operations resulted in 9% higher nitrogen use efficiency compared to one single operation (Table 5).

4. Discussion

Combined with other experimental factors, banded fertilization reduced *S. viridis* biomass compared to broadcast fertilization. Blackshaw et al. [40] recorded a similar reduction in biomass of this species. It is, therefore, suggested that banding fertilizers at a depth of 10 cm under spinach rows (or at a depth of 7.5 cm depth under seed) is an effective method for the management of grass weeds with a shallow, fibrous rooting system such as *S. viridis* [40]. Different results were obtained for *C. album*. The little impact of banded fertilization on *C. album* can be attributed to the deep taproot of this broadleaf species, which allows nutrient uptake from deeper soil layers [40]. Although the effects of the fertilization method may depend on weed species, it was observed that banded fertilization reduced total weed biomass. Because most annual weeds germinate in the upper soil layers, broadcast fertilizer incorporation boosts their emergence and growth. In contrast, precise application of fertilizers placement at a certain depth reduces weed seed germination and also limits the access of germinated seedlings to the added nutrients [41].

Regarding the effects of spinach intra-row spacing on weed biomass, it was found that narrow intra-row spacings (7 and 11 cm) resulted in significant reductions in *S. viridis* and *C. album* biomass and total weed biomass. An explanation might be that spinach establishment at narrow intra-row spacings accelerated crop canopy closure resulting in reductions of the light transmitted to the soil surface and the weeds growing beneath the crop canopy [27]. As a result, weed emergence and consequently weed biomass decreased with decreasing intra-row spacing, and these results are consistent with the corresponding findings of another recent study on spinach. In particular, De Cauwer et al. [9] also observed negative and strong linear regressions between weed biomass and intra-row spacing in two out of their three spinach cultivars studied. Reducing the distance between plants in the row is a common practice for weed suppression in a wide variety of minor crops and leafy vegetables [26]. Our results are also in agreement with recent studies conducted with poorly competitive legumes, whose growth rates are lower compared to spinach [28,29].

The above studies have also shown that increasing the number of mechanical weed control treatments is a recommended strategy to increase the efficacy of mechanical weed control. In the present study, such strong relationships were found as weed biomass decreased with the increasing number of weed control treatments. Two mechanical weed control treatments caused significant reductions in weed biomass compared to a single treatment. The explanation is that a single mechanical weed control treatment eliminated only the first flush of germinating weed seedlings; there were several weeds that escaped the treatment and continued to emerge and develop between crop rows. On the contrary, two mechanical weedings maximized the levels of weed control because the first treatment eliminated the first flush of early germinating weed seedlings, and the second operation controlled also the weed seedlings that escaped the first treatment [29]. Our results are in agreement with those of De Cauwer et al. [9] who found that the combination of pre-emergence and post-emergence mechanical weed control reduced weed biomass in spinach compared to a single treatment. These results are encouraging because optimizing mechanical weed control methods is a priority issue in spinach, where very few herbicides are approved for use [17]. Moreover, the performance of the cage weeder used for weed control between spinach rows in the present study was satisfactory. This is due to the fact that most weeds were between the 'white' or 'thread' (emergence) and 2-leaf growth stages at the time of treatment. Cage weeders work at very shallow depth and are, therefore, effective only on early germinating weed seedlings [33]. To control weeds at later growth stages between spinach rows, spring-tine harrows, and split-hoes are examples of the equipment that can be used [31,32]. In any case, research should continue and focus on both the efficacy and the selectivity of various mechanical weed control methods on spinach and other leafy vegetables.

Other findings of the current study were the cause-effect relationships established between weed and spinach biomass. The strong influence of weed competition on spinach productivity can be attributed to the low diversity in the weed community that infested this site. This situation is common in Greece, where the lack of crop rotation and sustainable farming practices has led to the development of low-diversity weed communities at several agricultural sites in the country. There are several examples from recent studies where strong and negative correlations between crop yield and weed biomass have been observed at such agricultural sites [29,42–44]. Weed diversity has recently been highlighted as an important factor determining the impact of weed interference on crop yield [45,46]. Meta-data analysis by Storkey and Neve [13] showed that yield losses exceed 60% when the weed community is less diversified and consists of only five species. In contrast, yield losses are expected to be less than 30% when the weed community is diversified and consists of 20 weed species. The increased competitiveness of low-diversity weed communities can be attributed to the fact that they are usually dominated by species that have similar resource requirements to crops in a given agricultural area [47]. As a result, crops suffer from competition from weeds that compete strongly for the same or similar resources and are present in very high densities in the field [13,48]. Combinations of sustainable weed management practices should be implemented at such sites aiming to promote weed species diversity and restore weed flora composition. Several recent reports highlight the importance of weed diversity in mitigating crop yield losses because diverse weed communities tend to be less competitive to crops [48–51]. In addition, several studies have also indicated that increasing species diversity in weed communities is a measure to prevent the establishment of noxious invasive weeds on agricultural lands [16,52–54].

Concerning banded fertilization, it increased spinach yields compared to broadcast fertilization. This is attributed to the better access of spinach plants to nutrients due to the precise placement of fertilizer below crop rows. Our results agree with those of other researchers who found that banding fertilizers under crop rows resulted in higher crop productivity in another leafy vegetable, namely lettuce (*Lactuca sativa* L.) [36,37]. The same authors also found that banded fertilization increased crop yield at various *C. album* densities and periods of competition from this particular weed species. A similar trend was observed in the current research where although fertilization had no effect on *C. album* biomass, total and commercial spinach biomass increased in the main plots that received banded fertilization. In adition, lower fertilization inputs (approximately 45%) and higher spinach yields resulted in improved nitrogen use efficiency for the crop. Our findings are consistent with other studies where alternative fertilization methods resulted in lower inputs, higher yields and improved nitrogen use efficiency for spinach [55].

As for the effects of the other factors on crop yield, strong relationships were observed either between intra-row spacing and spinach biomass or between the number of mechanical weed control treatments and spinach biomass. Spinach establishment at narrow intra-row spacing and two mechanical treatments resulted in lower weed biomass and consequently higher total and commercial biomass. These results are in line with those of other studies that also found a positive relationship between spinach stand density and crop yield, and between the number of weed control treatments and spinach biomass production [9]. These results are also in partial agreement with other studies where higher seeding rates and multiple mechanical weeding improved crop productivity compared to conventional agronomic practices [28,29].

5. Conclusions

Banded fertilization reduced weed biomass and fertilizer inputs and subsequently improved spinach yield and nitrogen use efficiency. Weed biomass decreased as spinach intra-row spacing decreased, and the number of mechanical weed control treatments increased. Reverse trends were observed in the total and commercial biomass of spinach. Both parameters increased when spinach intra-row spacing decreased, and the number of mechanical weed control treatments increased. Further research is needed to evaluate more cultural and non-chemical weed management practices in this crop where available herbicides are limited. In addition, competition from this low-diversity weed community affected the yield performance of spinach at a significant point. Our results validated that the loss of diversity in weed communities results in severe crop yield losses. Therefore, all sustainable weed management practices should be focused on increasing species diversity in the weed communities of such agricultural areas. As recently highlighted in several relevant studies, increasing the diversity of weed communities on agricultural land can be a measure to mitigate crop yield losses due to weed competition and prevent the establishment of noxious weed species.

Author Contributions: Conceptualization, I.G. and P.K.; methodology, I.G. and P.K.; software, N.A.; validation, I.G. and N.A.; formal analysis, I.G.; investigation, P.K.; resources, N.A.; data curation, I.G.; writing—original draft preparation, I.G.; writing—review and editing, P.K.; visualization, P.K. and N.A.; supervision, I.G.; project administration, P.K. All authors have read and agreed to the published version of the manuscript.

Funding: This research received no external funding.

Institutional Review Board Statement: Not applicable.

Informed Consent Statement: Not applicable.

Data Availability Statement: The data presented in this study are available on request from the corresponding author.

Acknowledgments: We thank the anonymous reviewers for their valuable recommendations that have contributed to the improvement of this manuscript.

Conflicts of Interest: The authors declare no conflict of interest.

References

1. Morelock, T.E.; Correll, J.C. Spinach. In *Vegetables I: Asteraceae, Brassicaceae, Chenopodicaceae, and Cucurbitaceae*, 1st ed.; Prohens, J., Nuez, F., Eds.; Springer: New York, NY, USA, 2008; Volume 1, pp. 189–218. [CrossRef]
2. Food and Agriculture Organization of the United Nations. FAOSTAT. Available online: http://faostat.fao.org (accessed on 14 October 2021).
3. Roberts, J.L.; Moreau, R. Functional properties of spinach (*Spinacia oleracea* L.) phytochemicals and bioactives. *Food Funct.* **2016**, *7*, 3337–3353. [CrossRef]
4. Bunea, A.; Andjelkovic, M.; Socaciu, C.; Bobis, O.; Neacsu, M.; Verhé, R.; Van Camp, J. Total and individual carotenoids and phenolic acids content in fresh, refrigerated and processed spinach (*Spinacia oleracea* L.). *Food Chem.* **2008**, *108*, 649–656. [CrossRef]
5. Cocetta, G.; Baldassarre, V.; Spinardi, A.; Ferrante, A. Effect of cutting on ascorbic acid oxidation and recycling in fresh-cut baby spinach (*Spinacia oleracea* L.) leaves. *Postharvest Biol. Technol.* **2014**, *88*, 8–16. [CrossRef]
6. Gil, M.I.; Garrido, Y. Leafy vegetables: Fresh and fresh-cut mature spinach. In *Controlled and Modified Atmospheres for Fresh and Fresh-Cut Produce*; Gil, M.I., Beaudry, R., Eds.; Academic Press: Cambridge, MA, USA, 2020; pp. 551–555. [CrossRef]
7. Ribera, A.; Bai, Y.; Wolters, A.M.A.; van Treuren, R.; Kik, C. A review on the genetic resources, domestication and breeding history of spinach (*Spinacia oleracea* L.). *Euphytica* **2020**, *216*, 48. [CrossRef]
8. Van der Vossen, H.A.M. Spinacia oleracea. In *Plant Resources of Tropical Africa 2: Vegetables*; Grubben, G.J.H., Denton, O.A., Eds.; Backhuys Publishers: Wageningen, The Netherlands, 2004; pp. 513–515. Available online: https://edepot.wur.nl/417517 (accessed on 14 October 2021).
9. De Cauwer, B.; Delanote, L.; Devos, M.; De Ryck, S.; Reheul, D. Optimisation of weed control in organic processing spinach (*Spinacia oleracea* L.): Impacts of cultivar, seeding rate, plant spacing and integrated weed management strategy. *Agronomy* **2021**, *11*, 53. [CrossRef]
10. Lati, R.N.; Rachuy, J.S.; Fennimore, S.A. Weed management in fresh market spinach (*Spinacia oleracea*) with phenmedipham and cycloate. *Weed Technol.* **2015**, *29*, 101–107. [CrossRef]
11. Wallace, R.W.; Phillips, A.L.; Hodges, J.C. Processing Spinach Response to selected herbicides for weed control, crop injury, and yield. *Weed Technol.* **2007**, *21*, 714–718. [CrossRef]
12. Légére, A.; Stevenson, F.C.; Benoit, D.L. Diversity and assembly of weed communities: Contrasting responses across cropping systems. *Weed Res.* **2005**, *45*, 303–315. [CrossRef]
13. Storkey, J.; Neve, P. What good is weed diversity? *Weed Res.* **2018**, *58*, 239–243. [CrossRef]
14. Travlos, I.S.; Cheimona, N.; Roussis, I.; Bilalis, D.J. Weed-species abundance and diversity indices in relation to tillage systems and fertilization. *Front. Environ. Sci.* **2018**, *6*, 11. [CrossRef]
15. Neve, P.; Busi, R.; Renton, M.; Vila-Aiub, M.M. Expanding the eco-evolutionary context of herbicide resistance research. *Pest Manag. Sci.* **2014**, *70*, 1385–1393. [CrossRef]
16. Perrino, E.V.; Calabrese, G. Endangered segetal species in southern Italy: Distribution, conservation status, trends, actions and ethnobotanical notes. *Genet. Resour. Crop Evol.* **2018**, *65*, 2107–2134. [CrossRef]
17. Wallace, R.W.; Miller, T.W.; Masabni, J.G. Sustainable weed control in vegetables. In *Weed Control*, 1st ed.; Korres, N.E., Burgos, N.R., Duke, S.O., Eds.; CRC Press: Boca Raton, FL, USA, 2018; pp. 404–424. Available online: https://www.taylorfrancis.com/chapters/edit/10.1201/9781315155913-21/sustainable-weed-control-vegetables-russell-wallace-timothy-miller-joseph-masabni (accessed on 13 November 2021).
18. Foppa Pedretti, E.; Boakye-Yiadom, K.A.; Valentini, E.; Ilari, A.; Duca, D. Life Cycle Assessment of spinach produced in Central and Southern Italy. *Sustainability* **2021**, *13*, 10001. [CrossRef]
19. Grigg, S. Spinach—A Global Expedition to Understand the Prevalence of Damping off and Practical Management Techniques. 2015. Available online: https://www.issinstitute.org.au/wp-content/uploads/2015/04/Report-Grigg-FINAL-LowRes.pdf (accessed on 14 November 2021).
20. Savvas, D.; Akoumianakis, K.; Karapanos, I.; Kontopoulou, C.K.; Ntatsi, G.; Liontakis, A.; Sintori, A.; Ropokis, A.; Akoumianakis, A. Recharging Greek Youth to Revitalize the Agriculture and Food Sector of the Greek Economy. Final Report—Sectoral Study 5. *Vegetables: Open-Field and Greenhouse Production*. 2015. Available online: https://www.generationag.org/assets/site/public/nodes/1019/1055-Vegetables_Open-Field_and_Greenhouse_Production.pdf (accessed on 14 November 2021).

21. Commission Implementing Regulation (EU) 2021/1449 of 3 September 2021 Amending Implementing Regulation (EU) No 540/2011 as Regards the Extension of the Approval Periods of the Active Substances 2-Phenylphenol (Including its Salts Such as the Sodium Salt), 8-Hydroxyquinoline, Amidosulfuron, Bifenox, Chlormequat, Chlorotoluron, Clofentezine, Clomazone, Cypermethrin, Daminozide, Deltamethrin, Dicamba, Difenoconazole, Diflufenican, Dimethachlor, Etofenprox, Fenoxaprop-P, Fenpropidin, Fludioxonil, Flufenacet, Fosthiazate, Indoxacarb, Lenacil, MCPA, MCPB, Nicosulfuron, Paraffin Oils, Paraffin Oil, Penconazole, Picloram, Propaquizafop, Prosulfocarb, Quizalofop-P-Ethyl, Quizalofop-P-Tefuryl, Sulphur, Tetraconazole, Tri-Allate, Triflusulfuron and Tritosulfuron (Text with EEA Relevance). Available online: https://eur-lex.europa.eu/legal-content/EN/TXT/PDF/?uri=CELEX:32021R1449&from=EN (accessed on 14 November 2021).
22. Lewis, K.A.; Tzilivakis, J.; Warner, D.; Green, A. An international database for pesticide risk assessments and management. *Hum. Ecol. Risk Assess.* **2016**, *22*, 1050–1064. [CrossRef]
23. Lati, R.N.; Mou, B.; Rachuy, J.S.; Fennimore, S.A. Light intensity is a main factor affecting fresh market spinach tolerance for phenmedipham. *Weed Sci.* **2016**, *64*, 146–153. [CrossRef]
24. Lati, R.N.; Mou, B.; Rachuy, J.S.; Fennimore, S.A. Evaluation of cycloate followed by evening two-leaf-stage phenmedipham application in fresh market spinach. *Weed Technol.* **2016**, *30*, 464–471. [CrossRef]
25. Heap, I. The International Herbicide-Resistant Weed Database. Available online: www.weedscience.org (accessed on 16 November 2021).
26. Pannacci, E.; Lattanzi, B.; Tei, F. Non-chemical weed management strategies in minor crops: A review. *Crop Prot.* **2017**, *96*, 44–58. [CrossRef]
27. Korres, N.E. Agronomic weed control: A trustworthy approach for sustainable weed management. In *Non-Chemical Weed Control*; Jabar, K., Chauhan, B.S., Eds.; Academic Press: Cambridge, MA, USA, 2018; pp. 97–114. [CrossRef]
28. Alba, O.S.; Syrovy, L.D.; Duddu, H.S.; Shirtliffe, S.J. Increased seeding rate and multiple methods of mechanical weed control reduce weed biomass in a poorly competitive organic crop. *Field Crops Res.* **2020**, *245*, 107648. [CrossRef]
29. Kanatas, P.J.; Gazoulis, I. The integration of increased seeding rates, mechanical weed control and herbicide application for weed management in chickpea (*Cicer arietinum* L.). *Phytoparasitica* **2021**, in press. [CrossRef]
30. Benaragama, D.; Shirtliffe, S.J. Integrating cultural and mechanical methods for additive weed control in organic systems. *Agron. J.* **2013**, *105*, 1728–1734. [CrossRef]
31. Peruzzi, A.; Ginanni, M.; Raffaelli, M.; Borelli, M. Physical weed control in organic spinach production. In Proceedings of the 6th EWRS Workshop on Physical and Cultural Weed Control, Lillehammer, Norway, 8–10 March 2004; Cloutier, D.C., Ascard, J., Eds.; pp. 15–23. Available online: https://citeseerx.ist.psu.edu/viewdoc/download?doi=10.1.1.622.7735&rep=rep1&type=pdf#page=22 (accessed on 9 October 2021).
32. Tei, F.; Stagnari, F.; Granier, A. Preliminary results on physical weed control in processing spinach. In Proceedings of the 5th EWRS Workshop on Physical and Cultural Weed Control, Pisa, Italy, 11–13 March 2002; Cloutier, D.C., Ed.; pp. 164–171. Available online: https://citeseerx.ist.psu.edu/viewdoc/download?doi=10.1.1.113.5384&rep=rep1&type=pdf#page=171 (accessed on 9 October 2021).
33. Cloutier, D.C.; Van der Weide, R.Y.; Peruzzi, A.; Leblanc, M.L. Mechanical weed management. In *Nonchemical Weed Management: Principles, Concepts and Technology*; Upadhyaya, M.K., Blackshaw, R.E., Eds.; CAB International: Wallingford, UK, 2007; pp. 111–134. Available online: https://edepot.wur.nl/116036 (accessed on 10 October 2021).
34. DiTomaso, J.M. Approaches for improving crop competitiveness through the manipulation of fertilization strategies. *Weed Sci.* **1995**, *43*, 491–497. [CrossRef]
35. Little, N.G.; DiTommaso, A.; Westbrook, A.S.; Ketterings, Q.M.; Mohler, C.L. Effects of fertility amendments on weed growth and weed-crop competition: A review. *Weed Sci.* **2021**, *69*, 132–146. [CrossRef]
36. Santos, B.M.; Dusky, J.A.; Stall, W.M.; Gilreath, J.P. Effects of phosphorus fertilization on common lambsquarters (*Chenopodium album*) duration of interference in lettuce (*Lactuca sativa*). *Weed Technol.* **2004**, *18*, 152–156. [CrossRef]
37. Santos, B.M.; Dusky, J.A.; Stall, W.M.; Gilreath, J.P. Influence of common lambsquarters (*Chenopodium album*) densities and phosphorus fertilization on lettuce. *Crop Prot.* **2004**, *23*, 173–176. [CrossRef]
38. Su, W.; Liu, B.; Liu, X.; Li, X.; Ren, T.; Cong, R.; Lu, J. Effect of depth of fertilizer banded-placement on growth, nutrient uptake and yield of oilseed rape (*Brassica napus* L.). *Eur. J. Agron.* **2015**, *62*, 38–45. [CrossRef]
39. Grouner. Grouner—Dischargers of Artificial Rain, Thessaloniki, Greece. 1938. Available online: http://www.grouner.gr/en/products.php?cat=8 (accessed on 15 November 2021).
40. Blackshaw, R.E.; Molnar, L.J.; Janzen, H.H. Nitrogen fertilizer timing and application method affect weed growth and competition with spring wheat. *Weed Sci.* **2004**, *52*, 614–622. [CrossRef]
41. Kaur, S.; Kaur, R.; Chauhan, B.S. Understanding crop-weed-fertilizer-water interactions and their implications for weed management in agricultural systems. *Crop Prot.* **2018**, *103*, 65–72. [CrossRef]
42. Kanatas, P.; Antonopoulos, N.; Gazoulis, I.; Travlos, I.S. Screening glyphosate-alternative weed control options in important perennial crops. *Weed Sci.* **2021**, *69*, 704–718. [CrossRef]
43. Kanatas, P.; Gazoulis, I.; Travlos, I. Irrigation timing as a practice of effective weed management in established alfalfa (*Medicago sativa* L.) crop. *Agronomy* **2021**, *11*, 550. [CrossRef]
44. Travlos, I.; Tsekoura, A.; Antonopoulos, N.; Kanatas, P.; Gazoulis, I. Novel sensor-based method (quick test) for the in-season rapid evaluation of herbicide efficacy under real field conditions in durum wheat. *Weed Sci.* **2021**, *69*, 147–160. [CrossRef]

45. Moss, S.R.; Storkey, J.; Cussans, J.W.; Perryman, S.A.M.; Hewitt, M.V. The Broadbalk long-term experiment at Rothamsted: What has it told us about weeds? *Weed Sci.* **2004**, *52*, 864–873. [CrossRef]
46. Pollnac, F.W.; Maxwell, B.D.; Menalled, F.D. Weed community characteristics and crop performance: A neighbourhood approach. *Weed Res.* **2009**, *49*, 242–250. [CrossRef]
47. Borgy, B.; Gaba, S.; Petit, S.; Reboud, X. Non-random distribution of weed species abundance in arable fields. *Weed Res.* **2012**, *52*, 383–389. [CrossRef]
48. Smith, R.G.; Mortensen, D.A.; Ryan, M.R. A new hypothesis for the functional role of diversity in mediating resource pools and weed-crop competition in agroecosystems. *Weed Res.* **2010**, *50*, 37–48. [CrossRef]
49. Adeux, G.; Vieren, E.; Carlesi, S.; Bàrberi, P.; Munier-Jolain, N.; Cordeau, S. Mitigating crop yield losses through weed diversity. *Nat. Sustain.* **2019**, *2*, 1018–1026. [CrossRef]
50. Cierjacks, A.; Pommeranz, M.; Schulz, K.; Almeida-Cortez, J. Is crop yield related to weed species diversity and biomass in coconut and banana fields of northeastern Brazil? *Agric. Ecosyst. Environ.* **2016**, *220*, 175–183. [CrossRef]
51. Ferrero, R.; Lima, M.; Davis, A.S.; Gonzalez-Andujar, J.L. Weed diversity affects soybean and maize yield in a long term experiment in Michigan, USA. *Front. Plant Sci.* **2017**, *8*, 236. [CrossRef]
52. Harker, K.N.; Clayton, G.W.; O'Donovan, J.T. Reducing agroecosystem vulnerability to weed invasion. In *Invasive Plants: Ecological and Agricultural Aspects*; Inderjit, S., Ed.; Birkhäuser: Basel, Switzerland, 2005; pp. 195–207.
53. Tracy, B.F.; Renne, I.J.; Gerrish, J.; Sanderson, M.A. Effects of plant diversity on invasion of weed species in experimental pasture communities. *Basic Appl. Ecol.* **2004**, *5*, 543–550. [CrossRef]
54. Wu, H.; Carrillo, J.; Ding, J. Invasion by alligator weed, *Alternanthera philoxeroides*, is associated with decreased species diversity across the latitudinal gradient in China. *J. Plant Ecol.* **2016**, *9*, 311–319. [CrossRef]
55. Canali, S.; Diacono, M.; Ciaccia, C.; Masetti, O.; Tittarelli, F.; Montemurro, F. Alternative strategies for nitrogen fertilization of overwinter processing spinach (*Spinacia oleracea* L.) in Southern Italy. *Eur. J. Agron.* **2014**, *54*, 47–53. [CrossRef]

diversity

MDPI

Article

High Differentiation among Populations of Green Foxtail, *Setaria viridis*, in Taiwan and Adjacent Islands Revealed by Microsatellite Markers

Wei-Hsun Hsieh [1], Yen-Chiun Chen [1], Hsien-Chun Liao [2], Yann-Rong Lin [1,3] and Chih-Hui Chen [2,*]

[1] Department of Agronomy, National Taiwan University, Taipei 10617, Taiwan; d06621103@ntu.edu.tw (W.-H.H.); wotan13233@gmail.com (Y.-C.C.); ylin@ntu.edu.tw or yann-rong.lin@worldveg.org (Y.-R.L.)
[2] Endemic Species Research Institute, Council of Agriculture, Nantou 552203, Taiwan; pan.darkwing@gmail.com
[3] Headquarter, World Vegetable Center, Shanhua, Tainan 741005, Taiwan
* Correspondence: chchen@tesri.gov.tw

Abstract: *Setaria viridis* (L.) Beauv., or green foxtail, is native to Eurasia and is the putative ancestor of foxtail millet. Due to the advantageous genetic characteristics of *S. viridis*, it is a model species for C4 plants. However, *S. viridis* has seriously spread to the agricultural system around the world because of its wide adaptability. This study is aimed to understand the distribution of *S. viridis* in Taiwan, and also to investigate the genetic diversity and relationships among different wild populations. A total of 141 *S. viridis* collected at 10 sites with sampling sizes ranging from 8 to 24 plants in Taiwan were analyzed by 13 highly polymorphic SSR markers, and 6.1 alleles per locus were detected in our study. The relationships of collected *S. viridis* mostly corresponded to its distribution in different parts of Taiwan revealed by PCoA and phylogenetic tree. Similarly, the results for population structure showed the significance of collecting site or geographical factors. Finally, the extent of gene flow was studied with the genetic differentiation (F_{ST}) and Nm values, and two *S. viridis* populations were found to significantly contain the existence of gene-flow events. In conclusion, *S. viridis* showed a pattern of low diversity and heterozygosity within a population, but high differentiation among populations because of its selfing attribute and the barriers of sea and mountain range for gene flow. In addition, the founder effect may be the other reason for this pattern of population genetic structure.

Keywords: green foxtail; *Setaria viridis*; weediness; genetic diversity; population genetic structure; gene flow

Citation: Hsieh, W.-H.; Chen, Y.-C.; Liao, H.-C.; Lin, Y.-R.; Chen, C.-H. High Differentiation among Populations of Green Foxtail, *Setaria viridis*, in Taiwan and Adjacent Islands Revealed by Microsatellite Markers. *Diversity* **2021**, *13*, 159. http://doi.org/10.3390/d13040159

Academic Editor: Mario A. Pagnotta

Received: 16 March 2021
Accepted: 1 April 2021
Published: 7 April 2021

1. Introduction

Setaria viridis (L.) Beauv., or green foxtail, belongs to the grass family Poaceae, and is native to Eurasia [1–3]. It is the putative ancestor of foxtail millet, *Setaria italica* (L.) Beauv., which was domesticated and selected by farmers in northern China about 10,500 years ago, based on archeological evidence [4–7]. Currently, foxtail millet is a minor crop cultivated in India, China and Taiwan for food, and in Europe for birdseed mainly [8]. The world production of millet was around 31 million tons in 2018, and foxtail millet is the largest crops among millets [9]. The cytogenetic study of GISH [10] and the phylogenetic relationship analyzed with chloroplast and nuclear genes [11,12] showed that the genome difference between *S. italica* and *S. viridis* was slightly distinguishable, which is concordant with the fact that few different morphological traits can be observed between them [5,13,14]. Furthermore, there was about 22% of SNP variation detected in both *S. viridis* and *S. italica* in the genome-wide data [15].

Due to its small stature, short lifecycle (6–8 weeks), small genome size (~510 Mb), diploidy ($2n$ = 18), self-pollination, efficient C4 photosynthesis and transformability, *S. viridis* has been recommended as a model species for the research of C4 plants, which also include switchgrass, maize and sorghum [1,16–19]. In addition, a new genomic resource of *S. viridis* was released by de novo assembly recently [15]. On the other hand, the whole-genome sequencing of foxtail millet has been released [16,20], and it makes *S. italica* important in model systems as well [17]. Recombinant inbred lines (RILs) of *S. italica* x *S. viridis*, accession A10, have been thoroughly investigated and applied in research on bioenergy, feed and forage of stock [16,17,19,21,22]. Moreover, A10 accession has been established and applied to QTL mapping and the construction of a genetic map [16,22].

The problem of invasive weeds caused losses of agriculture, overdose application of pesticide and more pollution to environments and eco-systems. *S. viridis* possesses the potential to adapt to various environments and habitats; consequently, it has invaded not only throughout the temperate, but in some subtropical and tropical regions of agricultural systems worldwide [8,23–26]. The variable biodiversity in phenotype and genotype allows *S. viridis* to invade, endure and colonize in different local environments [3]. The seeds can float with water flow for 10 days [27] or attach to animals or human clothing with the bristles on the panicles to disperse for a long distance [28–31]. Hence, it is considered as a serious weed for agriculture in many countries, such as Canada, the USA, Spain and Japan [4,30,32,33], and usually grows near riverbanks and lakes, and in roadsides, grain fields, wastelands and any other disturbed regions [34,35]. The invasiveness of *S. viridis* was detrimental to crop growth and production, with one reference of reported losses of 21% and 44% yields of wheat [36]. Furthermore, many other crops, like grain sorghum, rice and alfalfa, were also reported to be influenced by *S. viridis* significantly [28,29,37]. Despite being a noxious weed, *S. viridis* is extensively used as forage in Europe and as herbal medicine in some regions of Asia [38,39]. In North America, *S. viridis* is widely dispersed [30], but it does not tend to occupy high mountains and regions of lower latitudes [40]. Similarly, Jia et al. (2013) found that *S. viridis* was localized to northern latitude regions in China, suggesting that the local populations are more adapted to temperate climates [41].

Due to the interfile between *S. italica* and *S. viridis*, the occurrence of gene flow might make the weediness of *S. viridis* more powerful in nature [14,42]. For example, the giant green foxtail, *S. viridis* var. "major" (Gaud.) Posp., an intermediate type, is originally from the hybridization between wild *S. viridis* and cultivated *S. italica* [14,34,43]. In addition, *S. faberi* R. A. W. Herrma and *S. verticillata* (L.) P. Beauv., the tetraploid species (AABB genome), originated from a natural crossing between *S. adhaerans* (Forssk.) Chiov. and *S. viridis* [10]. On the contrary, it was also reported that reproductive barriers partially existed, resulting in a low level of introgression and gene flow in a wild-weed-crop complex [14,44–46]. It seems that the risk of gene flow between *S. italica* and *S. viridis* is low [47].

Until now, there were some studies about the genetics, genomics and evolution of *S. italica* published, but few about *S. viridis*. Related studies of *S. viridis* did not attract much attention, although it has been recommended as a model system for the studies of C4 photosynthesis, biofuels, and drought and salinity tolerance. The abundance of foxtail millet landraces has been observed in China and Taiwan [48–50]. The intrapopulation genetic diversity of weedy *Setaria* species (*S. viridis*, *S. faberi*, *S. parviflora* (Poir.) Kerguélen, *S. pumila* (Poir.) Roem. and Schult., *S. verticillata*) was quite low, but quite high in inter-population genetic diversity [40,51,52]. Interestingly, the low level of genetic diversity of *S. viridis* was observed in comparison with the other crop wild relatives, such as *Oryza rufipogon* and teosinte [26,53,54]. Instead, the cultivated *S. italica* displayed high diversity compared to sorghum and rice [55,56]. Overall, the genetic diversity and phylogenetic relationship was mostly assessed in the studies of foxtail millet [57–60], and only a few studies also included *S. viridis* collections [12,41,61]. In our opinion, these studies are not enough to deeply understand the genetic diversity and population structure of *S. viridis*, and many questions still remain unanswered. Importantly, that kind of information is crucial for germplasm conservation, genetic mapping, association studies and breeding programs [62,63].

The earliest described specimen of *S. viridis* in Taiwan is the one numbered 107,274 in the herbarium HAST, which was collected in 1916, and the earliest documentation in Taiwan was recorded in 1930 [64]. On the other hand, according to the archeological evidence, the cultivation of foxtail millet in Taiwan could be dated back to 5000 years ago [65]. Foxtail millet is one of the traditional foods for Taiwanese aborigines, and it also was used to brew alcoholic beverages because of the waxy property of Taiwanese landraces [48]. However, genomic or genetic diversity of *S. viridis* and its mechanism of success in becoming an aggressive weed in Taiwan are little understood so far. In this study, we hope to address the following issues: (i) how the pattern of the genetic diversity of *S. viridis* in Taiwan came about; (ii) the estimated population structure of genetic variation and genetic relationship of different *S. viridis* populations in Taiwan; and (iii) the pattern of genetic differentiation among *S. viridis* populations in different geographical regions of Taiwan.

2. Materials and Methods

2.1. Field Collections of S. viridis in Taiwan

The field survey covered Eastern and Western Taiwan, and several offshore islands belonging to the Taiwan Government as well. A total of 141 individuals of *S. viridis* were collected from 10 sites, including four on main island of Taiwan and six distributed on different offshore islands (Table 1). The individuals from the same collection site were regarded as a population. The sample size of every population ranged from 8 to 24. Fresh leaves were dried out using desiccant beads, then preserved at 4 °C. The longitudes and latitudes of all sites were recorded in WGS84 format.

Table 1. Information, including code, site, latitude, longitude and sample size of 10 collection sites in different parts of Taiwan.

Code	Site (Township/County)	Latitude	Longitude	Sample Size
#01	Tongluo, Miaoli	24°29′48.00″ N	120°47′17.00″ E	16
#02	Haiduan, Taitung	23°4′9.00″ N	121°9′33.00″ E	8
#03	Taitung, Taitung	22°43′36.00″ N	121°5′48.00″ E	16
#04	Ludao, Taitung	22°40′25.00″ N	121°28′17.00″ E	9
#05	Liujiao, Chiayi	23°30′56.00″ N	120°17′56.00″ E	10
#06	Qimei, Penghu	23°12′18.00″ N	119°25′37.00″ E	15
#07	Huxi, Penghu	23°35′40.00″ N	119°36′47.00″ E	11
#08	Jinning, Kinmen	24°27′20.00″ N	118°19′8.00″ E	16
#09	Jinsha, Kinmen	24°29′36.00″ N	118°24′44.00″ E	24
#10	Nangan, Lienchiang	26°8′45.00″ N	119°54′47.00″ E	16
Total				141

2.2. Genotyping by Microsatellite

The dried leaf of each sample was homogenized using tungsten carbide beads with Tissuelyser, then the genomic DNA was extracted by using the TPS method with modifications [66]. Twenty-five markers were used at first based on the PIC (polymorphism information content) value reported by Zhang et al. [67].

The forward and reversed primers were elongated with 5′ ACGACGTTGTAAAA 3′ and reversed 5′ CATTAAGTTCCCATTA 3′ sequences, respectively, to perform multiplex-ready PCR [68]. The PCR reaction mixture of each sample was 10 μL in total volume, containing 20 ng template DNA, 1 μL 1× IMMOLASE buffer (BIOLINE), 0.2 mM dNTP, 2.0 mM $MgCl_2$, 40 nM primers, 80 nM fluorophores (VIC, 6-FAM, NED, or PET), 0.05 μL IMMOLASE DNA polymerase (BIOLINE) and 4.55 μL ddH_2O. The amplification program was performed with the following steps: 95 °C 10 min, 20 cycles of 92 °C for 30 s, 63 °C for 90 s, 72 °C for 60 s, then 40 cycles of 92 °C for 15 s, 54 °C for 30 s and 72 °C for 60 s. The four PCR products were pooled together at a ratio of 2:3:4:6 (VIC:6-FAM:NED:PET) in volume, then we used GSLIZ600 as internal control. The size of the SSR fragments was detected by capillary electrophoresis. An elite variety of foxtail millet (*Setaria italica*) in

Taiwan, Taitung Number 8 (TT8), was also pooled into each plate to correct the size bias among different plates. The FSA (.fsa) files that contained fragment sizes and fluorescence intensities were analyzed using the "Fragman" v1.0.9 package in R software v4.0.2 [69,70]. After filtering out the markers with multiple peaks (multiallele), a set of 13 SSR markers distributed on nine chromosomes was selected and used for further analysis.

2.3. Genetic Diversity and Clustering Analysis

The genetic diversity of each marker and sampling site were evaluated with the parameters of allele numbers (Na), effective allele numbers (Ne), observed heterozygosity (Ho) and expected heterozygosity (He), which is also called gene diversity [71], using the "PopGenReport" v3.0.4 package in R software [72]. Additionally, the polymorphism information content (PIC) of each marker was calculated with the "polysat" v1.7.4 package in R software [73].

As for the clustering analysis of 141 *S. viridis* samples, the genetic distance between each pair of individuals was estimated by Euclidean distance using the "adegenet" v2.0.1 package in R software [74]. Subsequently, the distance matrix was used for a principal coordinate analysis (PCoA) using the "cmdscale" function and for constructing a neighbor-joining tree with the "ape" v5.4 package in R software [75]. The variety TT8 was included as an outgroup for phylogenetic tree.

The term "population structure" was defined as the genetic background of all 141 *S. viridis* individuals and estimated with STRUCTURE v2.3.4 using an admixture model [76]. The term "subpopulation" was defined as the clusters based on the optimum clustering number (*K* value). The *K* value was tested from 1 to 11 to find the best *K* for 141 *S. viridis* individuals. For each *K* value, burn-in period and Markov chain Monte Carlo (MCMC) were both set to 100,000× with 20 replications. The result of STRUCTURE was integrated with the "pophelper" v2.3.0 package in R software [77], and the optimum *K* value was determined by the value of delta *K* described in the Evanno method [78]. Finally, the results above were visualized using the "ggplot2" [79] and "ggtree" v3.11 packages in R software [80]. The individuals in the barplot of STRUCTURE were sorted by 10 collection sites, from #01 to #10. The different genetic background of each collection site based on structure analysis were represented by pie charts on the map of Taiwan.

2.4. F-Statistics of Populations at 10 Collection Sites

Wright's *F*-statistics, including F_{IS}, F_{ST} and F_{IT}, were used to summarize the population structure among the 10 collection sites [81]. The estimator of each parameters was described by Weir and Cockerham [82]. The F_{ST}-derived estimator, Nm value [83], was further used to evaluate the gene flow between each pair of collection sites under the island model theory. The formula used was $Nm = (1 - F_{ST})/4F_{ST}$ [81]. Finally, all of the *F*-statistics parameters were calculated in R software.

3. Results

3.1. Molecular Diversity of Microsatellite Markers from the Individuals of Green Foxtail in Taiwan

Thirteen SSR markers were used to assess the genetic diversity of 141 *S. viridis* individuals in Taiwan (Table 1). A total of 79 alleles were detected, with an average of 6.1 alleles per locus. The observed allele number ranged from 2 (SICAAS5005, SICAAS6052, SICAAS7090, SICAAS9121) to 16 (SICAAS1065). On the other hand, the effective allele number (Ne) was predicted when the number of alleles with equal frequency was assumed. In average, 3.558 effective alleles per locus were measured in all *S. viridis* populations. The marker SICAAS1015 showed the largest Ne (9.18), and SICAAS5005 showed the smallest Ne (1.26). The value of observed heterozygosity (Ho) ranged from 0.015 to 0.022 for all markers. Compared to the expected heterozygosity (He), the mean of Ho was only 0.0054, indicating the characteristic of high self-pollinated rate of *S. viridis*. In addition, nine out of 13 SSR markers displayed none of the observed heterozygosity in our study. Conversely, the values of expected heterozygosity (He) ranged from 0.21 (SICAAS5005) to 0.89

(SICAAS1015). Four markers, SICAAS1015, SICAAS1065, SICAAS2084 and SICAAS3090, exhibited more abundant gene diversity in our study. The PIC values ranged from 0.18 (SICAAS5005) to 0.88 (SICAAS1015) with a mean of 0.527, indicating that highly polymorphic SSR markers were used in our study. The marker SICAAS5005 showed the lowest PIC value, and provided little information for the analysis. The PIC values of seven markers were within the range of 0.25 to 0.5, displaying moderate polymorphic information. Finally, the PIC values of five markers were over 0.7, representing high polymorphism in our study (Table 2).

Table 2. Genetic diversity parameters of 13 SSR markers used to assess 141 *Setaria viridis* individuals.

Markers	Na	Ne	Ho	He	PIC
SICAAS1015	14	9.18	0.000	0.89	0.88
SICAAS1065	16	7.79	0.015	0.87	0.86
SICAAS2084	8	5.16	0.000	0.81	0.78
SICAAS3090	10	6.06	0.022	0.83	0.81
SICAAS5005	2	1.26	0.000	0.21	0.18
SICAAS5081	4	2.19	0.000	0.54	0.48
SICAAS6052	2	1.37	0.000	0.27	0.30
SICAAS7002	7	4.35	0.017	0.77	0.73
SICAAS7008	4	1.98	0.000	0.49	0.44
SICAAS7090	2	1.41	0.000	0.29	0.27
SICAAS8025	4	2.12	0.016	0.53	0.41
SICAAS9121	2	1.41	0.000	0.29	0.26
SICAAS9130	4	1.98	0.000	0.50	0.45
Average	6.1	3.558	0.0054	0.561	0.527

Na: observed allele number; Ne: effective allele number; Ho: observed heterozygosity; He: expected heterozygosity; PIC: polymorphic information content.

The genetic diversity indexes of *S. viridis* populations collected from 10 sites were revealed in this study (Table 3). Among all 10 collection sites, the number of observed alleles ranged from 13 (sites #01 and #02) to 29 (site #08), with a mean of 18.9. Similarly, the largest number of effective alleles was also detected at site #08, and the lowest was at both site #01 and #02. Overall, 15.619 effective alleles per site were found. The observed heterozygosity (Ho) was only detected at three sites (#03, #08 and #10) with very small values, and the other populations of *S. viridis* showed none of Ho. The mean of Ho was 0.005, which was expected due to the characteristics of self-pollination of *S. viridis*. As for the expected heterozygosity (He), it ranged from 0.03 (site #04) to 0.27 (site #08), and no He was observed at two collecting sites (#01 and #02).

Table 3. Genetic diversity index of *Setaria viridis* individuals in each collecting site.

Collecting Site	SS	Na	Ne	Ho	He	F_{IS}	F_{ST}	F_{IT}	Nm
#01	16	13	13.00	0.00	0.00	0.9896	0.4895	0.9947	0.26
#02	8	13	13.00	0.00	0.00	0.9901	0.4995	0.9950	0.25
#03	16	19	14.25	0.01	0.12	0.9841	0.3954	0.9904	0.38
#04	9	14	13.60	0.00	0.03	0.9904	0.4783	0.9950	0.27
#05	10	22	18.63	0.00	0.16	0.9921	0.3540	0.9949	0.46
#06	15	22	14.94	0.00	0.11	0.9905	0.4476	0.9948	0.31
#07	11	20	15.57	0.00	0.12	0.9915	0.3984	0.9949	0.38
#08	16	29	21.82	0.02	0.27	0.9723	0.2525	0.9793	0.74
#09	24	18	15.04	0.00	0.07	0.9901	0.4277	0.9943	0.33
#10	16	19	16.34	0.02	0.14	0.9653	0.3943	0.9790	0.38
Average		18.9	15.619	0.005	0.102				

SS: sample size; Na: observed allele number; Ne: effective allele number; Ho: observed heterozygosity; He: expected heterozygosity; F_{IS}: inbreeding coefficient of an individual relative to the subpopulation; F_{ST}: the degree of genetic differentiation between populations; F_{IT}": inbreeding coefficient of an individual relative to the total population; Nm: number of immigrants, the extent of gene flow.

3.2. Genetic Relationship of Green Foxtail Individuals Collected in Taiwan

The results of the principle coordinate analysis offer preliminary insight on the genetic relationship among 141 *S. viridis* individuals or 10 wild populations collected in Taiwan (Figure 1). The first and the second coordinates explained 23.00% and 14.90% of variability, respectively. First of all, sites #01 and #05, which are both in Western Taiwan, were grouped together. Secondly, three collecting sites, all in Taitung County (sites #02 and #03), including small island Ludao (site #4), had a slightly close relationship with each other. However, two individuals at site #03 were slightly away from the others. Interestingly, the *S. viridis* at site #06 and site #07 (Qimei and Huxi in Penghu, respectively) were significantly isolated from the other sites. However, one collection belonging to site #06 was closer to the individuals at sites #01 and #05. Finally, three outer collection sites (#08, #09 and #10) closer to mainland China showed an overlapped pattern with each other that indicated a closer relationship among them.

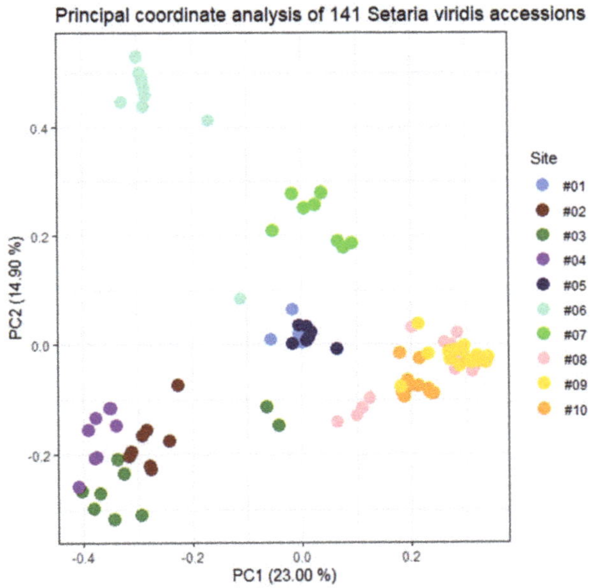

Figure 1. Principal coordinate analysis of the 141 *Setaria viridis* individuals in Taiwan. The first and the second coordinates explained 23.00 % and 14.90 % of variability, respectively. The 10 collecting sites are marked by different colors.

A similar pattern with more details is presented in the phylogenetic tree (Figure 2). Individuals collected in Taitung County, including sites #02, #03 and #04, were in Clade I. Broadly, *S. viridis* individuals collected at site #04 (Ludao) were closer to collections at site #03 (Taitung), however, several individuals collected at site #03 showed a complex and mixed relationship among these populations. Individuals from site #01 (Miaoli) and #05 (Chiayi), with clear geographical separation, were grouped in Clade II. Nevertheless, an individual from site #06 (Penghu) was grouped in this clade, indicating the probable movement behavior of *S. viridis* individuals could occur between Penghu and Chiayi occasionally. Besides, some individuals from site #08 were closer to site #09 (see Clade III), and the others were closer to site #10 instead (see Clade IV), suggesting the higher abundance of *S. viridis* at site #08 (Jinning, Kinmen). Furthermore, the individuals from site #09 (Jinsha, Kinmen) showed clear separation from the individuals from site #10 (Lienchang). Moreover, the individuals from site #09 can be further divided into two small clusters. The individuals from site #07 (Huxi, Penghu) were also isolated as Clade V. Similarly, most

S. viridis individuals from site #06 were obviously separated from the others (as Clade VI), but an exception was grouped in Clade II and closer to the individuals at site #05. This is consistent with the result of the PCoA (Figure 1).

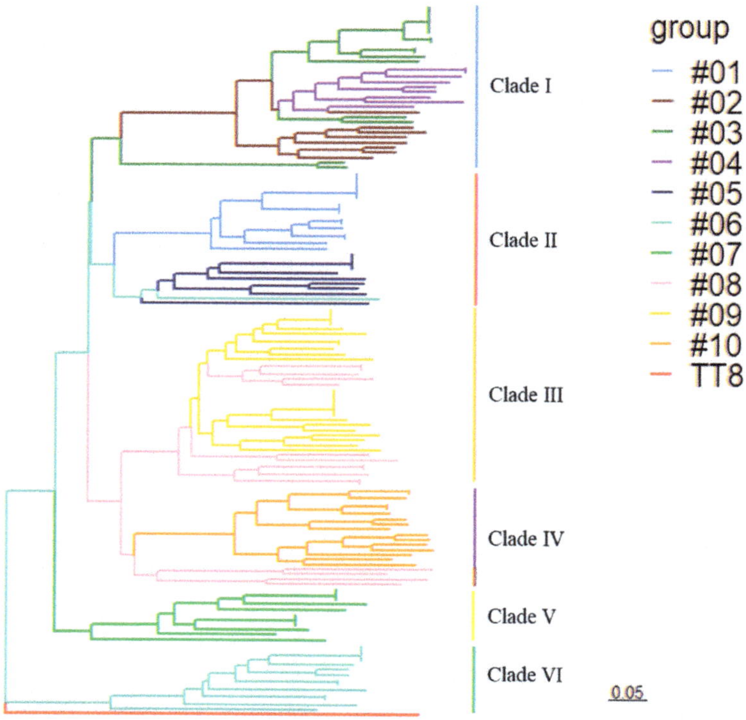

Figure 2. Neighbor-joining tree of the 141 *Setaria viridis* accessions in Taiwan. The color of each clade represents each individual collected from its own collecting site. The vertical line with different colors of every clade number corresponds to subpopulations that were inferred by the STRUCTURE analysis.

3.3. Population Structure and Geographic Distribution

The population structure of *S. viridis* collected from 10 sites was simulated by STRUCTURE (Figure 3a). The K value was inferred by the Evanno method, from $K = 1$ to $K = 11$ (Figure S1). The best K value was 10; however, 10 colors of different genetic backgrounds did not exactly represent the 10 collection sites (Figure 3b). In addition, the admixed individual is defined when the percentage of its own genetic background does not exceed at least 80%. The wild populations collected at sites #01, #02, #04 and #10 possessed pure genetic backgrounds with no admixture. In contrast, four individuals from site #05, three from #09, two from #06, one from #07 and most individuals from #08 were considered as admixture. Above all, the individuals collected from site #08 (Jinning, Kinmen) displayed the most complicated pattern (Figure 3b). The individuals from site #01 (Miaoli) and site #10 (Lienchiang) independently formed distinct subpopulations, called Pop 1 and Pop 7, respectively. Moreover, the individuals from sites #02, #03 and #04, in Eastern Taiwan (Taitung) were all grouped in Pop 2. Interestingly, two individuals from site #03 exceptionally formed another subgroup (Pop 10). The individuals from site #05 (Chiayi) were all grouped in Pop 3, mixed with another genetic background that also appeared in Pop 8 (in greenish blue), indicating potential gene flow or introgression. Surprisingly, two collecting sites of *S. viridis* from Penghu (sites #06 and #07) clearly formed two different subpopulations, Pop 4 for site #06 and Pop 5 for site #07 (see Figure 3a and Table 1). In addition, one individual collected at site #06 was grouped in Pop 3 instead of Pop 4, and showed high percentage of red background, which can be also noticed in the phylogenetic tree (Figure 2) and the PCoA (Figure 3). Finally, the individuals from sites #08 and #09 (Kinmen) were grouped in Pop 6, but displayed the mixed genetic background of Pop 8 and Pop 9.

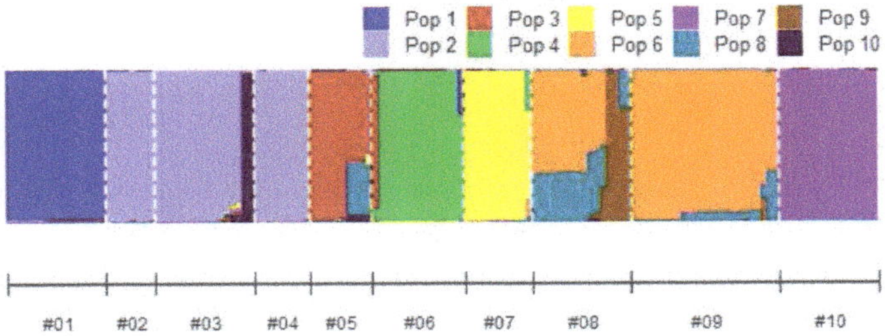

(a)

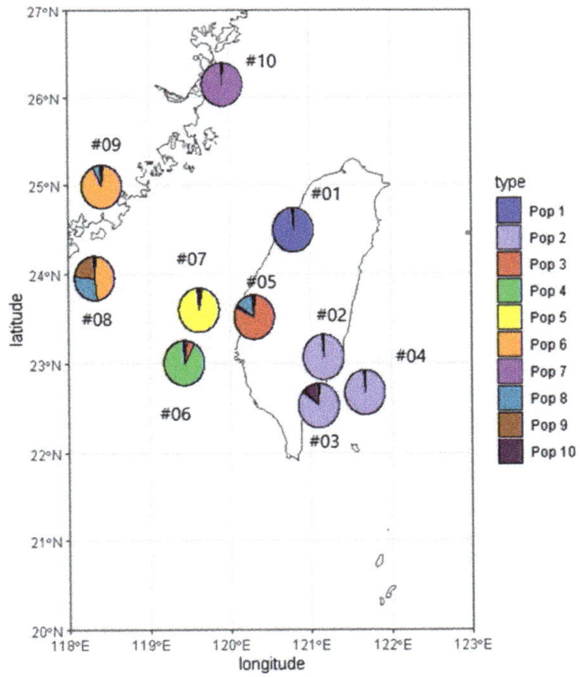

(b)

Figure 3. (**a**) STRUCTURE barplot of the 141 *Setaria viridis* individuals with *K* value = 10 inferred by the Evanno method. The order of results from STRUCTURE is sorted by 10 collection sites from #01 to #10. An individual is considered as the admixture when the genetic background is lower than 80% of its own percentage. (**b**) Geographic distribution of the 10 collection sites in Taiwan. The pie chart at each site represents the genetic proportion inferred by STRUCTURE when *K* = 10.

3.4. Genetic Differentiation among the Populations of S. viridis in Taiwan

F-statistics were also estimated in this study. All F_{IS} and F_{IT} values for the 10 sampling sites were all over 0.96. The lowest value of F_{IS} (0.9653) and F_{IT} (0.9790) were observed at site #10 (Lienchiang), an offshore island of Taiwan that is close to mainland China. The highest F_{IS} (0.9921) was detected at site #05 (Chiaiyi), and the highest F_{IT} (0.9950) was detected at sites #02 and #04 (Taitung). Overall, it was concluded that the wild *S. viridis* in Taiwan are highly self-pollinated within a population, which agreed with the result of low observed heterozygosity (Table 3). The extent of genetic differentiation between populations was estimated by F_{ST} value. We found high genetic differentiation among wild *S. viridis* populations in Taiwan. The populations at site #02 (Taitung) and site #01 (Miaoli) displayed the highest F_{ST} values, 0.4995 and 0.4895, respectively. Oppositely, site #08 (Kinmen) displayed a relatively smaller F_{ST} value (0.2525), suggesting that more gene flow events might take place in this area. Likewise, the number of immigrants (Nm) denotes the extent of gene flow, in which great differences of Nm value were found (0.25 to 0.74). Obviously, the highest extent of gene flow was detected at site #08, Kinmen (0.74), and smaller Nm values were observed at sites #01, #02 and #04, Taitung (0.25 to 0.27), which is in accordance with the results of genetic differentiation (Table 3).

3.5. Pairwise F_{ST} Index of Collecting Populations in Taiwan

Pairwise F_{ST} value, representing the extent of genetic differentiation between populations, was assessed for the 10 collection sites, and ranged from 0.14 to 0.62 (Table 4). The least extent of genetic differentiation was between sites #08 and #09 (Kinmen), indicating gene flow events might happen more frequently here. Next, the *S. viridis* populations collected at sites #03 (Taitung) and #04 (Ludao) also showed a low level of genetic differentiation (F_{ST} = 0.17). The F_{ST} values among sites #02, #03 and #04, in Taitung, were quite small in comparison with the others. On the contrary, the highest level of genetic differentiation was observed between site #07 (Huxi, Penghu) and site #04 (Ludao, Taitung), which are both small offshore islands off Taiwan's main island (Figure 3b). Next, three combinations from different collection sites showed the second-highest genetic differentiation (F_{ST} = 0.59): #07 (Penghu) and #02 (Taitung), #09 (Kinmen) and #06 (Penghu), and #10 (Lienchiang) and #06 (Penghu). In addition, two collection sites, #06 and #07 (Penghu), showed relatively high genetic differentiation with the other populations. On the contrary, the *S. viridis* population at site #08 (Jinning, Kinmen) exhibited the opposite result in our investigation. For example, the populations collected at sites #05 (Chiayi), #10 (Lienchiang) and #07 (Huxi, Penghu) separately showed the least genetic differentiation with site #08, suggesting that the composition of *S. viridis* at site #08 is more complex than the others. Finally, the population collected at site #01 (Miaoli) also showed the least genetic differentiation (F_{ST} = 0.36) with the population at site #05 (Chiayi), which also is located in Western Taiwan (Figure 3b).

Table 4. Pairwise F_{ST} value between subpopulations from the 10 collecting sites.

	#01	#02	#03	#04	#05	#06	#07	#08	#09
#02	0.47								
#03	0.54	0.22							
#04	0.55	0.28	0.17						
#05	0.36	0.47	0.46	0.51					
#06	0.57	0.52	0.56	0.52	0.50				
#07	0.58	0.59	0.58	0.62	0.50	0.50			
#08	0.41	0.38	0.42	0.44	0.32	0.49	0.40		
#09	0.52	0.49	0.55	0.55	0.42	0.59	0.50	0.14	
#10	0.51	0.52	0.55	0.58	0.46	0.59	0.54	0.32	0.45

4. Discussion

4.1. Molecular Diversity of Microsatellite from the Individuals of Green Foxtail in Taiwan

Simple sequence repeats (SSRs) are very useful for the investigation of population genetics and studies of evolution because they is easy to manipulate and contain abundant molecular variations and information [84–86]. In this study, 6.1 allele numbers were observed from 141 *S. viridis* individuals with 13 SSR markers. Despite the estimated PIC value, 0.527 on average, being lower than that in other related studies of *S. viridis* or *S. italica* [44,55,57], highly polymorphic markers were still used in our analysis (Table 2). The other related grass species of *S. viridis*, such as *Miscanthus sinensis* and switchgrass, showed different patterns. The naturalized populations of *M. sinensis* in the United States contained only 2.3 alleles per locus by 74 molecular markers, and the PIC value ranged from 0.2228 to 0.3030 [87]. Nevertheless, 8.7 alleles per locus were detected from 156 switchgrass individuals with 18 highly polymorphic markers [88]. Therefore, we thought that the quality of used markers was much more important than the quantity of markers. In our analysis, we preferred to select highly informative markers because it was reported that a small number of molecular markers with high quality can be enough for analysis of genetic variation [89]. Only four markers (SICAAS1065, SICAAS3090, SICAAS7002, SICAAS8025) displayed low extent of the observed heterozygosity (Ho), indicating that in Taiwan, *S. viridis* is a highly selfing species, which corresponds with previous studies [26,40]. In addition, a similar pattern of Ho value in terms of *S. viridis* at different collection sites was also observed (Table 3). On the other hand, importantly, the advantages of low heterozygosity (high rate of selfing) includes the ease of sequencing, and better analyzing and understanding of genetic and evolutionary mechanisms [90,91]. Besides, an effective allele number (Ne) based on the frequency of variants is an alternative measure of intrapopulation diversity [92]. Hence, it basically corresponds to the expected heterozygosity (Table 2). An average effective allele number of 3.558 can be seen in our *S. viridis* populations in terms of the assessment of 13 SSR markers. Interestingly, the Ne value of *S. viridis* in Taiwan was higher than that in mid/southern US, but much lower than the Asian populations [89].

For the purpose of better understanding the relationship between the genetic diversity and geographical distribution, we mainly focused on the genetic diversity of *S. viridis* populations from different parts of Taiwan (Table 3). The highest value of observed allele number (Na), effective allele numbers (Ne) and expected heterozygosity of *S. viridis* population were all detected at site #8 (Kinmen), indicating the highest genetic diversity of *S. viridis* was preserved in a population on Kinmen Island, near mainland China, which might be one of the origin of *S. viridis* (cp. [1,3]). On the contrary, the lowest genetic diversity was observed at site #1 (Miaoli) and site #2 (Taitung) in terms of the relatively low Na, Ne and He. The relatively low genetic diversity of intrapopulation of *S. viridis* was reported previously [40,51,52]. Wild sorghum populations collected from different regions in Africa showed high genetic diversity, He = 0.46 on average, assessed by six SSR markers [93]. Likewise, the *M. sinensis* population collected in the US showed a medium level of genetic diversity, with He values ranging from 0.2776 to 0.3738 [87]. However, in this study of *S. viridis* in Taiwan, the level of heterozygosity was considerably low in comparison with previous studies (cf. [41,89]).

4.2. Genetic Relationship and Structure of Green Foxtail Individuals among Different Parts of Taiwan

Basically, both the PCoA and phylogenetic analysis showed a similar pattern of results, which is accordance with geographical distribution and concordant with previous studies. The main island of Taiwan extends 394 km along its north–south oriented axis and a width of 140 km at its broadest. There are 268 peaks above 3000 m in elevation, and all of them are located within the so-called Central Range, which basically follows the axis of the island. No doubt this Central Range is a barrier for the gene flow between the east and west parts of Taiwan [94]. In addition to the isolation by the Taiwan Strait, the Central Range on the main island of Taiwan is also a barrier for gene flow. The *S. viridis* were also found to be grouped by regions or species in phylogenetic tree in a previous study [61]. Moreover, our results were concordant with the PCA pattern from the world-wide *S. viridis* panel in terms of SNP, PAV and structural variant genetic structure (cf. [15]). As a result, the genetic relationship may be influenced by geographical distribution for the same species [95]. In this study, some individuals from different populations clustered together in the same clade, displaying the mixed genetic branches of *S. viridis* collected at different sites in the phylogenetic tree (Figure 2). This might be the result of migration caused by human activities and long-distance animal movements. For example, there is routine transportation by ferry between Taitung City (site #03) and Ludao (site #04). The same ferry shuttle also ran between Chiayi (site #05) and Huxi (site #07). The simulated genetic backgrounds basically agreed with the administrative districts of Taiwan, except that Pop 4 and Pop 5 were two separate townships in Penghu (Figure 3a). Notably, four out of 10 *S. viridis* samples in Chiayi (site #05) showed admixed genetic background, which was also found at sites #08 and #09 in Kinmen. In addition, relatively low genetic differentiation between sites #05 and #08 was detected (Table 4), implying that some of the *S. viridis* in Chiayi might be taken from Kinmen because of human activities. Finally, pure and different genetic backgrounds at collection sites suggested that gene flow rarely happened among them, causing the pairwise genetic differentiation to be quite high. It was concluded that the *S. viridis* at these sites mainly reproduced by selfing, and it was hard for them to disperse by pollens or seeds of their own. Finally, the *S. viridis* individuals in the population at site #08 (Jinning, Kinmen) mixed in clade III and IV, showing the most complex genetic variation in our study in terms of the analysis of genetic-diversity parameters, population structure and genetic relationship (Table 3, Figure 3a).

4.3. Genetic Differentiation and Possible Gene Flow among the Populations of Different Collection Sites

F-statistics is one of the major measures to examine the differentiation between a subdivided population that deviates from the Hardy–Weinberg equilibrium. Briefly, F_{IS}, F_{IT} and F_{ST} represent the genetic diversity of different types of conditions that are individuals within the subpopulation, individuals within the total population and subpopulation within the total population, respectively [81,82]. In this study, low F_{IS} and F_{IT} values were observed at sites #08 (Kinmen) and #10 (Lienchiang), indicating that the inbreeding coefficient of *S. viridis* collected at sites #08 and #10 were relatively low. Consequently, the observed heterozygosity of *S. viridis* at these two sites were essentially high (Table 3). The extent of genetic differentiation (F_{ST}) among populations varied up to two times. *S. viridis* collected at site #08 showed the lowest genetic differentiation (F_{ST} = 0.2525), and the population at site #05 was ranked second (F_{ST} = 0.3540), indicating that gene flow in Kinmen and Chiayi might have occurred more frequently. Oppositely, the collected populations at site #01 (F_{ST} = 0.4895) and site #02 (F_{ST} = 0.4995) showed high extents of genetic differentiation, which might be due to less chance of interflow with other populations (Table 3). The higher Nm values at sites #05 and #08 were observed; however, the two admixed populations contained elevated heterozygosity (He), indicating gene flow might have taken place at these two sites recently [96,97]. Instead, the *S. viridis* at site #09 showed a relatively higher Nm value (0.33), but lower heterozygosity (0.07), which may be the result of ancient crossing, then followed by inbreeding with no cross-pollination [26]. It was reported that elevated diversity might be the result of multiple origins or larger populations [15]. The collection sites with no admixed individuals (sites #01, #02 and #04), exhibited a lower Nm value (<0.3), suggesting that less gene flow occurred before.

Pairwise F_{ST} varied dramatically, 0.14 to 0.62 (Table 4). The higher extent of genetic differentiation between the *S. viridis* populations was mostly observed at sites #06 or #07 compared with other collection sites. For example, the highest one (F_{ST} = 0.62) was detected between site #07 (Huxi, Penghu) and site #04 (Ludao, Taitung), two offshore islands of Taiwan. Likewise, some studies found that *S. viridis* both in Eurasia and North America displayed strong intracontinental differentiation [40,95]. In addition, the populations at site #09 (Kinmen) and #10 (Lienchiang) also showed the same extent of high genetic differentiation (F_{ST} = 0.59) as site #06. As a result, the *S. viridis* populations collected at sites #06 and #07 seemed to be more departed from the other populations in Taiwan, which agreed with the patterns of the PCoA and the phylogenetic tree (Figures 1 and 2). In fact, site #06 (Qimei) and site #07 (Huxi) belong to the Penghu archipelago, which is located in the middle of the Taiwan Strait (Table 1). One possible reason could be that the special climate of Penghu could have selected or filtered certain genotypes or genes, which is the same condition in North America (cf. [15,89]). Oppositely, the smallest F_{ST} was observed between sites #08 and #09 (F_{ST} = 0.14), followed by the ones between sites #03 and #04 (F_{ST} = 0.17), which are located in the same counties, Kinmen and Taitung, respectively. On the other hand, the collection sites in the same administrative district, such as sites #02, #03 and #04 (Taitung), and sites #08 and #09 (Kinmen), revealed less genetic differentiation than the others, indicating that isolation by distance or geographic barriers may affect the possibility of migration or exchange [98], such as the spread of seeds or pollens by wind and animals in nature. The other explanation might be that more frequent human activities take place within the neighboring area, which then increase the probability of artificial and unintentional movement of caryopses and individuals [99,100]. For example, there are routine ferry shuttles between the regions of sites #03 and #04.

Sites #05 and site #08 displayed relatively less genetic differentiation compared to other sites, which is congruent with their higher values of Nm (Table 3). Interestingly, the genetic distance between *S. viridis* and *S. italica* in the same area was lower than that between different areas of the same species, indicating the geographical factors was more important in shaping genetic differentiation than taxonomical factor between *S. viridis* and *S. italica* [95]. The F_{ST} value between *S. viridis* and *S. italica* (at the taxonomical level) was 0.05 to 0.41 in a previous study, which was lower than that of our study (at the geographical level) [56]. On the contrary, different races or species of rice showed significantly genetic differentiation (F_{ST} = 0.1166 to 0.42). The F_{ST} value of *Sorghum halepense* populations collected in the US were quite low, indicating not much differentiation among wild sorghum populations from different states in the US [101].

In conclusion, *S. viridis* showed a pattern of low diversity and heterozygosity within a population, but high differentiation among populations. This is accordant with a similar previous study of a Taiwanese coastal herbal plant, *Lysimachia mauritiana*, with allozyme markers [102]. In *L. mauritiana*, allozyme variation was very low, and *F*-statistics indicated an extremely high level of population differentiation, implying limited gene flow among populations. This pattern of population genetic structure probably resulted from severe genetic drift triggered by genetic bottlenecks, suggesting that Taiwanese populations were likely to be derived from four independent founder events [102]. Several selective forces could influence the population structure, genetic diversity and differentiation of *Setaria* grass were mentioned in [3]. For example, the decreased allele richness of *S. viridis* samples in North America was the result of the founder effect or genetic drift. A strong population structure of *S. viridis* had been presented in the US and Canadian collections because of different climatic zones or latitude [26,89,95]. Instead, the populations collected from China lacked an obvious population structure [41,89,103]. Surprisingly, the *S. viridis* populations in Taiwan, a small island, still displayed a strong population structure (Figure 3a). Up to 84 % of the *S. viridis* samples were considered as pure individual, indicating a high selfing rate and low extent of gene flow occurred among populations in Taiwan. However, most of admixed individuals were found at site #08, probably due to the short distance away from mainland China, resulting in a more complex genetic background than the other populations in Taiwan. The extent of genetic diversity was essentially low, but displayed strong population structure and differentiation [104–106], suggesting that not only normal, but specifically adapted genotypes still existed in the *S. viridis* populations in Taiwan (cp. [3]).

Investigation of the genetic structure, differentiation and related pattern of geographical distribution can help us to better understand the domestication and evolutionary mechanism [107]. Moreover, the *S. viridis* collections with abundant diversity and characteristics could be a valuable resource for breeding programs of foxtail millet, as is planned or is already underway globally for other crop wild relatives (CWR) for the purpose of enhancing genetic resources, including those at risk of extinction [108–110]. Finally, the related knowledge needs to be re-examined at a genome-wide level, and even include more geographic areas with larger collections to investigate subtler population structure and obtain a better diversity panel that can be applied in GWAS (Genome-wide association study) and breeding program in the future.

Supplementary Materials: The following are available online at https://www.mdpi.com/article/10.3390/d13040159/s1, Figure S1: Delta K value from $K = 2$ to $K = 10$ evaluated by the Evanno method.

Author Contributions: C.-H.C. and Y.-R.L. initiated this project and looked for financial support; W.-H.H. and Y.-R.L. designed the experiments. The experiment and data analysis were conducted by Y.-C.C., and H.-C.L. and C.-H.C. collected all the samples. Finally, W.-H.H. and C.-H.C. drafted and revised the manuscript. All authors have read and agreed to the final version of the manuscript.

Funding: This research was funded from the project 106-2313-B-002-034-MY3 by Ministry of Science and Technology, Taipei, Taiwan and also the project 106AS-11.9.2-EI-W1 from the Council of Agriculture, Executive Yuan, Taipei, Taiwan.

Institutional Review Board Statement: Not applicable.

Informed Consent Statement: Not applicable.

Data Availability Statement: The data is not publicly available owing to privacy concerns. The data is available on request from the corresponding author.

Acknowledgments: This work was supported by the project 106AS-11.9.2-EI-W1 from the Council of Agriculture, Executive Yuan, Taipei, Taiwan. The source of funding was from the project 106-2313-B-002-034-MY3 from the Ministry of Science and Technology, Taipei, Taiwan.

Conflicts of Interest: The authors declare no conflict of interest.

References

1. Brutnell, T.P.; Wang, L.; Swartwood, K.; Goldschmidt, A.; Jackson, D.; Zhu, X.G.; Kellogg, E.; Eck, J.V. *Setaria viridis*: A model for C4 photosynthesis. *Plant Cell* **2010**, *22*, 2537–2544. [CrossRef]
2. Chen, C.H.; Lin, C.Y.; Kuoh, C.S. Panicoideae, Paniceae. In *Grass Flora of Taiwan (2 of 3)*; Endemic Species Research Institute: Chichi, Taiwan, 2014; p. 158.
3. Dekker, J. The foxtail (*Setaria*) species-group. *Weed Sci.* **2003**, *51*, 641–656. [CrossRef]
4. Li, Y.; Wu, S.Z. Traditional maintenance and multiplication of foxtail millet (*Setaria italica* (L.) P. Beauv) landraces in China. *Euphytica* **1996**, *87*, 33–38. [CrossRef]
5. Li, H.W.; Li, C.H.; Pao, W.K. Cytological and genetical studies of the interspecific cross of the cultivated foxtail millet, *Setaria italica* (L.) Beauv., and the green foxtail millet, *S. viridis* L. *Agron. J.* **1945**, *9*, 32–54. [CrossRef]
6. Yang, X.; Wan, Z.; Perry, L.; Lu, H.; Wang, Q.; Zhao, C.; Li, J.; Xie, F.; Yu, J.; Cui, T.; et al. Early millet use in northern China. *Proc. Natl. Acad. Sci. USA* **2012**, *109*, 3726–3730. [CrossRef]
7. Barton, L.; Newsome, S.D.; Chen, F.; Wang, H.; Guilderson, T.P.; Bettinger, R.L. Agricultural origins and the isotopic identity of domestication in northern china. *Proc. Natl. Acad. Sci. USA* **2009**, *106*, 5523–5528. [CrossRef]
8. Austin, D.F. Foxtail millets (*Setaria*: Poaceae)—Abandoned food in two hemispheres. *Econ. Bot.* **2006**, *60*, 143–158. [CrossRef]
9. Food and Agriculture Organization of the United Nations. FAOSTAT Statistical Database. Available online: http://www.fao.org/faostat/en/#data/QC/visualize (accessed on 15 October 2020).
10. Benabdelmouna, A.; Shi, Y.; Abirached-Darmency, M.; Darmency, H. Genomic in situ hybridization (GISH) discriminates between the A and the B genomes in diploid and tetraploid *Setaria* species. *Genome* **2001**, *44*, 685–690. [CrossRef] [PubMed]
11. Giussani, L.M.; Cota-Sánchez, J.H.; Zuloaga, F.O.; Kellogg, E.A. A molecular phylogeny of the grass subfamily Panicoideae (Poaceae) shows multiple origins of C4 photosynthesis. *Am. J. Bot.* **2001**, *88*, 1993–2012. [CrossRef] [PubMed]
12. Doust, A.N.; Penly, A.M.; Jacobs, S.W.L.; Kellogg, E.A. Congruence, conflict, and polyploidization shown by nuclear and chloroplast markers in the monophyletic "bristle clade" (Paniceae, Panicoideae, Poaceae). *Syst. Bot.* **2007**, *32*, 531–544. [CrossRef]
13. Harlan, J.R.; de Wet, J.M.J. Towards a rational classification of cultivated plants. *Taxon* **1971**, *20*, 509–517. [CrossRef]
14. Darmency, H.; Zangre, G.R.; Pernes, J. The wild-weed-crop complex in *Setaria*: A hybridization study. *Genetica* **1987**, *75*, 103–107. [CrossRef]
15. Mamidi, S.; Healey, A.; Huang, P.U.; Grimwood, J.; Jenkins, J.; Barry, K.; Sreedasyam, A.; Shu, S.; Lovell, J.T.; Feldman, M.; et al. A genome resource for green millet *Setaria viridis* enables discovery of agronomically valuable loci. *Nat. Biotechnol.* **2020**, *38*, 1203–1210. [CrossRef]

16. Bennetzen, J.L.; Schmutz, J.; Wang, H.; Percifield, R.; Hawkins, J.; Pontaroli, A.C.; Estep, M.; Feng, L.; Vaughn, J.N.; Grimwood, J.; et al. Reference genome sequence of the model plant *Setaria*. *Nat. Biotechnol.* **2012**, *30*, 555–561. [CrossRef] [PubMed]
17. Brutnell, T.P.; Bennetzen, J.L.; Vogel, J.P. *Brachypodium distachyon* and *Setaria viridis*: Model genetic systems for the grasses. *Annu. Rev. Plant Biol.* **2015**, *66*, 465–485. [CrossRef] [PubMed]
18. Huang, P.; Shyu, C.; Coelho, C.P.; Cao, Y.; Brutnell, T.P. *Setaria viridis* as a model system to advance millet genetics and genomics. *Front. Plant Sci.* **2016**, *7*, 1781. [CrossRef] [PubMed]
19. Li, P.; Brutnell, T.P. *Setaria viridis* and *Setaria italica*, model genetic systems for the Panicoid grasses. *J. Exp. Bot.* **2011**, *62*, 3031–3037. [CrossRef] [PubMed]
20. Zhang, G.; Liu, X.; Quan, Z.; Cheng, S.; Xu, X.; Pan, S.; Xie, M.; Zeng, P.; Yue, Z.; Wang, W.; et al. Genome sequence of foxtail millet (*Setaria italica*) provides insights into grass evolution and biofuel potential. *Nat. Biotechnol.* **2012**, *30*, 549–554. [CrossRef]
21. Diao, X.; Schnable, J.; Bennetzen, J.L.; Li, J. Initiation of *Setaria* as a model plant. *Front Agric. Sci. Eng.* **2014**, *1*, 16–20.
22. Doust, A.N.; Kellogg, E.A.; Devos, K.M.; Bennetzen, J.L. Foxtail millet: A sequence-driven grass model system. *Plant Physiol.* **2009**, *149*, 137–141. [CrossRef]
23. Chen, S.L.; Phillips, S.M. *Setaria* P. Beauv. In *Flora of Chine*; Wu, Z.Y., Raven, P.R., Eds.; Science Press: Beijing, China, 2006; Volume 22, pp. 531–537.
24. Lorenzi, H.J.; Jeffery, L.S. *Weeds of the United States and Their Control*; Van Nostrand Reinhold: New York, NY, USA, 1987.
25. Defelice, M.S. Green Foxtail, *Setaria viridis* (L.) P. Beauv. *Weed Technol.* **2002**, *16*, 253–257. [CrossRef]
26. Huang, P.; Feldman, M.; Schröder, S.; Bahri, B.A.; Diao, X.; Zhi, H.; Estep, M.; Baxter, I.; Devos, K.M.; Kellogg, E.A. Population genetics of *Setaria viridis*, a new model system. *Mol. Ecol.* **2014**, *23*, 4912–4925. [CrossRef]
27. Ridley, H.N. *The Dispersal of Plants throughout the World*; Reeve: Ashford, UK, 1930; p. 744.
28. Douglas, B.J.; Morrison, I.N.; Thomas, A.G.; Maw, M.G. The Biology of Canadian Weeds. 70. *Setaria viridis* (L.) Beauv. *Can. J. Plant Sci.* **1985**, *65*, 669–690. [CrossRef]
29. Holm, L.G.; Plucknett, D.L.; Pancho, J.V.; Herberger, J.P. *The World's Worst Weeds: Distribution and Biology*; Krieger Publishing: Malabar, FL, USA, 1991; p. 609.
30. Holm, L.G.; Plucknett, D.L.; Pancho, J.V.; Herberger, J.P. *The World's Worst Weeds-Distribution and Biology*; The East-West Food Institute: Honolulu, HI, USA, 1977.
31. Wilson, R.G. Dissemination of weed seeds by surface irrigation water in Western Nebraska. *Weed Sci.* **1980**, *28*, 87–92. [CrossRef]
32. Frankton, C.; Mulligan, G.A. *Weeds of Canada*; Canada Department of Agriculture, Queen's Printer: Ottawa, ON, Canada, 1970; p. 948.
33. Perrino, E.V.; Calabrese, G. Endangered segetal species in southern Italy: Distribution, conservation status, trends, actions and ethnobotanical notes. *Genet. Resour. Crop Evol.* **2018**, *65*, 2107–2134. [CrossRef]
34. Rominger, J.M. *Taxonomy of Setaria (Graminaceae) in North America*; University of Illinois Press: Urbana, IL, USA, 1962.
35. Yatskievych, G. *Steyermarks's Flora of Missouri*; Missouri Botanical Garden Press: St. Louis, MO, USA, 1999; Volume 1, p. 991.
36. Blackshaw, R.E.; Stobbe, E.H.; Sturko, A.R.W. Effect of seeding dates and densities of green foxtail (*Setaria viridis*) on the growth and productivity of spring wheat (*Triticum aestivum*). *Weed Sci.* **1981**, *29*, 212–217. [CrossRef]
37. Gates, F.C. *Weeds in Kansas*; Kansas State Printing Plant: Topeka, KS, USA, 1941; p. 360.
38. Hunt, H.V.; Linden, M.V.; Liu, X.; Motuzaite-Matuzeviciute, G.; Colledge, S.; Jones, M.K. Millets across Eurasia: Chronology and context of early records of the genera *Panicum* and *Setaria* from archaeological sites in the Old World. *Veg. Hist. Archaeobot.* **2008**, *17*, 5–18. [CrossRef] [PubMed]
39. Duke, J.A. *Handbook of Edible Weeds*; CRC Press: Boca Raton, FL, USA, 1992; p. 246.
40. Wang, R.L.; Wendel, J.F.; Dekker, J.H. Weedy adaptation in *Setaria* spp. I. Isozyme analysis of genetic diversity and population genetic structure in *Setaria viridis*. *Am. J. Bot.* **1995**, *82*, 308–317. [CrossRef]
41. Jia, G.; Shi, S.; Wang, C.; Niu, Z.; Chai, Y.; Zhi, H.; Diao, X. Molecular diversity and population structure of Chinese green foxtail [*Setaria viridis* (L.) Beauv.] revealed by microsatellite analysis. *J. Exp. Bot.* **2013**, *64*, 3645–3656. [CrossRef]
42. Shi, Y.; Wang, T.; Li, Y.; Darmency, H. Impact of transgene inheritance on the mitigation of gene flow between crops and their wild relatives: The example of foxtail millet. *Genetics* **2008**, *180*, 969–975. [CrossRef]
43. Pohl, R.W. The genus *Setaria* in Iowa. Iowa State. *J. Sci.* **1951**, *25*, 501–508.
44. de Wet, J.M.J. The origin of weediness in plants. *Proc. Okla. Acad. Sci.* **1966**, *1966*, 14–17.
45. Harlan, J.R. The possible role of weed races in the evolution of cultivated plants. *Euphytica* **1965**, *14*, 173–176. [CrossRef]
46. Harlan, J.R.; de Wet, J.M.J.; Price, E.G. Comparative evolution of cereals. *Evolution* **1973**, *27*, 311–325. [CrossRef] [PubMed]
47. Wang, T.Y.; Zhao, Z.H.; Yan, H.B.; Li, Y.; Zhu, X.H.; Shi, Y.S.; Song, Y.C.; Ma, Z.Y.; Darmency, H. Gene flow from cultivated herbicide-resistant foxtail millet to its wild relatives: A basis for risk assessment of the release of transgenic millet. *Acta Agron. Sin.* **2001**, *27*, 681–687. (in Chinese).
48. Kuo, S.M.; Chen, Y.R.; Yin, S.Y.; Ba, Q.X.; Tsai, Y.C.; Kuo, W.H.J.; Lin, Y.R. Waxy allele diversification in foxtail millet (*Setaria italica*) landraces of Taiwan. *PLoS ONE* **2018**, *13*, e0210025. [CrossRef]
49. Hirano, R.; Naito, K.; Fukunaga, K.; Watanabe, K.N.; Ohsawa, R.; Kawase, M. Genetic structure of landraces in foxtail millet (*Setaria italica* (L.) P. Beauv.) revealed with transposon display and interpretation to crop evolution of foxtail millet. *Genome* **2011**, *54*, 498–506. [CrossRef]

50. Fukunaga, K.; Ichitani, K.; Kawase, M. Phylogenetic analysis of the rDNA intergenic spacer subrepeats and its implication for the domesticated history of foxtail millet, *Setaria italica*. *Theor. Appl. Genet.* **2006**, *113*, 261–269. [CrossRef]
51. Hamrick, J.L.; Godt, M.J.W. Allozyme diversity in plant species. In *Plant Population Genetics, Breeding and Genetic Resources*; Brown, A.H.D., Clegg, M.T., Kahler, A.L., Weir, B.S., Eds.; Sinauer: Sunderland, MA, USA, 1990; pp. 43–63.
52. Wang, R.L.; Wendel, J.F.; Dekker, J.H. Weedy adaptation in *Setaria* spp. II. Genetic diversity and population genetic structure in *S. glauca*, *S. geniculata*, and *S. faberii* (Poaceae). *Am. J. Bot.* **1995**, *82*, 1031–1039. [CrossRef]
53. Huang, X.; Kurata, N.; Wei, X.; Wang, Z.X.; Wang, A.; Zhao, Q.; Zhao, Y.; Liu, K.; Lu, H.; Li, W.; et al. A map of rice genome variation reveals the origin of cultivated rice. *Nature* **2012**, *490*, 497–501. [CrossRef]
54. Wright, S.I.; Bi, I.V.; Schröder, S.G.; Yamasaki, M.; Doebley, J.F.; McMullen, M.D.; Gaut, B.S. The effects of artificial selection on the maize genome. *Science* **2005**, *308*, 1310–1314. [CrossRef] [PubMed]
55. Liu, Z.; Bai, G.; Zhang, D.; Znu, C.; Xia, X.; Cheng, Z.; Shi, Z. Genetic diversity and population structure of elite foxtail millet (*Setaria italica* (L.) P. Beauv.) germplasm in China. *Crop Sci.* **2011**, *51*, 1655–1663. [CrossRef]
56. Wang, C.; Chen, J.; Zhi, H.; Yang, L.; Li, W.; Wang, Y.; Li, H.; Zhao, B.; Chen, M.; Diao, X. Population genetics of foxtail millet and its wild ancestor. *BMC Genet.* **2010**, *11*, 90. [CrossRef]
57. Wang, C.; Jia, G.; Zhi, H.; Niu, Z.; Chai, Y.; Li, W.; Wang, Y.; Li, H.; Lu, P.; Zhao, B.; et al. Genetic diversity and population structure of Chinese foxtail millet [*Setaria italica* (L.) Beauv.] landraces. *G3 (Bethesda)* **2012**, *2*, 769–777. [CrossRef]
58. Gupta, S.; Kumari, K.; Muthamilarasan, M.; Parida, S.K.; Prasad, M. Population structure and association mapping of yield contributing agronomic traits in foxtail millet. *Plant Cell Rep.* **2014**, *33*, 881–893. [CrossRef]
59. Vetriventhan, M.; Upadhyaya, H.D.; Anandakumar, C.R.; Senthilvel, S.; Varshney, R.K.; Parzies, H.K. Population structure and linkage disequilibrium of ICRISAT foxtail millet (*Setaria italica* (L.) P. Beauv.) core collection. *Euphytica* **2014**, *196*, 423–435. [CrossRef]
60. Fukunaga, K.; Wang, Z.; Kato, K.; Kawase, M. Geographical variation of nuclear genome RFLPs and genetic differentiation in foxtail millet, *Setaria italica* (L.) P. Beauv. *Genet. Resour. Crop Evol.* **2002**, *49*, 95–101. [CrossRef]
61. Li, W.; Zhi, H.; Wang, Y.; Li, H.; Diao, X. Assessment of genetic relationship of foxtail millet with its wild ancestor and close relatives by ISSR markers. *J. Integr. Agric.* **2012**, *11*, 556–566. [CrossRef]
62. Kim, S.; Plagnol, V.; Hu, T.T.; Toomajian, C.; Clark, R.M.; Ossowski, S.; Ecker, J.R.; Weigel, D.; Nordborg, M. Recombination and linkage disequilibrium in *Arabidopsis thaliana*. *Nat. Genet.* **2007**, *39*, 1151–1155. [CrossRef] [PubMed]
63. Brachi, B.; Morris, G.P.; Borevitz, J.O. Genome-wide association studies in plants: The missing heritability is in the field. *Genome Biol.* **2011**, *12*, 232. [CrossRef]
64. Honda, M. *Monographia Poacearum Japonicarum, Bambusoideis Exclusis (Journal of the Faculty of Science: Section III, Botany)*; Imperial University of Tokyo: Tokyo, Japan, 1930; Volume 3, pp. 1–484.
65. Tsang, C.H.; Li, K.T.; Hsu, T.F.; Tsai, Y.C.; Fang, P.H.; Hsing, Y.I.C. Broomcorn and foxtail millet were cultivated in Taiwan about 5000 years ago. *Bot. Stud.* **2017**, *58*, 3. [CrossRef] [PubMed]
66. Thomson, D.; Dietzgen, R.G. Detection of DNA and RNA plant viruses by PCR and RT-PCR using a rapid virus release protocol without tissue homogenization. *J. Virol. Methods* **1995**, *54*, 85–95. [CrossRef]
67. Zhang, S.; Tang, C.; Zhao, Q.; Li, J.; Yang, L.; Qie, L.; Fan, X.; Li, L.; Zhang, N.; Zhao, M.; et al. Development of highly polymorphic simple sequence repeat markers using genome-wide microsatellite variant analysis in Foxtail millet [*Setaria italica* (L.) P. Beauv.]. *BMC Genom.* **2014**, *15*, 78. [CrossRef]
68. Hayden, M.J.; Nguyen, T.M.; Waterman, A.; Chalmers, K.J. Multiplex-Ready PCR: A new method for multiplexed SSR and SNP genotyping. *BMC Genom.* **2008**, *9*, 80. [CrossRef]
69. Covarrubias-Pazaran, G.; Diaz-Garcia, L.; Schlautman, B.; Salazar, W.; Zalapa, J. Fragman: An R package for fragment analysis. *BMC Genet.* **2016**, *17*, 62. [CrossRef] [PubMed]
70. R Core Team. *R: A language and Environment for Statistical Computing*; R Foundation for Statistical Computing: Vienna, Austria, 2019; Available online: https://www.R-project.org/ (accessed on 10 November 2019).
71. Nei, M. Analysis of gene diversity in subdivided populations. *Proc. Nat. Acad. Sci. USA* **1973**, *70*, 3321–3323. [CrossRef]
72. Adamack, A.T.; Gruber, B. Popgen Report: Simplifying basic population genetic analyses in R. *Methods Ecol. Evol.* **2014**, *5*, 384–387. [CrossRef]
73. Clark, L.V.; Jasieniuk, M. POLYSAT: An R package for polyploid microsatellite analysis. *Mol. Ecol. Resour.* **2011**, *11*, 562–566. [CrossRef]
74. Jombart, T. Adegenet: A R package for the multivariate analysis of genetic markers. *Bioinformatics* **2008**, *24*, 1403–1405. [CrossRef] [PubMed]
75. Saitou, N.; Nei, M. The neighbor-joining method: A new method for reconstructing phylogenetic trees. *Mol. Biol. Evol.* **1987**, *4*, 406–425.
76. Pritchard, J.K.; Stephens, M.; Donnelly, P. Inference of population structure using multilocus genotype data. *Genetics* **2000**, *155*, 945–959. [PubMed]
77. Francis, R.M. pophelper: An R package and web app to analyze and visualize population structure. *Mol. Ecol. Resour.* **2017**, *17*, 27–32. [CrossRef]
78. Evanno, G.; Regnaut, S.; Goudet, J. Detecting the number of clusters of individuals using the software Structure: A simulation study. *Mol. Ecol.* **2005**, *14*, 2611–2620. [CrossRef]

79. Wickham, H. *ggplot2: Elegant Graphics for Data Analysis*; Springer Publishing Company: New York, NY, USA, 2009.
80. Yu, G.; Smith, D.K.; Zhu, H.; Guan, Y.; Lam, T.T.Y. ggtree: An R package for visualization and annotation of phylogenetic trees with their covariates and other associated data. *Methods Ecol. Evol.* **2017**, *8*, 28–36. [CrossRef]
81. Wright, S. The genetical structure of populations. *Ann. Eugen.* **1951**, *15*, 323–354. [CrossRef] [PubMed]
82. Weir, B.S.; Cockerham, C.C. Estimating *F*-statistics for the analysis of population structure. *Evolution* **1984**, *38*, 1358–1370.
83. Wright, S. Isolation by distance. *Genetics* **1943**, *28*, 114–138.
84. Matsuoka, Y.; Mitchell, S.E.; Kresovich, S.; Goodman, M.; Doebley, J. Microsatellites in Zea-variability, patterns of mutations, and use for evolutionary studies. *Theor. Appl. Genet.* **2002**, *104*, 436–450. [CrossRef]
85. Tautz, D.; Renz, M. Simple sequences are ubiquitous components of eukaryotic genomes. *Nucleic Acids Res.* **1984**, *12*, 4127–4138. [CrossRef]
86. Kumar, P.; Gupta, V.K.; Misra, A.K.; Modi, D.R.; Pandey, B.K. Potential of molecular markers in plant biotechnology. *Plant OMIC* **2009**, *2*, 141–162.
87. Zhao, Y.; Basak, S.; Fleener, C.E.; Egnin, M.; Sacks, E.J.; Prakash, C.S.; He, G. Genetic diversity of *Miscanthus sinensis* in US naturalized populations. *GCB Bioenergy* **2017**, *9*, 965–972. [CrossRef]
88. Ecker, G.; Zalapa, J.; Auer, C. Switchgrass (*Panicum virgatum* L.) genotypes differ between coastal sites and inland road corridors in the Northeastern US. *PLoS ONE* **2015**, *10*, e0130414. [CrossRef] [PubMed]
89. Schröder, S.; Bahri, B.A.; Eudy, D.M.; Layton, D.J.; Kellogg, E.A.; Devos, K.M. Genetic diversity and origin of North American green foxtail [*Setaria viridis* (L.) Beauv.] accessions. *Genet. Resour. Crop Evol.* **2017**, *64*, 367–378. [CrossRef]
90. Fang, X.; Dong, K.; Wang, X.; Liu, T.; He, J.; Ren, R.; Zhang, L.; Liu, R.; Liu, X.; Li, M.; et al. A high density genetic map and QTL for agronomic and yield traits in Foxtail millet [*Setaria italica* (L.) P. Beauv.]. *BMC Genom.* **2016**, *17*, 336. [CrossRef]
91. Gupta, S.; Kumari, K.; Sahu, P.P.; Vidapu, S.; Prasad, M. Sequence-based novel genomic microsatellite markers for robust genotyping purposes in foxtail millet [*Setaria italica* (L.) P. Beauv]. *Plant Cell Rep.* **2012**, *31*, 323–337. [CrossRef]
92. Nei, M. Estimation of average heterozygosity and genetic distance from a small number of individuals. *Genetics* **1978**, *89*, 583–590.
93. Muraya, M.M.; de Villiers, S.; Parzies, H.K.; Mutegi, E.; Sagnard, F.; Kanyenji, B.M.; Kiambi, D.; Geiger, H.H. Genetic structure and diversity of wild sorghum populations (*Sorghum* spp.) from different eco-geographical regions of Kenya. *Theor. Appl. Genet.* **2011**, *123*, 571–583. [CrossRef]
94. Hsieh, C.F.; Shen, C.F. Introduction to the flora of Taiwan, 1: Geography, geology, climate, and soils. In *Flora of Taiwan*, 2nd ed.; Department of Botany, National Taiwan University: Taipei, Taiwan, 1994; Volume 1, pp. 1–3.
95. Jusuf, M.; Pernes, J. Genetic variability of foxtail millet (*Setaria italica* P. Beauv.). *Theor. Appl. Genet.* **1985**, *71*, 385–391. [CrossRef]
96. Slatkin, M.; Barton, N.M. A comparison of three indirect methods for estimating average levels of gene flow. *Evolution* **1989**, *43*, 1349–1368. [CrossRef]
97. Nei, M. *Molecular Evolutionary Genetics*; Columbia University Press: New York, NY, USA, 1987.
98. Upadhyaya, H.D.; Dwivedi, S.L.; Baum, M.; Varshney, R.K.; Udupa, S.M.; Gowda, C.L.L. Genetic structure, diversity, and allelic richness in composite collection and reference set in chickpea (*Cicer arietinum* L.). *BMC Plant Biol.* **2008**, *8*, 106. [CrossRef] [PubMed]
99. Nguyen, V.E.; Pernès, J. Genetic diversity of foxtail millet (*S. italica*). In *Genetic Differenciation and Dispersal in Plants*; Jacquard, P., Ed.; NATO ASI series; Springer: Berlin/Heidelberg, Germany, 1985; Volume G5, pp. 113–128.
100. Ohadi, S.; Littlejohn, M.; Mesgaran, M.; Rooney, W.; Bagavathiannan, M. Surveying the spatial distribution of feral sorghum (*Sorghum bicolor* L.) and its sympatry with johnsongrass (*S. halepense*) in South Texas. *PLoS ONE* **2018**, *13*, e0195511.
101. Sezen, U.U.; Barney, J.N.; Atwater, D.Z.; Pederson, G.A.; Pederson, J.F.J.; Chandler, M.; Cox, T.S.; Cox, S.; Dotray, P.; Kopec, D.; et al. Multi-Phase US Spread and Habitat Switching of a Post-Columbian Invasive, *Sorghum halepense*. *PLoS ONE* **2016**, *11*. [CrossRef]
102. Kono, Y.; Chung, K.F.; Chen, C.H.; Hoshi, Y.; Setoguchi, H.; Chou, C.H.; Oginuma, K.; Peng, C.I. Intraspecific karyotypic polymorphism is highly concordant with allozyme variation in *Lysimachia mauritiana* (Primulaceae: Myrsinoideae) in Taiwan: Implications for the colonization history and dispersal patterns of coastal plants. *Ann. Bot.* **2012**, *110*, 1119–1135. [CrossRef] [PubMed]
103. Le Thierry d'Ennequin, M.; Panaud, O.; Toupance, B.; Sarr, A. Assessment of genetic relationships between *Setaria italica* and its wild relative *S. viridis* using AFLP markers. *Theor. Appl. Genet.* **2000**, *100*, 1061–1066. [CrossRef]
104. Barrett, S.C.H.; Shore, J.S. Isozyme variation in colonizing plants. In *Isozymes in Plant Biology*; Soltis, D.E., Soltis, P.S., Eds.; Dioscorides: Portland, OR, USA, 1989; pp. 106–126.
105. Barrett, S.C.H. Genetics and evolution of agricultural weeds. In *Weed Management in Agro-Ecosystems: Ecological Approach*; Altieri, M., Liebman, M., Eds.; CRC Press: Boca Raton, FL, USA, 1988; pp. 58–75.
106. Rice, K.; Jain, S.K. Plant population genetics and evolution in disturbed environments. In *The Ecology of Natural Disturbance and Patch Dynamics*; Pickett, S.T.A., White, P.A., Eds.; Academic Press: New York, NY, USA, 1985; pp. 287–303.
107. Hartl, D.L.; Clark, A.G. *Principles of Population Genetics*; Sinauer Associates: Sunderland, MA, USA, 2007.
108. Perrino, E.V.; Wagensommer, R.P. Crop Wild Relatives (CWR) Priority in Italy: Distribution, Ecology, In Situ and Ex Situ Conservation and Expected Actions. *Sustainability* **2021**, *13*, 1682. [CrossRef]

109. Castañeda-Álvarez, N.; Khoury, C.; Achicanoy, H.; Bernau, V.; Dempewolf, H.; Eastwood, R.J.; Guarino, L.; Harker, R.H.; Jarvis, A.; Maxted, N.; et al. Global conservation priorities for crop wild relatives. *Nat. Plants* **2016**, *2*, 16022. [CrossRef]
110. Jarvis, D.I.; Brown, A.H.; Cuong, P.H.; Collado-Panduro, L.; Latournerie-Moreno, L.; Gyawali, S.; Tanto, T.; Sawadogo, M.; Mar, I.; Sadiki, M.; et al. A global perspective of the richness and evenness of traditional crop-variety diversity maintained by farming communities. *Proc. Natl. Acad. Sci. USA* **2008**, *105*, 5326–5331. [CrossRef]

diversity

MDPI

Article

Detection of Target-Site Herbicide Resistance in the Common Ragweed: Nucleotide Polymorphism Genotyping by Targeted Amplicon Sequencing

Barbara Kutasy [1], Zoltán Farkas [2], Balázs Kolics [1], Kincső Decsi [1], Géza Hegedűs [3], Judit Kovács [4], János Taller [1], Zoltán Tóth [2], Nikoletta Kálmán [5], Gabriella Kazinczi [6] and Eszter Virág [1,*]

1 Department of Plant Science and Biotechnology, Georgikon Campus, Szent István University, 7 Festetics Str., 8360 Keszthely, Hungary; kutasybarbara@gmail.com (B.K.); bkolics@gmail.com (B.K.); decsikincso@gmail.com (K.D.); taller.janos@szie.hu (J.T.)
2 Department of Crop Production and Land Use, Georgikon Campus, Szent István University, 7 Festetics Str., 8360 Keszthely, Hungary; farkas.zoltan@szie.hu (Z.F.); toth.zoltan@szie.hu (Z.T.)
3 Department of Economic Methodology, Georgikon Campus, Szent István University, 16 Deák Ferenc Str., 8360 Keszthely, Hungary; hegedus@educomat.hu
4 Gregor Mendel Institute of Molecular Plant Biology, Austrian Academy of Sciences, Vienna Biocenter, Dr. Bohr-Gasse 3, 1030 Vienna, Austria; judit.kovacs@gmi.oeaw.ac.at
5 Department of Biochemistry and Medical Chemistry, University of Pécs Medical School, 12 Szigeti Str., 7624 Pécs, Hungary; knikoletta@gmail.com
6 Institute of Plant Science, Kaposvár Campus, Szent István University, 40 Guba S. Str., 7400 Kaposvár, Hungary; pacseszakne.kazinczi.gabriella@szie.hu
* Correspondence: virag.eszter.andrea@szie.hu

Citation: Kutasy, B.; Farkas, Z.; Kolics, B.; Decsi, K.; Hegedűs, G.; Kovács, J.; Taller, J.; Tóth, Z.; Kálmán, N.; Kazinczi, G.; et al. Detection of Target-Site Herbicide Resistance in the Common Ragweed: Nucleotide Polymorphism Genotyping by Targeted Amplicon Sequencing. *Diversity* **2021**, *13*, 118. https://doi.org/10.3390/d13030118

Academic Editor: Mario A. Pagnotta

Received: 29 January 2021
Accepted: 6 March 2021
Published: 10 March 2021

Abstract: Background: The spread of herbicide-resistance *Ambrosia artemisiifolia* threatens not only the production of agricultural crops, but also the composition of weed communities. The reduction of their spread would positively affect the biodiversity and beneficial weed communities in the arable habitats. Detection of resistant populations would help to reduce herbicide exposure which may contribute to the development of sustainable agroecosystems. Methods: This study focuses on the application of target-site resistance (TSR) diagnostic of *A. artemisiifolia* caused by different herbicides. We used targeted amplicon sequencing (TAS) on Illumina Miseq platform to detect amino acid changes in herbicide target enzymes of resistant and wild-type plants. Results: 16 mutation points of four enzymes targeted by four herbicide groups, such as Photosystem II (PSII), Acetohydroxyacid synthase (AHAS), 5-enolpyruvylshikimate 3-phosphate synthase (EPSPS) and protoporphyrinogen IX oxidase (PPO) inhibitors have been identified in common ragweed populations, so far. All the 16 mutation points were analyzed and identified. Out of these, two mutations were detected in resistant biotypes. Conclusions: The applied next-generation sequencing-targeted amplicon sequencing (NGS-TAS) method on *A. artemisiifolia* resistant and wild-type populations enable TSR detection of large sample numbers in a single reaction. The NGS-TAS provides information about the evolved herbicide resistance that supports the integrated weed control through the reduction of herbicide exposure which may preserve ecological properties in agroecosystems.

Keywords: target-site herbicide resistance; *ahas*; *als*; *psbA*; *epsps*; targeted amplicon sequencing; *Ambrosia artemisiifolia*

1. Introduction

Extensive weed management in agricultural fields is a major threat to agroecosystem biodiversity as weed communities play a key role in maintaining heterogeneity. The total number of weed species in agricultural fields has increased since the 1950s. Although this has resulted in a more diverse composition of agricultural weeds, it has negatively affected landscape structure and land-use intensity. As a result, crop intensification methods, such as herbicide treatment, have become critical to control weed communities to improve agricultural crop production [1]. It is important not to endanger beneficial weed species while using weed control. To define the correct management of invasive species, local environmental drivers have to be identified due to their severe effects on individual performance [2].

Common ragweed (*Ambrosia artemisiifolia* L.: Asteraceae, *A. artemisiifolia)* is a highly adaptive, invasive weed and has a negative impact on biodiversity [3,4]. Its population can be found not only in agricultural habitats where herbicides are normally used but also in field margins and paths, abandoned fields, and forest edges that could be affected by herbicide or intensified mowing [5,6]. Common ragweed has spread from North America to habitats in temperate climates such as Eurasia, Asia and Australia [7,8], and has become distributed in all countries of Southern and Eastern Europe [9–12]. In North America large common ragweed populations have extensively arisen along roadsides and arable fields as well, which was recently observed mainly in Eastern Canada [13].

Many weed scientists deliberate on the knowledge and practice gap of the perspectives and approaches in weed ecology and management [14–16]. Due to the complexity of the problem, it requires transdisciplinary. Transdisciplinary approaches contribute to the most promising technological improvements managing weed control and agroecosystem diversity and reduce the potential unwanted environmental impacts of weed management. The negative correlation between weed diversity and crop yield loss has been reported [17]. As a consequence of the diverse weed management and crop production systems, the more adaptable weedy plants may replace the invasive species and the given ecosystems benefit [15]. Therefore, a well-thought herbicide use strategy should be effective and reasonable. The weed thresholds are required to promote the usage of optimal herbicides to protect crop yields and arable habitats [18].

To improve weed management and avoid ecological damages, monitoring the different herbicide-resistance in a population seems to be a great solution to find the optimal chemical crop protection. In this study, we focus on the subject of the common ragweed chemical control. Our primary goal is to achieve the controlled eradication of ragweed or at least to maintain its low moderate level while minimizing the ecological damage caused by the applied herbicide.

Common ragweed has evolved different herbicide resistance modes, making it particularly problematic to control [19]. Herbicide resistance can be categorized into two categories based on mode of action: target-site resistance (TSR) and non-target-site resistance (NTSR) [20–22]. Here, we focus on TSR resistance, which can be caused by mutation or polymorphism that results in either increased abundance of the target protein or a structural change in the target protein that reduces affinity to the herbicide [23,24]. Structural changes are generally due one or several amino acid substitutions [25–29]. The accumulation of multiple amino acid substitutions in a target protein has been known as multiple target-site resistance (MTSR) which triggers an increased resistance level of individual plants [22,30,31]. MTSR has been detected in different weeds conferring resistance to Photosystem II (PSII) and Acetohydroxyacid synthase (AHAS) inhibitors [32–34]. Resistant genotypes proliferate in areas where herbicides are used for weed control. In order to effectively combat herbicide-resistant common ragweed, effective methods to measure the prevalence of TSR and MTSR within target populations are needed.

Four herbicide mode-of-action groups are commonly used to control *A. artemisiifolia*. These are: (i) triazines and ureas (Photosystem II (PSII) inhibitors), (ii) sulfonylureas, imidazolinones, sulfonylaminocarbonyl-triazolinones, pyrimidinylthiobenzoates and triazolopyrimidines (AHAS inhibitors), (iii) glyphosate (5-enolpyruvylshikimate 3-phosphate synthase (EPSPS) inhibitors) and (iv) diphenylethers, N-phenylphthalimides, aryl triazolinones, phenylpyrazoles, pyrimidinediones (protoporphyrinogen IX oxidase (PPO) inhibitors).

PS II inhibitors bind to the D1 protein of the photosystem II (EC: 1.10.3.9) complex, blocking electron transport and stopping CO_2 fixation. The D1 protein is encoded by the chloroplast *psbA* gene. The first documented herbicide resistant population of *A. artemisiifolia* was discovered in 1976 and was resistant to the PSII-inhibitor atrazine (a triazines) [35]. Resistance was conferred by a point mutation in the *psbA* gene resulting in a serine-to-glycine substitution, which alters D1 protein conformation, causing reduced herbicide target interaction [36,37]. To detect the causal mutation (Ser264Gly) in common ragweed, a bidirectional allele-specific PCR (Bi-PASA) method was developed [38]. Resistant biotypes and populations have also been reported against linuron (belonging to ureas) in eastern Canada [39,40]. TSR, conferred by the *psbA* mutation Val219Ile, was detected after linuron treatment using TaqMan single nucleotide polymorphism (SNP) genotyping and Sanger sequencing [41]. Moreover, three additional mutations at the site of action (Ala251Val, Phe255Ile, and Asn266Thr) have been confirmed in populations of other species, each with a unique cross resistance pattern [42].

AHAS inhibitors are among the most used common ragweed inhibitors worldwide. AHAS (EC 4.1.3.18 also known as acetolactate synthase, ALS) catalyses the first common step in the pathway for synthesis of the branched-chain amino acids leucine, isoleucine, and valine in plants [43]. AHAS inhibitors have low mammalian toxicity, are broad spectrum and highly effective in low doses. However, resistance to AHAS inhibitors tends to arise rapidly in weed populations. Multiple studies have reported that AHAS inhibitor-resistant weed species are more numerous than weed species resistant to other herbicide groups [44,45]. Many AHAS inhibitor-resistant common ragweed populations have been reported. Cross resistance in these populations is common, with resistant individuals displaying resistance to additional herbicides, including cloransulam-methyl (triazolopyrimidines), chlorimuron (sulfonylureas), and imazaquin (imidazolinones) [46]. Resistance in common ragweed against AHAS inhibitors is conferred by the substitution Trp574Leu [47]. Although eight different AHAS amino acid substitutions confer resistance to AHAS inhibitors in other species, only Trp574Leu has been reported in common ragweed [20,48].

Several polymorphisms in the *epsps* gene of many plant species confer resistance to glyphosate. Resistance is conferred by mutations at Pro106 and Thr102, and recently an example of MTSR was reported in *Amaranthus hybridus* that contained the triple-substitution Thr102Ile, Ala103Val and Pro106Ser [49–52]. Glyphosate resistance in *A. artemisiifolia* was first reported in Missouri and Arkansas in 2004 but the mechanism of the resistance has not yet been determined [53].

PPO inhibitors have been used widely to control weeds in different crop cultures since the 1960s. PPO (EC 1.3.3.4) is a key enzyme in the tetrapyrrole biosynthetic pathway that produces heme and chlorophyll in plastids and mitochondria [54]. Two different nuclear-encoded PPO isozymes are in plants: plastidal PPO1 encoded by the gene *ppx1* and mitochondrial PPO2 encoded by the *ppx2* gene [55]. The weed reported to evolve resistance to PPO inhibitors was *Amaranthus tuberculatus*, which contained a glycine deletion at position 210 (ΔG210) in PPX2 (GenBank accession: DQ386114) [56]. In *A. artemisiifolia*, an Arg98Leu substitution in the *ppx2* gene contributes to PPO resistance [57].

Multi-resistant *A. artemisiifolia* populations can be found worldwide, containing multiple resistance-conferring mutations in genes encoding PSII, AHAS, EPSPS, and PPO [46]. The investigation of specific mutations conferring herbicide resistance in populations is essential for the selection of effective herbicides for weed control. Targeted amplicon sequencing (TAS) is an effective approach that takes advantage of NGS (next generation sequencing) to detect specific mutations. To sequence large numbers of targeted gene regions, NGS is using indexed primers to label samples that are expansively used for population genetics and for genome-wide genotyping in plants with ultralow SNP densities [58].

This study presents an NGS-TAS method for TSR identification in *A. artemisiifolia*. In order to discover multiple polymorphisms in *psbA*, *ahas*, *epsps*, and *ppx2* that confer resistance, we analysed sensitive biotypes, as well as imazethapyr-resistant (imidazolinones, AHAS inhibitor) and linuron-resistant (ureas, PSII inhibitor) biotypes. This method may contribute to the sustainable maintenance of biological integrity in agroecosystems by screening herbicide resistant common ragweed populations.

2. Materials and Methods

2.1. Plant Materials and DNA Extraction

Leaves of 20 susceptible *A. artemisiifolia* plants were collected from an agricultural field in West-Transdanubian region of Hungary (46°44′54.7″ N 17°14′07.7″ E). Seeds of linuron resistant biotypes were collected in a field border of Ste-Clotilde Legault Quebec, Canada (45°07′14″ N 73°38′19″ W) and seeds of imazethapyr resistant biotypes were collected in a soybean field of Mirabel Quebec, Canada (45°39′16.94″ N 74°05′45.31″ W). Seedlots were sampled from sites where ragweed was tested and identified as imazethapyr and linuron resistant. Resistance seeds were germinated in petri dishes at 30 °C/20 °C 16 h photoperiod and were planted in pots and were growing in greenhouse. Leaves of 20-20 planted individuals were further collected. Samples were frozen in liquid nitrogen immediately after collection and stored at −80 °C until DNA and RNA analysis.

Total DNA from leaf tissues was isolated according to the protocol Doyle, J. J. and Doyle, J. L. [59]. The DNA quantity and purity were assessed on NanoDrop2000 (Thermo Scientific, Waltham, MA, USA). For the experiments, DNA of resistant and sensitive biotypes were used according to the following classification: (i) bulked sample of 20 sensitive genotypes deriving from Hungarian sampling site; (ii) bulked sample of 20 imazethapyr (imidazolinones) resistant genotypes collected from Mirabel sampling site (AMI) and (iii) bulked sample of 20 linuron (ureas) resistant genotypes collected from Ste-Clotilde Legault sampling site (AMU). All individuals, both susceptible and resistant genotypes, were analyzed by Sanger method.

2.2. Identification of A. artemisiifolia Herbicide Target Enzyme mRNAs

In order to identify in silico coding sequences (cds) of herbicide target enzymes of each herbicide group, we used the transcript dataset of *A. artemisiifolia*, deposited in the National Centre for Biotechnology Information's (NCBI) Transcriptome Shotgun Assembly (TSA) dataset under the accession GEZL01000001 [60]. Reference genes were selected based on taxonomic relationship from closely related species and were downloaded from the UniProt database: *A. artemisiifolia*, *psbA* (B5MF75); *Xanthium* sp., *ahas* (Q41716); *Cichorium intybus*, *ppx1* (Q9SPL6); *Amaranthus palmeri*, *ppx2* (A0A291B3V5); *Erigeron canadensis*, *epsps* (G4U4J5) [61]. The identity of reference genes and identified genes was 96–99% [62]. The investigated sequences were deposited in the NCBI GenBank database (Accession numbers see below).

2.3. Intron Analysis, Amplification and Cloning of A. artemisiifolia Herbicide Target Enzyme Genes

Multiple primer pairs were designed in order to cover all possible mutation points along the whole *psbA* and *ahas* gene sequences by using Primer3Plus [63] (Figures S1 and S2).

In the case of intron containing *epsps* gene, the exon-intron bounds were predicted based on complete gene sequences of closely related species by using ClustalW aligning the sequences of *Conyza canadensis* (AY545667.1) and *Amaranthus palmeri* (FJ861242.1 and FJ861243.1) [64] (Figure S3).

In order to validate polymorphisms that characterize herbicide resistance, the imazethapyr, linuron resistant, and sensitive bulked samples were amplified. The PCR amplifications was performed by using Dream Taq DNA polymerase (Thermo Fisher Scientific, Waltham, MA, USA) The PCR products were separated on 1.5% agarose gel (Promega, Madison, WI, USA) and were purified using NucleoSpin Gel and PCR Clean-up system (Macharey-Nagel GmbH & Co, Düren, Germany). The pGEM-T Easy Vector System kit (A1360 Promega, Madison, WI, USA) and JM109 Competent Cells were used to clone PCR products [65]. Sanger sequencing of the cloned fragments was performed with ABI 3130xl Genetic Analyzer (Applied Biosystems, Carlsbad, CA, USA). Sequenced fragments of herbicide target genes were aligned to predicted gene sequences certifying amino acid substitutions in the mutation points.

2.4. PCR Amplification for NGS-TAS Experiments

The KAPA HiFi HotStart PCR Kit (Kapa Biosystems, Wilmington, NC, USA) was used for amplification of target fragments. In NGS-TAS experiments of investigated regions of *psbA*, *ahas*, *epsps*, and *ppx2* genes, primers were designed by using Primer3Plus (Figure S7).

The PCR amplifications were performed and the PCR products were separated on 1.5% agarose gel (Promega, Madison, USA). PCR products were controlled with Agilent 2100 Bioanalyzer (Agilent Technologies, Waldbronn, Germany) using Agilent DNA 1000 Kit and were purified using NucleoSpin Gel and PCR Clean-up system (Macharey-Nagel GmbH &Co, Düren, Germany).

2.5. NGS-TAS Experiments

For NGS-TAS experiments, resistant samples were used and grouped with the following nomenclature.

Sample AMI was named based on the imidazolinone (imazethapyr) resistant bulked DNA of *A. artemisiifolia* and sample AMU was named based on the urea (linuron) resistant bulked DNA of *A. artemisiifolia*. Numbers 1–3 were three biological repeats of the same bulked DNA (AMI1-3, AMU1-3). The concentrations of PCR products were measured by NanoDrop2000 (Thermo Scientific, Waltham, USA) and fragments for each sample were diluted relative to the PCR product with the lowest concentration. AMI1control and AMU1control contained the not diluted samples of the first biological repeats. In the AMI and AMU groups bulked DNA of 20 individuals were used.

In order to obtain long fragments that cover all the mutation points in each gene, fragments were sequenced on MiSeq 550 platform obtaining 300 bp paired-end reads detailed as follows: locus-specific PCR products were purified using 1.0 volume KAPA PureBeads (F. Hoffmann-La Roche, Basel, Switzerland) according to the manufacturer's protocol. The concentration of eluted DNA was measured using a Qubit 3.0 Fluorometer with Qubit dsDNA HS Assay Kit (Thermo Fisher Scientific, Waltham, USA). Index PCR reactions (20 μL each) were set up by using 20 ng of purified template in 6 μL, 2 μL Nextera XT Index kit v2 Primers (N7xx & S5xx) (Illumina, Inc. San Diego, CA, USA), and 10 μL of 2xKAPA Hifi Hot Start Ready Mix (F. Hoffmann-La Roche, Switzerland). The used primer and IlluminaNextera adapter sequences are detailed in Figure S7. PCR cycling parameters for index PCRs were as follows: initial denaturation at 95 °C for 3 min; 8 cycles at 95 °C for 30 s, 55 °C for 30 s, 72 °C for 30 s; final extension at 72 °C for 5 min. PCR products were purified using 1.0 volume KAPA PureBeads and eluted in 20 μL of 10 mM Tris-HCl pH 8. The product libraries were quantified and qualified by using High Sensitivity D1000 ScreenTape on TapeStation 2200 instrument (Agilent Technologies, Santa Clara, CA, USA) (Figure S11). Equimolar concentrations of libraries were pooled, diluted to 4 nM, and combined with other sample pools to gain the desired sequencing depth. Sequencing was carried out using Illumina MiSeq platform and 600-cycle Reagent Kit v3 (Illumina, Inc. San Diego, CA, USA). Samples were demultiplexed and adapter-trimmed by using MiSeq Control Software.

2.6. Bioinformatics Analysis

Pear software [66] was used to merge paired-end reads with minimum overlap size 10 bp, without discard uncalled bases, and with 30 parallel threads in processing (arguments –v 10 –u 1 –j 30). All the samples contained the reads amplified by the 7 primer pairs covering all the mutation points. Accordingly, the reads belonging to each fragment were counted in each sample (Figure S12). Usearch software [67] was applied to collect identical sequences for resistance polymorphism. Usearch was used in three steps: (i) Quality filter with fastq_filter command with fastq_maxee 1.0 and fastq_minlen 160 parameters; (ii) Singleton filter with fastq_uniques command; (iii) Grouping with cluster_otus parameter with minsize 10. Groups were shown the similarity threshold 99% representing the different amplicon products with nucleotide polymorphism for the investigated gene.

3. Results

In order to perform NGS-TAS analysis, it would be necessary to know cds and gene sequences of herbicide target enzymes. However, no whole genome sequence data of *A. artemisiifolia* yet exists. To design correct NGS-TAS primer pairs, it is essential to know at least the sequences at the intron-exon boundaries close to the mutation points. Accordingly, gene sequences and point mutations of the herbicide target enzymes were determined in sensitive genotypes by using in silico NGS transcriptome analysis (NGS-TA), SS Sanger sequencing, and NGS-TAS (Figure 1).

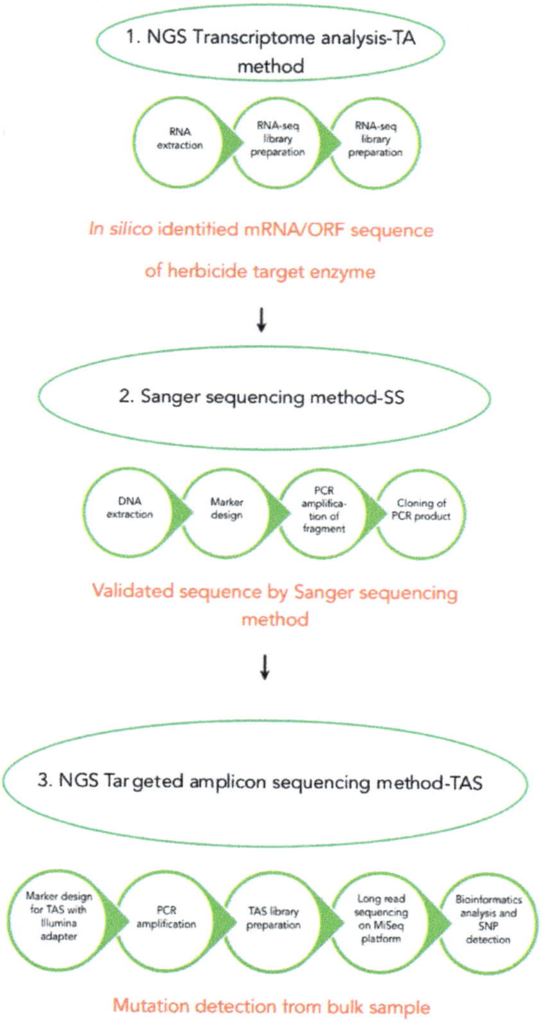

Figure 1. Flow chart of target-site resistance (TSR) diagnostic method using the following steps: (i) Next generation sequencing (NGS) Transcriptome analysis (NGS-TA). Open reading frames of target enzymes were identified based on transcriptome database; (ii) Sanger sequencing method (SS). Specific primers were designed covering the mutation points, fragments were amplified by PCR, PCR products were cloned and sequenced; (iii) Next-generation sequencing-Targeted amplicon sequencing (NGS-TAS) method. PCR products were sequenced on Illumina MiSeq550 platform resulting 300 bp paired-end reads that were sorted based on primer motifs, assembled, and grouped.

3.1. Identification of A. artemisiifolia Herbicide Target Enzyme cds and Genes

In silico analysis of investigated herbicide target enzyme cds was performed with BLAST+. The accession numbers of predicted GOIs are MT425203 (*psbA*), MK096760 (*ahas*), MK096765 (*epsps*), and MK096762 (*ppx2*). Open reading frame (ORF) sequences were predicted using NCBI ORF finder [68]. The predicted sequence of the *psbA* gene consisted of one exon with a total length of 1062bp, as previously described (Table 1) [37]. The *psbA* gene was amplified with 2 primer pairs covering the resistance mutation points as described before (Figure S1, GenBank accession MT879746).

The structure of the *ahas* gene contained one exon with a total length of 1965bp. The whole *ahas* gene was amplified with five primer pairs covering the resistance mutation points as described before (Figures S2 and S4, GenBank accession MT415954) [20,47].

The structure of the *epsps* gene contained 8 exons and 7 introns with a total length of 3539bp. A similar structure was described in *Amaranthus palmeri* (JX564536.1) [64]. The *epsps* gene was amplified with eight primer pairs covering the resistance mutation points as described before (Figures S3 and S4, GenBank accessions MT415955 (exon1-intron1), MT409110 (intron1-exon8)) [32].

The first published weed species to evolve resistance to PPO-inhibiting herbicides was *Amaranthus tuberculatus* [56]. The *ppx2* gene (GenBank accession DQ386118) from the resistant biotype of *A. tuberculatus* contained a glycine deletion at position 210 (ΔG210) that was shown to confer resistance. This deletion in *A. artemisiifolia* has not been published, yet, although the substitution Arg98Leu was discovered to contribute to PPO-inhibitor resistance *A. artemisiifolia* [57]. So far, no complete gene sequence exists of the whole *ppx2* gene in plants covering all of the introns and exons. Therefore, identifying the complete sequence of *ppx2* gene is not possible only using Sanger sequencing. By using the NGS-TAS, a 324 bp fragment of *ppx2* gene was amplified with one primer pair covering the resistance mutation point Arg98 [55]. Electropherogram of Sanger sequenced PCR product revealed two product lengths. The differences between the two fragments were 13 bp localised in one intron (Figures S4 and S5, GenBank accession MT879747, MT879748).

In order to obtain accurate information about the investigated enzymes, in silico predicted cds of Illumina data and the Sanger sequenced data of the appropriate PCR products were compared. The two technologies showed slight variability (1–2% differences) in amino acid sequences across *ahas* and *epsps*, however these discrepancies were not at mutation points. There was no difference in sequencing results of the *psbA* gene between the two methods (Figure S6).

3.2. NGS-TAS Experiments of AMI and AMU Biological Repeats

In NGS-TAS experiments specific regions of *psbA*, *ahas*, *epsps*, and *ppx2* genes were amplified by different primers (Figure 2, Figures S9 and S10). Sample AMI was named based on the imidazolinone (imazethapyr) resistant bulked DNA of *A. artemisiifolia* and sample AMU was named based on the urea (linuron) resistant bulked DNA of *A. artemisiifolia*. Both AMI and AMU contained three replicates.

Table 1. Detailed information of gene structures and coding sequences (cds) of *psbA*, *ahas*, *epsps*, and *ppx2* genes.

	cds							Complete Gene				
Gene Name	**Length (bp)**	**Lenght (aa)**	**Similarity (%)**	**Reference ID NCBI GeneBank**	**Contig ID NCBI GeneBank GEZL01000001**	**NCBI Accession**	**GC Content (%)**	**Length (bp)**	**Intron Number**	**Intron Position from … to**	**NCBI Accession**	**GC Content (%)**
psbA	1062	353	99%	*A. artemisiifolia* AB427162.1	TR92155|c0_g1_i1	MT425203	41.7	1062	-	-	MT879746	41
ahas	1965	654	95%	*Xanthium* sp. AAA74913.1	TR49503|c0_g3_i1 TR49503|c0_g8_i1	MK096760	48.1	1965	-	-	MT415954	48.3
epsps	1539	512	96%	*Helianthus annuus* XP_022017499.1	TR44247|c0_g1_i1	MK096765	45	3539	7	316 … 1311 1557 … 1832 1987 … 2114 2330 … 2410 2529 … 2656 2868 … 3006 3069 … 3400	MT415955 MT409110	34 37.5
ppx2	1476	491	95%	*Helianthus annuus* XP_021982414.1	TR33881|c0_g1_i1	MK096762	42.8	Partial 325	1	56–276	MT879747 MT879748	34.3 33.1

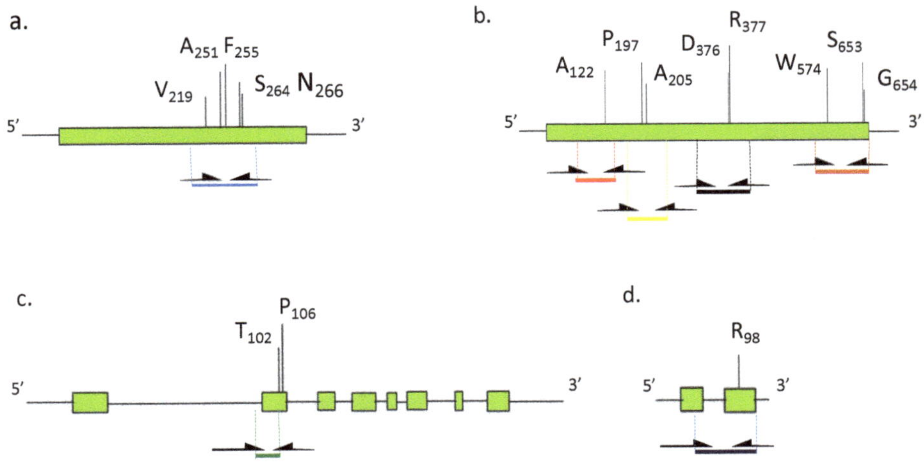

Figure 2. The schematic representation of (**a**), *psbA*; (**b**), *ahas*; (**c**), *epsps*, and (**d**), *ppx2* genes in *A. artemisiifolia* indicating regions amplified by different primers. Vertical grey lines indicate amino acid positions that contain substitutions in herbicide resistant plants. The number after the amino acid refers to the amino acid position. Green boxes indicate exons and grey lines between boxes indicate introns. Colour lines show positions of primer pairs.

Each sample contained the PCR products of *psbA*, *ahas*, *epsps*, and *ppx2* genes. The control of AMI1-3 group was labeled as AMI1control and contained PCR products of AMI1. Similarly, AMU1control served as a control for AMU1-3 group and contained PCR products of AMU1. Both in AMI and AMU samples, the concentrations of the PCR products of the four genes were measured in each replicate, respectively. Afterwards, the lowest concentration was used as a base for further dilution of the other PCR products to avoid error in sequencing analysis (Table 2, Figure S8). In NGS-TAS analyses the samples were sequenced in four lanes. Based on the ratio of concentration/contig number of the samples, we found that all concentrations of AMI and AMU samples were optimal for the diagnostic procedure.

3.3. *Bioinformatics Analysis of AMI and AMU Groups*

Samples with different concentrations showed different distributions in contig number (Table 2). As a result of amplicon sequencing the number of Illumina MiSeq 300 bp reads were 276,462 (forward), 296,722 (reverse) in case of AMI and 316,465 (forward), 369,748 (reverse) in case of AMU samples. After quality filtering dropped reads were 0.7% on average of AMI samples and 4.7% on average of AMU samples. PEAR software was used to assemble cleaned reads of which unpaired ones were filtered out with the average ratio 5.45 and 9.05% of AMI and AMU samples, respectively. Numbers of filtered and assembled reads were 255,357–287,197. In order to separate the investigated gene fragments we clustered the contigs (assembled reads) into groups based on appropriate primer motifs by using Usearch software. During clustering singletons were dropped (unique groups with one sequence). Average ratios of the survived contigs were 69.8% and 69.7%. The group number was 18 and 12 in AMI and AMU samples (Figure S12).

Table 2. Concentrations of PCR product (cc: ng/µL) and numbers of contig in each group in imidazolinone (imazethapyr) resistant bulked samples (AMI) and urea (linuron) resistant bulked samples (AMU).

Fragments	Length of Fragment with Adapter (bp)	AMI								AMU							
		1 Sample		2 Sample		3 Sample		1 Control Sample		1 Sample		2 Sample		3 Sample		1 Control Sample	
		cc	Number of Contig	cc	Number of Contig	cc	Number of Contig	cc	Number of Contig	cc	Number of Contig	cc	Number of Contig	cc	Number of Contig	cc	Number of Reads
psbA	377		26,989		34,594		30,560	72.1	32,592		28,106		28,523		30,492	48.3	40,178
ahas 1	288		10,009		25,817		11,802	60.3	9129		8066		21,401		12,709	24.3	11,426
ahas 2	273		45,613		32,641		45,887	50.3	34,158		43,411		47,525		50,635	43.4	52,289
ahas 3	319	50.3	18,488	32.8	15,106	40.8	23,639	74.3	20,210	24.3	25,580	19.1	21,738	19.9	19,612	37	23,140
ahas 4	453		12,563		13,668		11,955	88.6	17,567		13,133		15,549		15,854	37.1	12,032
epsps	310		21,543		34,053		29,779	51.9	16,599		18,103		10,839		17,476	31.7	15,737
ppx2	390		11,580		9280		26,659	58.8	10,546		15,205		10,108		9346	37.5	15,583

Fragments in each sample contain different numbers of contigs. Primer sorted groups contain min 8066 max 52,289 contigs (Table 2). Undiluted AMI1control and AMU1 control samples showed an equable read distribution in each fragment, therefore dilution of samples in the same concentration is recommended. Confirmation of a fragment with such a large number of reads provides an opportunity for large-sample analysis.

3.4. Detection of Mutation Points in Resistant Biotypes

To be able to identify the mutations causing resistance in *A. artemisiifolia*, bulked and amplified DNA sequences of imazethapyr and linuron resistant samples were further analysed by NGS-TAS that covered 16 known resistance mutations sites in *A. artemisiifolia*. In the AMI group, sequence alignment analysis showed that a single nucleotide in the *ahas* gene was changed from TGG to TTG and resulted in a Trp574Leu substitution conferring herbicide resistance [47] (Figure 3). Other amino acid substitutions in the D1 protein, EPSPS and PPX2 enzymes at known resistance-conferring mutations sites were not found. Based on these data, the investigated population was susceptible to PSII, EPSP synthase and PPO inhibitors (Figure 3). In the AMU group, sequence alignment analysis showed that a single triplet was changed (GTA to ATA) in the *psbA* gene and resulted in a resistance-conferring Val219Ile substitution [41] (Figure 3). Other amino acid substitutions in AHAS, EPSPS, and PPX2 enzymes at known resistance-conferring mutations sites were not found. Based on this data the investigated biotype was susceptible to AHAS, EPSP synthase, and PPO inhibitors in terms of TSR (Figure 3).

Figure 3. Amino acid mutations in investigated gene products. Imidazolinone and urea resistant bulked samples were analyzed by NGS-TAS that covered 16 known resistance-conferring mutations sites in *A. artemisiifolia*. Marking: green, possible mutation sites; red, mutant amino acids. DNA sequencing revealed that an inferred leucine for tryptophan substitution at amino acid position 574 in AHAS enzyme was responsible for the imazethapyr herbicide resistance. Amino acid substitution at the position 219 (valine/isoleucine) in *psbA* gene product, D1 protein, was responsible for the linuron resistance. Amino acid substitutions in EPSPS and PPX2 enzymes at known resistance-conferring mutations sites were not found.

4. Discussion

Chemical weed control is one of the most common ways to fight against weeds in agricultural areas, giving a high selection pressure for resistant biotypes. After herbicide treatment, the rapid diagnosis of resistance mutations present in surviving individuals using would inform and improve the effectiveness of the next herbicide treatment. In order to maintain sustainability in crop fields and beneficial weed communities that are required for natural ecosystems, the monitoring of resistant common ragweed populations between treatments is important. In this study, we present a molecular biology-based approach (NGS-TAS) to detect herbicide resistant common ragweed populations in arable habitats [69,70].

It is important that only a few herbicides with new mechanism of action (MOA) have been released since the 1980s. The increasing number of herbicide resistant weeds and the lack of discovery of new MOAs make it difficult or even impossible to use existing herbicides effectively. Some studies predict that these problems will endanger the sustainability of weed control [16,71]. Therefore, it is essential to inform farmers about integrated and resistant weed management strategies that are less harmful to the environment [16].

Determining TSR would largely improve the weed management strategy of arable lands. Field sampling of populations and testing them by NGS-TAS method would guide farmers to select the proper herbicides thereby decreasing the impact of agricultural chemical inputs on the populations of crop fields. Thereby, unnecessary and ineffective chemical usage could be avoided [16].

In this study, we describe an effective diagnostic process using NGS-TAS to get information about evolved TSR in *A. artemisiifolia* populations. In NGS-TAS, TSR is determined using SNP genotyping following genome-wide genotyping. We demonstrate that NGS-TAS analysis is a method that can monitor TSR against four different enzymes targeted by different herbicides in *A. artemisiifolia* simultaneously. In order to analyse SNPs as part of the resistance mechanism of *A. artemisiifolia*, coding sequences and complete genes of D1 protein, AHAS and EPSPS were identified. However, the sequence of PPX2 only partially was identified at the region of interest as its length and genomic data is unknown. Mutations and polymorphisms in these four proteins that had previously been reported to confer resistance were specifically investigated. The developed NGS-TAS markers can identify 16 amino acid substitutions of the investigated common ragweed genes among which 4 (*psbA*: Val219, Ser264, *ahas*: Trp 574 and *ppx2*:Arg98) were proven. As a result, the investigated samples characterized by imidazolinone (imazethapyr) and urea (linuron) TSR were proved to carry mutation points at Trp574Leu and Val219Ile in the *ahas* and *psbA* genes, respectively. Although no MTSR existed in the studied resistant samples (AMI, AMU), the NGS-TAS method can be used to detect multiple herbicide resistance, which needs to be confirmed by specific MTSR samples.

NGS technology can examine a large number of samples simultaneously using fragment or sample-specific indexed primers. Therefore, amplicon primers were provided that are suitable for discriminating mutation points. In this study, we demonstrated the trial of a pooled sample evolved mutations points of herbicide target genes supporting resistance of *A. artemisiifolia* populations and biotypes.

5. Conclusions

Invasive weed species appeared to be the bane of biodiversity, ecosystem services, and food security. Integrated weed management practices provide several comprehensive solutions for weed control to reduce weed coverage in agricultural fields. Unfortunately, many weed management approaches used nowadays are highly unsustainable. Moreover, many studies report on the rapid evolution of herbicide-resistant weed populations. This complex evolution of herbicide-resistant weeds makes it difficult to use sustainable herbicide technology [20]. Therefore, applying weed management tools for invasive plants are often ineffective at producing long-term benefits [72]. In order to map evolved resistance in populations, different herbicide resistance mechanisms have to be investigated, such as TSR.

To summarize, NGS-TAS method is a powerful approach that was used for the detection of amino acid polymorphism and mutations that induce TSR. Therefore, it can improve effective weed control taking into account biological integrity and sustainability in agroecosystems by selecting the optimal and effective herbicide usage in cropping systems.

Supplementary Materials: The following are available online at https://www.mdpi.com/1424-2818/13/3/118/s1, Figures S1–S12: Primers, gel pictures and electropherograms of Sanger sequencing. Primers, gel pictures, electropherograms and concentrations of amplicon sequencing. Bioinformatics analysis.

Author Contributions: Conceptualization, B.K. (Barbara Kutasy), B.K. (Balázs Kolics); methodology, B.K. (Barbara Kutasy), Z.F., K.D., B.K. (Balázs Kolics), N.K.; software, G.H.; validation, B.K. (Barbara Kutasy), E.V. and G.H. formal analysis, B.K. (Barbara Kutasy); investigation, B.K. (Barbara Kutasy), E.V.; resources B.K. (Barbara Kutasy); data curation, E.V.; writing—original draft preparation, B.K. (Barbara Kutasy), E.V.; writing—review and editing, E.V., J.K., G.K.; visualization B.K. (Barbara Kutasy), E.V. and J.K.; supervision, E.V.; project administration, B.K. (Barbara Kutasy), E.V.; funding acquisition, G.K., J.T. and Z.T. All authors have read and agreed to the published version of the manuscript.

Funding: The work was supported by the EFOP-3.6.3-VEKOP-16-2017-00008 and the GINOP-2.3.4-15-2016-00005 projects. The project is co-financed by the European Union and the European Regional Development Fund. This work/research was supported by the Hungarian Government and the European Union, with the co-funding of the European Regional Development Fund in the frame of Széchenyi 2020 Programme GINOP-2.3.2-15-2016-00054 project.

Institutional Review Board Statement: Not applicable.

Informed Consent Statement: Not applicable.

Data Availability Statement: Not applicable.

Acknowledgments: We express our thanks to Diane Lyse Benoit, Agriculture and Agri-Food Canada, Horticulture Research and Development Centre, Quebec, to release available the resistant *Ambrosia artemisiifolia* seeds (biological samples) to our experiments. The work was supported by the EFOP-3.6.3-VEKOP-16-2017-00008 and the GINOP-2.3.4-15-2016-00005 projects. We are indebted in the financial supporting to TKP2020-IKA-12 project. The project is co-financed by the European Union and the European Regional Development Fund. This work/research was supported by the Hungarian Government and the European Union, with the co-funding of the European Regional Development Fund in the frame of Széchenyi 2020 Programme GINOP-2.3.2-15-2016-00054 project. We are grateful to the editor and to two anonymous reviewers for valuable comments that have helped to improve the manuscript.

Conflicts of Interest: The authors declare no conflict of interest. The funders had no role in the design of the study; in the collection, analyses, or interpretation of data; in the writing of the manuscript, or in the decision to publish the results.

References

1. Baessler, C.; Klotz, S. Effects of changes in agricultural land-use on landscape structure and arable weed vegetation over the last 50 years. *Agric. Ecosyst. Environ.* **2006**, *115*, 43–50. [CrossRef]
2. Lommen, S.T.E.; Hallmann, C.A.; Jongejans, E.; Chauvel, B.; Leitsch-Vitalos, M.; Aleksanyan, A.; Tóth, P.; Preda, C.; Šćepanović, M.; Onen, H.; et al. Explaining variability in the production of seed and allergenic pollen by invasive *Ambrosia artemisiifolia* across Europe. *Biol. Invasions* **2018**, *20*, 1475–1491. [CrossRef]
3. Pinke, G.; Király, G.; Barina, Z.; Mesterházy, Á.; Balogh, L.; Csiky, J.; Schmotzer, A.; Molnár, A.V.; Pál, R.W. Assessment of endangered synanthropic plants of Hungary with special attention to arable weeds. *Plant Biosyst. Int. J. Deal. All Asp. Plant Biol.* **2011**, *145*, 426–435. [CrossRef]
4. Alberternst, B.; Nawrath, S.; Starfinger, U. Biodiversity impacts of common ragweed. *Julius-Kühn-Archiv* **2016**. [CrossRef]
5. Essl, F.; Biró, K.; Brandes, D.; Broennimann, O.; Bullock, J.M.; Chapman, D.S.; Chauvel, B.; Dullinger, S.; Fumanal, B.; Guisan, A.; et al. Biological Flora of the British Isles: *Ambrosia artemisiifolia*. *J. Ecol.* **2015**, *103*, 1069–1098. [CrossRef]

6. Fumanal, B.; Girod, C.; Fried, G.; Bretagnolle, F.; Chauvel, B. Can the large ecological amplitude of *Ambrosia artemisiifolia* explain its invasive success in France? *Weed Res.* **2008**, *48*, 349–359. [CrossRef]
7. Bassett, I.J.; Crompton, C.W. The biology of Canadian weeds.: 11. *Ambrosia artemisiifolia* L. and *A. psilostachya* DC. *Can. J. Plant Sci.* **1975**, *55*, 463–476. [CrossRef]
8. Zhu, J.; Wang, J.; DiTommaso, A.; Zhang, C.; Zheng, G.; Liang, W.; Islam, F.; Yang, C.; Chen, X.; Zhou, W. Weed research status, challenges, and opportunities in China. *Crop. Prot.* **2020**, *134*, 104449. [CrossRef]
9. Kiss, L.; Beres, I. Anthropogenic factors behind the recent population expansion of common ragweed (*Ambrosia artemisiifolia* L.) in Eastern Europe: Is there a correlation with political transitions? *J. Biogeogr.* **2006**, *33*, 2156–2157. [CrossRef]
10. Csontos, P.; Vitalos, M.; Barina, Z.; Kiss, L. Early distribution and spread of *Ambrosia artemisiifolia* in Central and Eastern Europe. *Bot. Helv.* **2010**, *120*, 75–78. [CrossRef]
11. Chauvel, B.; Dessaint, F.; Cardinal-Legrand, C.; Bretagnolle, F. The historical spread of *Ambrosia artemisiifolia* L. in France from herbarium records. *J. Biogeogr.* **2006**, *33*, 665–673. [CrossRef]
12. Hamaoui-Laguel, L.; Vautard, R.; Liu, L.; Solmon, F.; Viovy, N.; Khvorostyanov, D.; Essl, F.; Chuine, I.; Colette, A.; Semenov, M.A.; et al. Effects of climate change and seed dispersal on airborne ragweed pollen loads in Europe. *Nat. Clim. Chang.* **2015**, *5*, 766–771. [CrossRef]
13. Simard, M.-J.; Benoit, D.L. Distribution and abundance of an allergenic weed, common ragweed (*Ambrosia artemisiifolia* L.), in rural settings of southern Quebec, Canada. *Can. J. Plant Sci.* **2010**, *90*, 549–557. [CrossRef]
14. Müller-Schärer, H.; Sun, Y.; Chauvel, B.; Karrer, G.; Kazinczi, G.; Kudsk, P.; Oude, A.L.; Schaffner, U.; Skjoth, C.; Smith, M.; et al. Cross-fertilizing weed science and plant invasion science to improve efficient management: A European challenge. *Basic Appl. Ecol.* **2018**, *33*, 1–13. [CrossRef]
15. Neve, P.; Barney, J.N.; Buckley, Y.; Cousens, R.D.; Graham, S.; Jordan, N.R.; Lawton-Rauh, A.; Liebman, M.; Mesgaran, M.B.; Schut, M.; et al. Reviewing research priorities in weed ecology, evolution and management: A horizon scan. *Weed Res.* **2018**, *58*, 250–258. [CrossRef]
16. Westwood, J.H.; Charudattan, R.; Duke, S.O.; Fennimore, S.A.; Marrone, P.; Slaughter, D.C.; Swanton, C.; Zollinger, R. Weed Management in 2050: Perspectives on the Future of Weed Science. *Weed Sci.* **2018**, *66*, 275–285. [CrossRef]
17. Moss, S.R.; Storkey, J.; Cussans, J.W.; Perryman, S.A.M.; Hewitt, M.V. Symposium the Broadbalk long-term experiment at Rothamsted: What has it told us about weeds? *Weed Sci.* **2004**, *52*, 864–873. [CrossRef]
18. Coble, H.D. Future Directions and Needs for Weed Science Research. *Weed Technol.* **1994**, *8*, 410–412. [CrossRef]
19. Kazinczi, G.; Novák, R.; Pathy, Z.; Béres, I. Common ragweed (*Ambrosia artemisiifolia* L.): A review with special regards to the results in Hungary. III. Resistant biotypes, control methods and authority arrangements. *Herbologia* **2008**, *9*, 119–144.
20. Powles, S.B.; Yu, Q. Evolution in Action: Plants Resistant to Herbicides. *Annu. Rev. Plant Biol.* **2010**, *61*, 317–347. [CrossRef] [PubMed]
21. Harker, K.N.; O'Donovan, J.T. Recent Weed Control, Weed Management, and Integrated Weed Management. *Weed Technol.* **2013**, *27*, 1–11. [CrossRef]
22. Délye, C.; Jasieniuk, M.; Le Corre, V. Deciphering the evolution of herbicide resistance in weeds. *Trends Genet.* **2013**, *29*, 649–658. [CrossRef]
23. Vila-Aiub, M.M.; Neve, P.; Powles, S.B. Fitness costs associated with evolved herbicide resistance alleles in plants. *New Phytol.* **2009**, *184*, 751–767. [CrossRef] [PubMed]
24. Powles, S.B.; Preston, C. Evolved Glyphosate Resistance in Plants: Biochemical and Genetic Basis of Resistance. *Weed Technol.* **2006**, *20*, 282–289. [CrossRef]
25. Bernasconi, P.; Woodworth, A.R.; Rosen, B.A.; Subramanian, M.V.; Siehl, D.L. A Naturally Occurring Point Mutation Confers Broad Range Tolerance to Herbicides That Target Acetolactate Synthase. *J. Biol. Chem.* **1995**, *270*, 17381–17385. [CrossRef] [PubMed]
26. Han, H.; Yu, Q.; Purba, E.; Li, M.; Walsh, M.; Friesen, S.; Powles, S.B. A novel amino acid substitution Ala-122-Tyr in ALS confers high-level and broad resistance across ALS-inhibiting herbicides. *Pest Manag. Sci.* **2012**, *68*, 1164–1170. [CrossRef] [PubMed]
27. Li, J.; Li, M.; Gao, X.; Fang, F. A novel amino acid substitution Trp574Arg in acetolactate synthase (ALS) confers broad resistance to ALS-inhibiting herbicides in crabgrass (*Digitaria sanguinalis*). *Pest Manag. Sci.* **2017**, *73*, 2538–2543. [CrossRef]
28. McNaughton, K.E.; Letarte, J.; Lee, E.A.; Tardif, F.J. Mutations inALSconfer herbicide resistance in redroot pigweed (*Amaranthus retroflexus*) and Powell amaranth (*Amaranthus powellii*). *Weed Sci.* **2005**, *53*, 17–22. [CrossRef]
29. Riar, D.S.; Norsworthy, J.K.; Srivastava, V.; Nandula, V.; Bond, J.A.; Scott, R.C. Physiological and Molecular Basis of Acetolactate Synthase-Inhibiting Herbicide Resistance in Barnyardgrass (*Echinochloa crus-galli*). *J. Agric. Food Chem.* **2013**, *61*, 278–289. [CrossRef] [PubMed]
30. Warwick, S.I.; Xu, R.; Sauder, C.; Beckie, H.J. Acetolactate Synthase Target-Site Mutations and Single Nucleotide Polymorphism Genotyping in ALS-Resistant Kochia (*Kochia scoparia*). *Weed Sci.* **2008**, *56*, 797–806. [CrossRef]
31. Yu, Q.; Jalaludin, A.; Han, H.; Chen, M.; Sammons, R.D.; Powles, S.B. Evolution of a Double Amino Acid Substitution in the 5-Enolpyruvylshikimate-3-Phosphate Synthase in Eleusine indica Conferring High-Level Glyphosate Resistance. *Plant Physiol.* **2015**, *167*, 1440–1447. [CrossRef] [PubMed]
32. Brosnan, J.T.; Vargas, J.J.; Breeden, G.K.; Grier, L.; Aponte, R.A.; Tresch, S.; Laforest, M. A new amino acid substitution (Ala-205-Phe) in acetolactate synthase (ALS) confers broad spectrum resistance to ALS-inhibiting herbicides. *Planta* **2016**, *243*, 149–159. [CrossRef] [PubMed]

33. Diebold, R.S.; McNaughton, K.E.; Lee, E.A.; Tardif, F.J. Multiple resistance to imazethapyr and atrazine in Powell amaranth (*Amaranthus powellii*). *Weed Sci.* **2003**, *51*, 312–318. [CrossRef]
34. Liu, M.; Hulting, A.G.; A Mallory-Smith, C. Characterization of multiple-herbicide-resistant Italian ryegrass (*Lolium perenne* spp. multiflorum). *Pest Manag. Sci.* **2014**, *70*, 1145–1150. [CrossRef]
35. Stephenson, G.R.; Dykstra, M.D.; McLaren, R.D.; Hamill, A.S. Agronomic Practices Influencing Triazine-Resistant Weed Distribution in Ontario. *Weed Technol.* **1990**, *4*, 199–207. [CrossRef]
36. Alfonso, M.; Pueyo, J.J.; Gaddour, K.; Etienne, A.L.; Kirilovsky, D.; Picorel, R. Induced New Mutation of D1 Serine-268 in Soybean Photosynthetic Cell Cultures Produced Atrazine Resistance, Increased Stability of S2QB- and S3QB-States, and Increased Sensitivity to Light Stress. *Plant Physiol.* **1996**, *112*, 1499–1508. [CrossRef]
37. Cseh, A.; Cernák, I.; Taller, J. Molecular characterization of atrazine resistance in common ragweed (*Ambrosia artemisiifolia* L.). *J. Appl. Genet.* **2009**, *50*, 321–327. [CrossRef] [PubMed]
38. Mátyás, K.K.; Taller, J.; Cseh, A.; Poczai, P.; Cernák, I. Development of a simple PCR-based assay for the identification of triazine resistance in the noxious plant common ragweed (*Ambrosia artemisiifolia*) and its applicability in higher plants. *Biotechnol. Lett.* **2011**, *33*, 2509–2515. [CrossRef] [PubMed]
39. Saint-Louis, S.; DiTommaso, A.; Watson, A.K. A Common Ragweed (*Ambrosia artemisiifolia*) Biotype in Southwestern Québec Resistant to Linuron. *Weed Technol.* **2005**, *19*, 737–743. [CrossRef]
40. Beckie, H.J.; Hall, L.; Tardif, F.; Blackshaw, R. Herbicide resistance in Canada—Where are we today. In *Integrated Weed Management: Explore the Potential*; Expert Committee on Weeds: Sainte-Anne-de-Bellevue, QC, Canada, 2001; pp. 1–36.
41. Simard, M.-J.; Laforest, M.; Soufiane, B.; Benoit, D.L.; Tardif, F. Linuron resistant common ragweed (*Ambrosia artemisiifolia*) populations in Québec carrot fields: Presence and distribution of Target site and non-target site resistant biotypes. *Can. J. Plant Sci.* **2017**, *98*, 345–352. [CrossRef]
42. Beckie, H.J.; Tardif, F.J. Herbicide cross resistance in weeds. *Crop. Prot.* **2012**, *35*, 15–28. [CrossRef]
43. Duggleby, R.G.; Pang, S.S. Acetohydroxyacid synthase. *J. Biochem. Mol. Biol.* **2000**, *33*, 1–36.
44. Tranel, P.J.; Wright, T.R. Resistance of weeds to ALS-inhibiting herbicides: What have we learned? *Weed Sci.* **2002**, *50*, 700–712. [CrossRef]
45. Yu, Q.; Powles, S.B. Resistance to AHAS inhibitor herbicides: Current understanding. *Pest Manag. Sci.* **2014**, *70*, 1340–1350. [CrossRef] [PubMed]
46. Heap, I. The International Survey of Herbicide Resistant Weeds. 2021. Available online: www.weedscience.org (accessed on 10 February 2021).
47. Patzoldt, W.L.; Tranel, P.J.; Alexander, A.L.; Schmitzer, P.R. A common ragweed population resistant to cloransulam-methyl. *Weed Sci.* **2001**, *49*, 485–490. [CrossRef]
48. Massa, D.; Krenz, B.; Gerhards, R. Target-site resistance to ALS-inhibiting herbicides in Apera spica-venti populations is conferred by documented and previously unknown mutations. *Weed Res.* **2011**, *51*, 294–303. [CrossRef]
49. Ng, C.H.; Wickneswari, R.; Salmijah, S.; Teng, Y.T.; Ismail, B.S. Gene polymorphisms in glyphosate-resistant and -susceptible biotypes of *Eleusine indica* from Malaysia. *Weed Res.* **2003**, *43*, 108–115. [CrossRef]
50. Lee, L.J.; Ngim, J. A first report of glyphosate-resistant goosegrass (*Eleusine indica* (L) Gaertn) in Malaysia. *Pest Manag. Sci. Former. Pestic. Sci.* **2000**, *56*, 336–339. [CrossRef]
51. Baerson, S.R.; Rodriguez, D.J.; Tran, M.; Feng, Y.; Biest, N.A.; Dill, G.M. Glyphosate-Resistant Goosegrass. Identification of a Mutation in the Target Enzyme 5-Enolpyruvylshikimate-3-Phosphate Synthase. *Plant Physiol.* **2002**, *129*, 1265–1275. [CrossRef]
52. Perotti, V.E.; Larran, A.S.; E Palmieri, V.; Martinatto, A.K.; E Alvarez, C.; Tuesca, D.; Permingeat, H.R. A novel triple amino acid substitution in the EPSPS found in a high-level glyphosate-resistant *Amaranthus hybridus* population from Argentina. *Pest Manag. Sci.* **2019**, *75*, 1242–1251. [CrossRef]
53. Brewer, C.E.; Oliver, L.R. Confirmation and Resistance Mechanisms in Glyphosate-Resistant Common Ragweed (*Ambrosia artemisiifolia*) in Arkansas. *Weed Sci.* **2009**, *57*, 567–573. [CrossRef]
54. Smith, A.G.; Marsh, O.; Elder, G.H. Investigation of the subcellular location of the tetrapyrrole-biosynthesis enzyme coproporphyrinogen oxidase in higher plants. *Biochem. J.* **1993**, *292*, 503–508. [CrossRef] [PubMed]
55. Lermontova, I.; Kruse, E.; Mock, H.-P.; Grimm, B. Cloning and characterization of a plastidal and a mitochondrial isoform of tobacco protoporphyrinogen IX oxidase. *Proc. Natl. Acad. Sci. USA* **1997**, *94*, 8895–8900. [CrossRef]
56. Patzoldt, W.L.; Hager, A.G.; McCormick, J.S.; Tranel, P.J. A codon deletion confers resistance to herbicides inhibiting protoporphyrinogen oxidase. *Proc. Natl. Acad. Sci. USA* **2006**, *103*, 12329–12334. [CrossRef]
57. Rousonelos, S.L.; Lee, R.M.; Moreira, M.S.; Vangessel, M.J.; Tranel, P.J. Characterization of a Common Ragweed (*Ambrosia artemisiifolia*) Population Resistant to ALS- and PPO-Inhibiting Herbicides. *Weed Sci.* **2012**, *60*, 335–344. [CrossRef]
58. Shirasawa, K.; Kuwata, C.; Watanabe, M.; Fukami, M.; Hirakawa, H.; Isobe, S. Target Amplicon Sequencing for Genotyping Genome-Wide Single Nucleotide Polymorphisms Identified by Whole-Genome Resequencing in Peanut. *Plant Genome* **2016**, *9*, 1–8. [CrossRef] [PubMed]
59. Doyle, J.J.; Doyle, J.L. A rapid DNA isolation procedure for small quantities of fresh leaf tissue. *Phytochem. Bull.* **1987**, *19*, 11–15.
60. Virág, E.; Hegedűs, G.; Barta, E.; Nagy, E.; Mátyás, K.; Kolics, B.; Taller, J. Illumina Sequencing of Common (Short) Ragweed (*Ambrosia artemisiifolia* L.) Reproductive Organs and Leaves. *Front. Plant Sci.* **2016**, *7*, 1506. [CrossRef] [PubMed]

61. Altschul, S.F.; Gish, W.; Miller, W.; Myers, E.W.; Lipman, D.J. Basic local alignment search tool. *J. Mol. Biol.* **1990**, *215*, 403–410. [CrossRef]
62. Thompson, J.D.; Higgins, D.G.; Gibson, T.J. CLUSTAL W: Improving the sensitivity of progressive multiple sequence alignment through sequence weighting, position-specific gap penalties and weight matrix choice. *Nucleic Acids Res.* **1994**, *22*, 4673–4680. [CrossRef]
63. Untergasser, A.; Nijveen, H.; Rao, X.; Bisseling, T.; Geurts, R.; Leunissen, J.A.M. Primer3Plus, an enhanced web interface to Primer3. *Nucleic Acids Res.* **2007**, *35*, W71–W74. [CrossRef]
64. Gaines, T.A.; Wright, A.A.; Molin, W.T.; Lorentz, L.; Riggins, C.W.; Tranel, P.J.; Beffa, R.; Westra, P.; Powles, S.B. Identification of Genetic Elements Associated with EPSPS Gene Amplification. *PLoS ONE* **2013**, *8*, e65819. [CrossRef]
65. Marcus, L.; Hartnett, J.; Storts, D.R. The pGEM®-T and pGEM®-T Easy Vector Systems. *Promega Notes Mag.* **1996**, *58*, 36.
66. Zhang, J.; Kobert, K.; Flouri, T.; Stamatakis, A. PEAR: A fast and accurate Illumina Paired-End reAd mergeR. *Bioinformatics* **2014**, *30*, 614–620. [CrossRef] [PubMed]
67. Edgar, R.C. Search and clustering orders of magnitude faster than BLAST. *Bioinformatics* **2010**, *26*, 2460–2461. [CrossRef] [PubMed]
68. Rombel, I.T.; Sykes, K.F.; Rayner, S.; Johnston, S.A. ORF-FINDER: A vector for high-throughput gene identification. *Gene* **2002**, *282*, 33–41. [CrossRef]
69. Kniss, A.R. Genetically Engineered Herbicide-Resistant Crops and Herbicide-Resistant Weed Evolution in the United States. *Weed Sci.* **2017**, *66*, 260–273. [CrossRef]
70. Gaines, T.A.; Zhang, W.; Wang, D.; Bukun, B.; Chisholm, S.T.; Shaner, D.L.; Nissen, S.J.; Patzoldt, W.L.; Tranel, P.J.; Culpepper, A.S.; et al. Gene amplification confers glyphosate resistance in *Amaranthus palmeri*. *Proc. Natl. Acad. Sci. USA* **2009**, *107*, 1029–1034. [CrossRef] [PubMed]
71. Bajwa, A.A.; Mahajan, G.; Chauhan, B.S. Nonconventional Weed Management Strategies for Modern Agriculture. *Weed Sci.* **2015**, *63*, 723–747. [CrossRef]
72. Pearson, D.E.; Ortega, Y.K.; Runyon, J.B.; Butler, J.L. Secondary invasion: The bane of weed management. *Biol. Conserv.* **2016**, *197*, 8–17. [CrossRef]

diversity

MDPI

Article

Do Weeds Hinder the Establishment of Native Plants on a Reclaimed North American Boreal Mine Site?

Kaitlyn E. Trepanier *, Brea Burton and Bradley D. Pinno *

Department of Renewable Resources, University of Alberta, 442 Earth Sciences Building, Edmonton, AB T6G 2E3, Canada; brea@ualberta.ca
* Correspondence: ktrepani@ualberta.ca (K.E.T.); bpinno@ualberta.ca (B.D.P.)

Abstract: The majority of plant diversity in the boreal forest of northern Alberta, Canada is comprised of native understory plant species that are continuously facing competition from other species, including both undesirable native and weedy species. In oil sands mine reclamation, cover soils rich in organic matter are used to cap overburden materials. The aim of this study is to understand the role of weeds on different reclamation cover soils (forest floor-mineral mix and peat-mineral mix) and determine if they hinder the establishment of the native plant community. This study was conducted four growing seasons after site establishment in June 2019. At that time, both soil types had approximately 45% total cover, had 21 species per plot, and were composed of mainly native vegetation. Competition from non-native forbs (11% average cover, mainly *Sonchus arvensis* and *Melilotus alba*) did not seem to impact the development of the native vegetation community on either soil type given the high cover and richness of native forbs. However, native graminoids (predominantly *Calamagrostis canadensis*) were associated with reduced native forb cover and richness at graminoid cover greater than 17%. Overall, non-native forbs appeared to have little impact on the native forb community on either soil type while native graminoids had a negative influence. We suggest that the classification of what is considered an undesirable weedy species should be evaluated in the context of ecosystem management goals rather than simply the presence of non-native species.

Keywords: reclamation; boreal; competition; plant community; weedy species

Citation: Trepanier, K.E.; Burton, B.; Pinno, B.D. Do Weeds Hinder the Establishment of Native Plants on a Reclaimed North American Boreal Mine Site?. *Diversity* **2021**, *13*, 76. https://doi.org/10.3390/d13020076

Academic Editors: Robert Pal and Michael Wink

Received: 11 December 2020
Accepted: 7 February 2021
Published: 12 February 2021

1. Introduction

Weed invasion on mine sites post-reclamation is an ongoing management issue worldwide, particularly if weeds dominate to the point of altering the desired plant community. For example, the introduced forb *Tripleurospermum perforatum* (scentless chamomile) is establishing and dominating on reclaimed well sites in Alberta [1,2], *Andropogon gayanus* (Gamba grass) is invading reclaimed mine sites in northern Australia [3], and the encroachment of a non-native shrub, *Elaeagnus umbellate*, is dominating coal mining reclamation sites across the eastern United States [4]. These species, and many others around the world [5,6], inhibit revegetation of desirable species post-disturbance and are considered detrimental to the reclamation goal of a native plant community. Weeds have the potential to alter the successional trajectory of a reclamation site and hinder the establishment of a native plant community [1,2,7]. Weeds typically reproduce prolifically, spread rapidly, are disturbance-adapted [8], and outcompete other plants via utilization of vital resources required for successful establishment and growth; they may also migrate outwards into the surrounding landscape [9]. Consequently, in this context of post-mining land use a "weed" is any undesirable plant species, native or introduced, that interferes with ecological, economic, and management objectives for revegetation post-disturbance and may prevent sites from meeting certification requirements [7].

"Weedy" species, including both native and non-native types in the North American boreal region, are predominantly disturbance-adapted forbs and perennial grasses capable of thriving in a variety of growing conditions [10,11]. Many of the boreal weeds may be acting as temporary placeholders in the early stage of vegetation community development and can potentially be outcompeted over time. A study done in Alberta's oil sands region on a reclaimed upland forest suggests that within early successional communities, weedy species can act as pioneers and may be facilitating native revegetation long-term [12]. However, facilitative interactions from non-native plants on the native vegetation community are scarce and generally occur in high stress environments [13–16]. In many cases, non-native species in higher stress environments can be crucial in the recovery of optimal soil conditions which may be uninhabitable by native trees or understory species [12,16]. Through facilitation, weedy species have been shown to create better soil conditions and structure due to their ability to efficiently capture limited resources and in turn enhance aboveground and belowground biomass, leading to the addition of organic matter to the soil [12,15,16]. This positive interaction has been seen in a weedy graminoid species, *Stipa tenacissim,* in the degraded steppes of Spain, where it has directly enhanced seedling survival through the alteration of microclimate, soil fertility, and moisture availability [16]. However, it is not clear if competition or facilitation processes are dominating on our site.

Anthropogenic activities, particularly oil sands mining, represent a high severity disturbance that both removes the vegetation and modifies the soil [7,10,11,17]. Current reclamation operations in the oil sands mining region of northern Alberta include constructing landforms from overburden or tailings and then capping these with cover soil rich in organic matter. The reclamation cover soils used in northern Alberta are either upland-derived forest floor-mineral mix (FFMM) or peatland-derived peat-mineral mix (PMM). Due to the high stress environments created post-mining, the direction and interactions that will occur within the novel vegetation communities are relatively unknown. The interactions between weedy and native species are, in part, largely dependent on soil quality and type. FFMM often displays higher initial species diversity than PMM, resulting in more potential competitive and facilitative interactions. Cover and diversity dynamics on PMM, on the other hand, likely have a more neutral relationship, as there is initially more growing space available.

Weed research tends to be focused on the impact of non-native weedy species, but in the boreal region there is evidence that native graminoid species pose a risk to the ecosystem, tree, and plant community. However, these may not be a concern to land managers as they are native species. Further research is needed, therefore, to understand and help evaluate the differences in the plant community in response to various levels of competition from both native and non-native species on reclamation sites. The goal of this project is to gain a greater understanding of the role that weedy species, both native and non-native, play on reclaimed sites and determine if weeds hinder the establishment of the vegetation community, in particular native forbs. Alternatively, the weeds may simply be utilizing available resources with no long-term impacts on plant community development. The specific research questions include: Are weeds impeding the successful establishment of a native plant community? What are the impacts of weeds on the different soil types? Which species are acting as weeds?

2. Materials and Methods

This study was conducted on a reclamation site located at an oil sands mine approximately 75 km northwest of Fort McMurray, Alberta, Canada (57.3377° N 111.7552° W) within the boreal mixedwood ecoregion [18]. This area experiences annual precipitation of 419 mm with an annual average temperature of 1 °C [19]. Reclamation occurred in 2015 on an overburden-based tailings dyke extending over 19 ha. The reclamation site was capped with two distinct cover soils, upland-derived FFMM and peatland-derived PMM, to a depth of 0.25 m. FFMM is an upland forest-based cover soil that incorporates the seedbank rich organic forest floor layer with underlying mineral soil resulting in high levels of initial plant diversity and biomass [10,11]. PMM is derived from peatlands and is a mixture of lowland peat deposits and underlying mineral soil (ratio of 60:40 peat-to-soil volume) [17]. PMM has high organic matter content, water holding capacity, and surface roughness, making it ideal for tree and woody species establishment [20,21].

Plant community composition was measured in June 2019 at 40 (20 FFMM and 20 PMM) plots. Within each plot, four 1 m^2 quadrats were averaged to assess the percent cover and richness of all vascular plants identified to species-level to a minimum of 0.5% cover [22–25]. Bare soil (%) was calculated as 100—total cover. Species were also grouped into functional groups: native forbs, non-native forbs, graminoids (grasses, rushes, and sedges), shrubs, and trees [22–25]. All graminoids, shrubs, and trees were native species. Species nomenclature follows USDA [26]. The main competitive native graminoid species on reclamation sites in Alberta is *Calamagrostis canadensis* [27–30]. *Calamagrostis canadensis* is an aggressive rhizomatous grass that grows to be 60–120 cm tall, is highly competitive for light and nutrient resources, and establishes earlier in the growing season than other plant species [23,29,30]. The large amounts of dense aboveground litter and belowground root mass produced by *Calamagrostis canadensis* negatively influence light availability, surface access, and soil temperature, which can prevent understory plant and tree regeneration [31,32].

General soil characteristics were measured in all plots. Volumetric soil water content (FieldScout TDR 300, Spectrum Technologies Inc., Aurora, IL, USA), soil temperature (Fisher Scientific digital thermometers), and pH (400-m, Field Scout, Spectrum Technologies Inc., Aurora, IL, USA) were measured within a short period (3–5 days) to minimize variation [33]. Soil measurements were not performed within 24 h after a rain event [33,34]. Volumetric water content (VWC) was collected from three random locations within the inner plot to acquire an average reading. Soil temperature and pH readings were collected at the plot center; the thermometer and probes were left in the ground to stabilize for 2–3 min before the reading was taken. All soil measurements were taken at a depth of 12 cm. VWC was higher by 10.5% on PMM compared to FFMM, while pH and soil temperature were similar between soil types (Table 1). All reclamation topsoil must meet chemical quality guidelines for pH, electrical conductivity, and other variables as outlined in [17].

Table 1. Study site characteristics for showing average soil temperature, VWC, pH, (standard deviation), and species richness for both soil types (FFMM n = 20; PMM n = 20). *p*-values for soil temperature are based on ANOVA and all other *p*-values are based on permANOVA.

		FFMM	PMM	*p*
Soil	Soil Temperature (°C)	14.1 (2.2)	15.4 (1.4)	0.090
	VWC (%)	16.5 (4.4)	27.0 (6.4)	<0.001
	pH	6.9 (0.7)	6.5 (0.9)	0.176
Average Species Richness Per 1 m^2 Quadrat	Tree	1	2	<0.001
	Native Forb	10	9	0.118
	Non-Native Forb	3	4	0.270
	Graminoid	5	5	1
	Shrub	2	1	0.980

All data analysis was done using R software (version R.3.1.1, R Core Team 2019). The main response variable of interest is native forb cover in relation to soil type and competing weed cover. Species cover, volumetric water content, and pH data had non-normal distribution based on the Shapiro–Wilk test and residual plots and could not be normalized; therefore, non-parametric tests were completed. To determine the differences between functional group cover, richness (native forbs, non-native, graminoids, shrubs, and trees), pH, VWC, and tree regeneration by soil type, a two-way permutational analysis of variance (permANOVA) (significance level of 0.05) was completed using the *lmPerm* package (version 2.1.0) [35]. ANOVA was used to compare soil temperature between soil types. Spearman correlation was used to compare the cover of non-native forbs and graminoids to native forb cover and richness on both soil types. Negative correlations are interpreted to indicate competition, while positive correlations indicate facilitation or no direct interaction. Regression tree analysis using the mvpart package (Version 1.6–2) [36] was used to evaluate the potential thresholds of weedy, graminoid, and individual species cover that would be potentially limiting on native forb cover.

3. Results

Four growing seasons after establishment, species richness (average = 21 species per 1 m^2 quadrat, $p = 1$) and total cover (average = 45%, $p = 1$) were similar between FFMM and PMM. The largest component of the plant community in terms of both richness and cover was from native forbs on both soil types (Table 1, Figure 1). Native forb cover was similar on both soil types ($p = 0.408$) with an average cover of 21.3% on FFMM and 18.1% on PMM. Shrub cover was similar between soil types ($p = 0.076$) with an average cover of 4.5% on FFMM and 1.8% on PMM (Figure 1). Non-native forb cover ($p < 0.001$) was higher on PMM, which had a cover of 10.9%, compared to FFMM, which had an average cover of 4.3% (Figure 1). Tree cover ($p < 0.001$) was higher on PMM, which had an average cover of 4.7%, while FFMM had 0.6% (Figure 1). Graminoid cover was higher on the FFMM ($p = 0.034$), which had an average cover of 14.4%, while PMM had 9.9%.

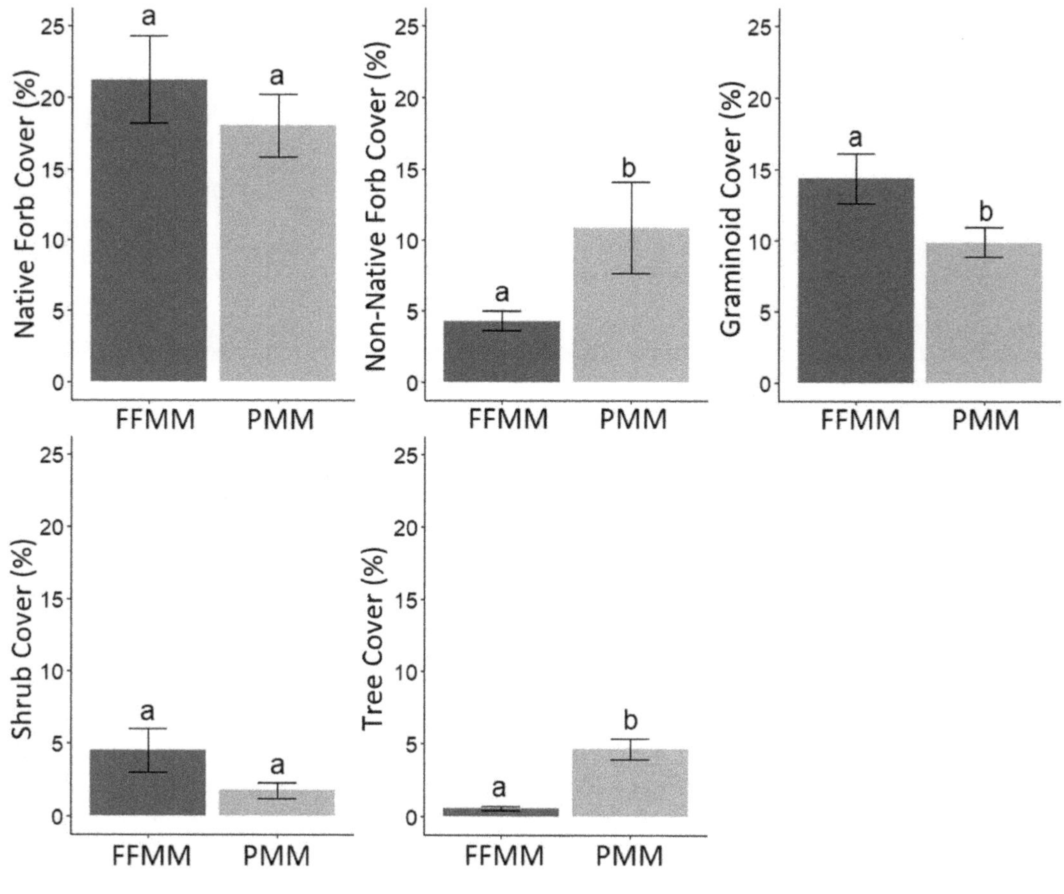

Figure 1. Average percent cover by functional group comparison of forest floor-mineral mix (FFMM) and peat-mineral mix (PMM) soil types with standard error bars and significant letters based on permutational analysis of variance (permANOVA) (FFMM n = 20; PMM n = 20).

On FFMM, the species with the highest cover were *Calamagrostis canadensis* (native grass), *Chamerion angustifolium* (native forb), *Rubus idaeus* (native shrub), and *Sonchus arvensis* (non-native forb) (Table 2). On PMM, the species with the highest cover were *Chamerion angustifolium*, *Calamagrostis canadensis*, *Equisetum arvense* (native forb), and *Melilotus alba* (non-native forb) (Table 2). *Calamagrostis canadensis* accounted for the largest proportion of graminoid cover on both soil types; 47% of graminoid cover was *Calamagrostis* on FFMM and 53% on PMM. *Sonchus arvensis* accounted for 60% of non-native forb cover on FFMM and 34% on PMM, while *Melilotus alba* accounted for 14% of non-native cover on FFMM and 40% on PMM. These three weedy species were clearly the dominant weedy species on both soil types.

Table 2. Four species with the highest average percent cover per plot found within each functional group and soil type showing cover (%).

	FFMM		PMM	
Functional Group	**Species**	**Average Cover Per Plot (%)**	**Species**	**Average Cover Per Plot (%)**
Non-Native Forbs	*Sonchus arvensis*	2.6	*Melilotus alba*	4.4
	Melilotus alba	<1.0	*Sonchus arvensis*	3.7
	Taraxacum officinale	<1.0	*Crepis tectorum*	1.5
	Crepis tectorum	<1.0	*Taraxacum officinale*	<1.0
Native Forbs	*Chamerion angustifolium*	6.2	*Chamerion angustifolium*	6.5
	Equisetum arvense	3.9	*Equisetum arvense*	5.8
	Lathyrus ochroleucus	1.9	*Achillea millefolium*	1.2
	Fragaria virginiana	1.7	*Lathyrus ochroleucus*	<1.0
Graminoids	*Calamagrostis canadensis*	6.8	*Calamagrostis canadensis*	5.3
	Poa pratensis	2.7	*Agropyron trachycaulum*	1.0
	Agropyron trachycaulum	2.5	*Carex* spp.	<1.0
	Leymus innovatus	1.3	*Poa pratensis*	<1.0
Shrubs	*Rubus idaeus*	3.7	*Salix* spp.	1.0
	Rosa acicularis	<1.0	*Rubus idaeus*	<1.0
	Cornus sericea	<1.0	*Rosa acicularis*	<1.0
	Ribes oxyacanthoides	<1.0	*Shepherdia canadensis*	<1.0

On both soil types, neither native forb cover (rho = −0.307, $p = 0.056$) nor richness (FFMM rho = −0.131, $p = 0.582$; PMM rho = −0.044, $p = 0.854$) showed a significant relationship related to non-native forb cover. Similarly, no relationships occurred between native forb cover and the two main non-native forb species, *Sonchus arvensis* (FFMM rho = −0.278, $p = 0.235$; PMM rho = −0.223, $p = 0.344$) and *Melilotus alba* (FFMM rho = −0.149, $p = 0.530$; PMM rho = −0.166, $p = 0.483$). There was also no relationship between non-native forb cover and either tree (FFMM rho = 0.298 $p = 0.202$; PMM rho = −0.070 $p = 0.768$) or shrub (FFMM rho = 0.040 $p = 0.868$; PMM rho = −0.071 $p = 0.767$) cover.

In contrast to non-native forbs, the relationship between graminoid cover and native forb cover showed a clear threshold relationship with reduced cover of native forbs when graminoid cover was greater than 17.4% (Figure 2b). Native forb richness, however, was not impacted by graminoid cover on either soil type (graminoid FFMM rho = 0.104, $p = 0.663$; PMM rho = 0.278, $p = 0.235$). There was no relationship between graminoid cover and either tree (FFMM rho = −0.043, $p = 0.857$; PMM rho = −0.027, $p = 0.910$) or shrub (FFMM rho = 0.0928, $p = 0.697$; PMM rho = −0.027, $p = 0.513$) cover on either soil type. These relationships were unaffected by soil pH, temperature, and VWC at the plot level.

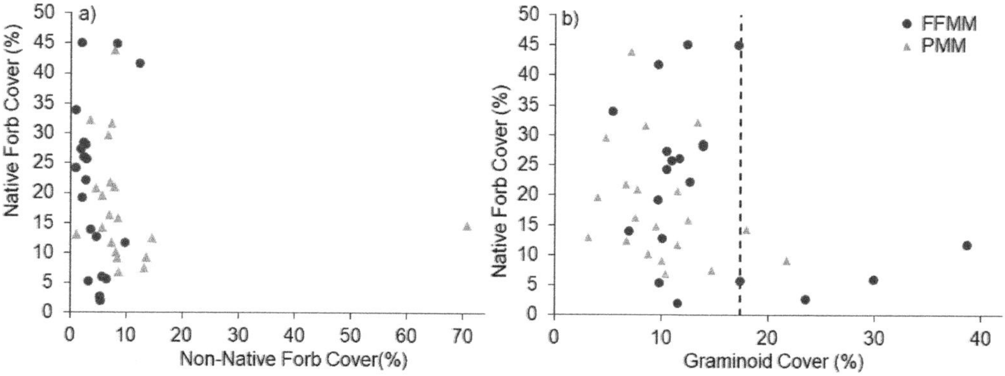

Figure 2. Scatter plot between native forb cover (%), (**a**) non-native forbs, and (**b**) graminoid cover (%) with a threshold created from a regression tree analysis (threshold for graminoid cover ≥17.4% n = 6, <17.4% n = 34).

4. Discussion

There is little evidence of competition among functional groups or of non-native forbs hindering the development of native forbs. On the other hand, there is also no evidence of facilitation with non-native species enhancing the establishment of native forbs such as has been hypothesized for the nitrogen-fixing *Melilotus* [37–40]. Based on the results of this study, *Melilotus alba* does not appear to be a factor in understory growth, displaying little to no impact on abundance or diversity of native species on either soil type; however, this may also be due to lifecycle, as *Melilotus spp.* are biennials [28,39]. Overall, it appears that non-native forbs could simply be acting as placeholders or niche fillers alongside other pioneer species within the native plant community and may not impede the establishment and development of other species [38,41]. This random vegetation community structure is common in early successional communities [39,42] and is evidenced in our study by the abundance of available seedbeds (55% bare soil on average) and growing space. Competition may play a larger role in structuring the vegetation community once all available growing space is occupied, but competition from non-native forbs does not appear to be negatively influencing the plant community at this stage.

On oil sands mine sites that utilized either FFMM or PMM as cover soil, *Sonchus arvensis* (Perennial sow-thistle, non-native perennial forb), *Melilotus alba* (White sweet clover, non-native biennial forb), and *Calamagrostis canadensis* (Bluejoint reedgrass, native perennial grass) have been identified as potentially having the most impact on the desired native forb community post-disturbance [1,2,27]. Listed on Alberta's noxious weed list [20], *Sonchus arvensis* is 40–200 cm tall and forms large, aggressive root networks [22,43] capable of capitalizing on belowground resources, which can result in an aboveground competitive advantage deleteriously affecting the native plant community development and diversity [27]. *Melilotus alba* establishes quickly, grows to 250 cm tall, spreads rapidly, and acts as a nitrogen-fixer [22,28,44], making it potentially useful as a cover crop. However, some of these same characteristics enable *Melilotus alba* to become weedy and difficult to control [28,44]. On a reclaimed coal mine in Alberta utilizing *Melilotus* spp. as a cover crop, findings show the establishment of the native understory was not affected and may have been facilitated via repression of other weeds [28]. Therefore, whether or not *Melilous alba* should be considered a weed is dependent on many factors.

Sonchus arvensis, the other main non-native species found in our study, is listed on Alberta's noxious weed list [45], meaning that it is legally required to be controlled to prevent the spread. However, based on our results *Sonchus arvensis* showed no evidence of competition with native forbs four years after site establishment and has less than 5% cover on both soil types. Species like *Sonchus arvensis,* and many other weeds, have high initial cover after reclamation and then decrease in abundance within a few years. This is because they often originate from the seedbank, particularly on FFMM, and have rapid expansion [28,35,46]. However, it appears that this initially high cover of *Sonchus arvensis* does not inhibit the growth and development of native forbs and is a low risk to the long-term ecosystem development on the site [2].

Although all graminoids, including the dominant *Calamagrostis canadensis*, present on our site are native species and are not considered noxious weeds, they appear to have a greater potential to reduce native forb cover [47,48]. Graminoids reduced cover of native forbs on both FFMM and PMM at graminoid cover levels greater than 17%. In other studies, the abundance of native forb species was reduced by 10% in the presence of native perennial graminoids, in particular *Calamagrostis canadensis*, and the growth and establishment of native trees were substantially reduced [25,43]. Graminoids have the potential to impede other native plants through a decrease in soil temperature, reduced light availability, and snow press damage due to the dense leaf litter [31,32]. A recent study from northeastern Alberta found that perennial grasses showed the greatest impact on tree growth and competition dynamics in this system [49]. We see a similar relationship in our study with higher graminoid cover and lower tree cover on FFMM, with the opposite being true on PMM.

The future of weeds on this site is still relatively unknown. On some mine reclamation sites, non-native forbs can still be a large problem, showing 95% cover more than 20 years post reclamation [50]. In other studies, weeds average only 10% cover after 20 years [21]. In our study, non-native forbs were around 10% cover after four years. Non-native species still have the potential to drive ecosystem development when native species richness is low and non-native forb cover is high [21]. In other cases, such as in our study, when non-native cover is low and native species richness is high, non-native forbs are not likely to have a large impact in structuring the vegetation community [51].

The future role of non-native forbs is generally dependent on the development of an overstory canopy [51–53]. As long as trees can develop into the initial community on reclamation sites, the long-term competitive relationships will likely be similar to what is found in nearby natural ecosystems [46,48]. In both mature natural stands and post-disturbance stands, the tree canopy has a strong influence on the altering of the understory plant community, on soil nutrients, and on light availability [46,48]. With the presence of an overstory canopy, the understory will be less likely to be invaded by weedy species due to the barrier for wind-dispersal and heavy shading [50,52]. However, as in natural forests, graminoids are capable of remaining in the understory canopy throughout the entire rotation [29].

We expected the weed relationships to vary by soil type, but we did not see this trend. This indicates that the vegetation communities on both soil types are behaving in a similar manner. The initial vegetation community on this site was different between soil types due to the differences in the seedbank [46]. Within a few years, however, other factors such as seed rain alter the vegetation community on both soil types [46,53]. It also appears that even at this early successional stage, the competitive processes between plant functional groups are similar between soil types; in the future, we expect the successional patterns to be similar as well.

5. Management Implications

From a management perspective, if weeds are not posing a problem by reducing desirable plant cover or altering successional pathways, they can likely be monitored rather than eradicated, resulting in saved time and money on weed control. Instead of focusing on the presence and subsequent control of a particular weed species, reclamation requirements and policies may need to focus more on ecosystem health and productivity outcomes, such as the cover and richness of a desirable native plant community. This approach will better account for "novel" ecosystems, which are a more realistic and economically achievable outcome [51]. A viable weed management option for our site appears to be simply allowing the desirable plant community to develop over time with no further interventions required. However, on sites where native graminoids appear to be acting as a more serious weedy competitor than non-native forbs, more active interventions may be needed.

Sonchus arvensis, *Melilotus alba*, and *Calamagrostis canadensis* on this reclamation site established through the seedbank. Determining the pre-salvage abundance of these species would allow managers to identify the probability of reclamation sites having higher weed pressures and potentially prevent them from establishing on site at all [2,29]. Although competition did not seem to have a large influence in structuring the vegetation community in our study, caution must be used to extrapolate to other reclamation sites and successional stages. This study was a short-term experiment during the summer season of 2019 and was only one time period on a successional stage. In addition, our study was intended to identify the actual impact of weeds on native forbs on a specific site, and we believe we have accomplished this goal using relatively simple plant community metrics, specifically cover and richness. However, future studies using more sophisticated metrics such as plant volume and competition indices may be able to better quantify plant interactions [54]. Of course, all reclamation sites are unique; therefore, continued monitoring is needed to gain a better understanding of the long-term implications of different weed management techniques on different site types.

Author Contributions: Conceptualization, B.B., B.D.P., and K.E.T.; methodology, B.B., B.D.P., and K.E.T.; validation, B.B., B.D.P., and K.E.T.; formal analysis, K.E.T.; investigation, B.B., B.D.P., and K.E.T.; data curation, B.B., B.D.P., and K.E.T.; writing—original draft preparation, K.E.T., B.B., and B.D.P.; writing—review and editing, K.E.T., B.B., and B.D.P.; visualization, K.E.T. and B.D.P.; supervision, B.D.P. and K.E.T.; funding acquisition, B.D.P. All authors have read and agreed to the published version of the manuscript.

Funding: This research was funded by Canadian Natural Resources Limited.

Institutional Review Board Statement: Not applicable.

Informed Consent Statement: Not applicable.

Data Availability Statement: The data presented in this study are available on request from the corresponding author. The data are not publicly available due to privacy concerns.

Acknowledgments: We would like to thank Sofia Toledo and Mathew Anderson for help with lab and field work. Thank you, Sebastian Dietrich for the comments.

Conflicts of Interest: The authors declare no conflict of interest.

References

1. Cole, D.E.; King, J.R.; Oyarzun, D.A.; Dietzler, T.H.; McClay, A. Experiences with invasive plant management and ecology in Alberta. *Can. J. Plant Sci.* **2007**, *87*, 1013–1022. [CrossRef]
2. Small, C.; Degenhardt, D.; Drozdowski, B.; Thacker, S.; Powter, C.; Schoonmaker, A.; Schreiber, S. *Optimizing Weed Control for Progressive Reclamation: Literature Review*; Canada's Oil Sands Innovation Alliance: Calgary, AB, Canada, 2018; 48p.
3. Luck, L.; Bellairs, S.M.; Rossiter-Rachor, N.A. Residual herbicide treatments reduce *Andropogon gayanus* (Gamba Grass) recruitment for mine site restoration in northern Australia. *Ecol. Manag. Restor.* **2019**, *20*, 214–221. [CrossRef]
4. Franke, M.E.; Zipper, C.; Barney, J.N. Native hardwood tree seedling establishment following invasive autumn-olive (*elaeagnus umbellata*) removal on a reclaimed coal mine. *Invasive Plant Sci. Manag.* **2018**, *11*, 155–161. [CrossRef]
5. Holmes, P.M. Shrubland restoration following woody alien invasion and mining: Effects of topsoil depth, seed source, and fertilizer addition. *Restor. Ecol.* **2001**, *9*, 71–84. [CrossRef]
6. González-Muñoz, N.; Costa-Tenorio, M.; Espigares, T. Invasion of alien *Acacia dealbata* on Spanish Quercus robur forests: Impact on soils and vegetation. *For. Ecol. Manag.* **2012**, *269*, 214–221. [CrossRef]
7. Alberta Environment. *Guidelines for Reclamation to Forest Vegetation in the Athabasca Oil Sands Region*, 2nd ed.; Terrestrial Subgroup of the Reclamation Working Group of the Cumulative Environmental Management Association: Fort McMurray, AB, Canada, 2010.
8. Baker, H.G. The evolution of weeds. *Annu. Rev. Ecol. Syst.* **1974**, *5*, 1–24. [CrossRef]
9. Espeland, E.; Perkins, L. Weed establishment and persistence after water pipeline installation and reclamation in the mixed grass prairie of Western North Dakota. *Ecol. Restor.* **2017**, *35*, 303–310. [CrossRef]

10. Errington, R.C.; Pinno, B.D. Early successional plant community dynamics on a reclaimed oil sands mine in comparison with natural boreal forest communities. *Ecoscience* **2015**, *22*, 133–144. [CrossRef]
11. Mackenzie, D.D.; Naeth, M.A. The role of the forest soil propagule bank in assisted natural recovery after oil sands mining. *Restor. Ecol.* **2009**, *18*, 418–427. [CrossRef]
12. Gingras-Hill, T.; Nwaishi, F.C.; Macrae, M.; Price, J.S.; Petrone, R.M. Ecohydrological functioning of an upland undergoing reclamation on post-mining landscape of the Athabasca oil sands region, Canada. *Ecohydrology* **2018**, *11*, 1–12. [CrossRef]
13. Cavieres, L.; Arroyo, M.T.K.; Peñaloza, A.; Molina-Montenegro, M.; Torres, C. Nurse effect of Bolax gummifera cushion plants in the alpine vegetation of the Chilean Patagonian Andes. *J. Veg. Sci.* **2002**, *13*, 547–554. [CrossRef]
14. Bruno, J.F.; Stachowicz, J.J.; Bertness, M.D. Inclusion of facilitation into ecological theory. *Trends Ecol. Evol.* **2003**, *18*, 119–125. [CrossRef]
15. Jordan, N.; Larson, D.L.; Huerd, S.C. Soil modification by invasive plants: Effects on native and invasive species of mixed-grass prairies. *Biol. Invasions* **2007**, *10*, 177–190. [CrossRef]
16. Maestre, F.T.; Bautista, S.; Cortina, J.; Bellot, J. Potential for using facilitation by grasses to establish shrubs on a semiarid degraded steppe. *Ecol. Appl.* **2001**, *11*, 1641–1655. [CrossRef]
17. Alberta Environment and Water. *Best Management Practices for Conservation of Reclamation Materials in the Mineable Oil Sands Region of Alberta*; Province of Alberta: Edmonton, AB, Canada, 2012; pp. 1–161.
18. Natural Regions Committee. *Natural Regions and Subregions of Alberta*; Natural Regions Committee: Edmonton, AB, Canada, 2006.
19. Canadian Climate Normals 1981–2010 Station Data. Available online: https://climate.weather.gc.ca/climate_normals/results_1981_2010_e.html?searchType=stnProv&lstProvince=AB&txtCentralLatMin=0&txtCentralLatSec=0&txtCentralLongMin=0&txtCentralLongSec=0&stnID=2704&dispBack=0 (accessed on 22 April 2020).
20. Pinno, B.D.; Errington, R.C. Maximizing natural trembling aspen seedling establishment on a reclaimed boreal oil sands site. *Ecol. Restor.* **2015**, *33*, 43–50. [CrossRef]
21. Pinno, B.D.; Hawkes, V.C. Temporal trends of ecosystem development on different site types in reclaimed boreal forests. *Forests* **2015**, *6*, 2109–2124. [CrossRef]
22. Tannas, K.E. *Common Plants of the Western Rangelands—Volume 3: Forbs*, 1st ed.; Alberta Agriculture, Food and Rural Development: Lethbridge, AB, Canada, 2004.
23. Tannas, K.E. *Common Plants of the Western Rangelands—Volume 1: Grasses and Grass-Like Species*, 1st ed.; Alberta Agriculture, Food and Rural Development: Lethbridge, AB, Canada, 2001.
24. Moss, E.H.; Packer, J. *Flora of Alberta*, 2nd ed.; University of Toronto Press: Toronto, ON, Canada, 1994.
25. Tannas, K.E. *Common Plants of the Western Rangelands—Volume 2: Trees and Shrubs*, 1st ed.; Alberta Agriculture, Food and Rural Development: Lethbridge, AB, Canada, 2003.
26. United States Department of Agriculture. Available online: https://plants.sc.egov.usda.gov/adv_search.html (accessed on 7 December 2019).
27. de Bortoli, L.A.; Pinno, B.D.; Mackenzie, M.D.; Li, E.H. Plant community composition and tree seedling establishment in response to seeding and weeding treatments on different reclamation cover soils. *Can. J. For. Res.* **2019**, *49*, 836–843. [CrossRef]
28. Macdonald, S.E.; Snively, A.E.K.; Fair, J.M.; Landhausser, S.M. Early trajectories of forest understory development on reclamation sites: Influence of forest floor placement and a cover crop. *Restor. Ecol.* **2015**, *23*, 698–706. [CrossRef]
29. Lieffers, V.J.; Macdonald, S.E.; Hogg, E.H. Ecology of and control strategies for *Calamagrostis canadensis* in boreal forest sites. *Can. J. For. Res.* **1993**, *23*, 2070–2077. [CrossRef]
30. Macdonald, S.E.; Lieffers, V.J. Rhizome plasticity and clonal foraging of *Calamagrostis canadensis* in response to habitat heterogeneity. *J. Ecol.* **1993**, *81*, 769–776. [CrossRef]
31. Landhäusser, S.M.; Mulak, T.L.; Lieffers, V.J. The effect of roots and litter of Calamagrostis canadensis on root sucker regeneration of Populus tremuloides. *Forestry* **2007**, *80*, 481–488. [CrossRef]
32. Landhausser, S.M.; Stadt, K.J.; Lieffers, V.J. Screening for control of a forest weed: Early competition between three replacement species and Calamagrostis canadensis of *Picea glauca*. *J. Appl. Ecol.* **1996**, *33*, 1517. [CrossRef]
33. Macdonald, S.E.; Fenniak, T.E. Understory plant communities of boreal mixedwood forests in western Canada: Natural patterns and response to variable-retention harvesting. *For. Ecol. Manag.* **2007**, *242*, 34–48. [CrossRef]
34. Pinno, B.D.; Sherr, I.; Errington, R.C.; Shea, K. Islands – Soil Patches and Plant Community Dynamics on a New Oil Sands Reclamation Design. *J. Am. Soc. Min. Reclam.* **2016**, *5*, 28–44. [CrossRef]
35. Package 'lmperm'. Available online: https://cran.r-project.org/web/packages/lmPerm/lmPerm.pdf (accessed on 7 December 2019).
36. Package 'mvpart'. Available online: https://mran.microsoft.com/snapshot/2014-12-11/web/packages/mvpart/mvpart.pdf (accessed on 7 December 2019).
37. Foster, B.L.; Tilman, D. Dynamic and static views of succession: Testing the descriptive power of the chronosequence approach. *Plant Ecol.* **2000**, *146*, 1–10. [CrossRef]
38. Alday, J.G.; Pallavicini, Y.; Marrs, R.H.; Martínez-Ruiz, C. Functional groups and dispersal strategies as guides for predicting vegetation dynamics on reclaimed mines. *Plant Ecol.* **2011**, *212*, 1759–1775. [CrossRef]
39. Spellman, K.V.; Schneller, L.C.; Mulder, C.P.H.; Carlson, M.L. Effects of non-native *Melilotus albus* on pollination and reproduction in two boreal shrubs. *Oecologia* **2015**, *179*, 495–507. [CrossRef] [PubMed]

40. Tilman, D.; Wedin, D. Dynamics of nitrogen competition between successional grasses. *Ecology* **1991**, *72*, 1038–1049. [CrossRef]
41. MacDougall, A.S.; Turkington, R. Are invasive species the drivers or passengers of change in degraded ecosystems? *Ecology* **2005**, *86*, 42–55. [CrossRef]
42. Dhar, A.; Comeau, P.G.; Karst, J.; Pinno, B.D.; Chang, S.X.; Naeth, A.M.; Vassov, R.; Bampfylde, C. Plant community development following reclamation of oil sands mine sites in the boreal forest: A review. *Environ. Rev.* **2018**, *298*, 1–13. [CrossRef]
43. Plants Database: *Sonchus arvensis*, L. Field Sowthistle. Available online: https://plants.usda.gov/core/profile?symbol=SOAR2 (accessed on 7 December 2019).
44. Turkington, R.A.; Cavers, P.B.; Rempel, E. The biology of Canadian weeds. 29. *Melilotus alba* Desr. and *M. offtcinalis* (L.) Lam. ROY. *Can. J. Planr Sci.* **1978**, *58*, 523–537. [CrossRef]
45. Provincially Regulated Weeds. Available online: https://www.alberta.ca/provincially-regulated-weeds.aspx (accessed on 7 December 2019).
46. Trepanier, K.E.; Pinno, B.D.; Errington, R.C. Dominant drivers of plant community assembly vary by soil type and time in reclaimed forests. *Plant Ecol.* **2020**. [CrossRef]
47. Brown, C.S.; Bugg, R.L. Effects of established perennial grasses on introduction of native forbs in California. *Restor. Ecol.* **2001**, *9*, 38–48. [CrossRef]
48. Landhäusser, S.M.; Lieffers, V.J. Growth of *Populus tremuloides* in association with *Calamagrostis canadensis*. *Can. J. For. Res.* **1998**, *28*, 396–401. [CrossRef]
49. Henkel-Johnson, D.; Macdonald, S.E.; Bork, E.W.; Thomas, B.R. Influence of weed composition, abundance, and spatial proximity on growth in young hybrid poplar plantations. *For. Ecol. Manag.* **2016**, *362*, 55–68. [CrossRef]
50. Strong, W.L. Vegetation development on reclaimed lands in the Coal Valley Mine of western Alberta, Canada. *Can. J. Bot.* **2000**, *78*, 110–118. [CrossRef]
51. Audet, P.; Pinno, B.D.; Thiffault, E. Reclamation of boreal forest after oil sands mining: Anticipating novel challenges in novel environments. *Can. J. For. Res.* **2015**, *45*, 364–371. [CrossRef]
52. Hart, S.A.; Chen, H.Y.H. Understory vegetation dynamics of North American boreal forests. *CRC. Crit. Rev. Plant Sci.* **2006**, *25*, 381–397. [CrossRef]
53. Cole, R.J.; Holl, K.D.; Zahawi, R.A. Seed rain under tree islands planted to restore degraded lands in a tropical agricultural landscape. *Ecol. Appl.* **2010**, *20*, 1255–1269. [CrossRef] [PubMed]
54. Weigelt, A.; Jolliffe, P. Indices of plant competition. *J. Ecol.* **2003**, *91*, 707–720. [CrossRef]

diversity

MDPI

Article

Ecology and Diversity of Angiosperm Parasites and Their Host Plants along Elevation Gradient in Al-Baha Region, Saudi Arabia

Sami Asir Al-Robai

Department of Biology, Faculty of Science, Al-Baha University, Al-Baha 1988, Saudi Arabia; salrobai@bu.edu.com

Abstract: The ecology and diversity of flowering parasitic plants and their hosts are poorly investigated and usually ignored in Saudi Arabian plant communities. Therefore, this work aimed at assessing the ecology and diversity of parasitic plants and their hosts along an elevation gradient in the Al-Baha region (1300–2400 m.a.s.l.). Different quantitative vegetation parameters were applied to analyze the collected data. Eight parasitic plants from six genera and four families were identified along the gradient, with 67% of them being zoochorously dispersed species. They accounted for approximately 23.5% (8 out of 34) of those found throughout Saudi Arabia. Perennials, stem hemi-parasites, and biregional taxa accounted for around 62.5% of the total parasites, whereas indigenous species accounted for 75%. The dominant family of parasitic species was Loranthaceae (50%), and *Phragmanthera austroarabica* A.G.Mill. & J.A.Nyberg was the most important species (IVI = 107.28). *Orobanche cernua* Loefl. and *Loranthella deflersii* (Tiegh.) S.Blanco & C.E.Wetzel were restricted to the dry zone (low elevation) only, while the other parasites were distributed across the surveyed region. Twenty-three host plants were identified throughout the study region. About 83% of them were phanerophytes and bioregional plants, with 91% being perennial species. The prevalent host plant family across all sites was Fabaceae, with *Nicotina glauca* Graham being the most important host species (IVI = 32.44%). *P. austroarabica* and *Plicosepalus curviflorus* Tiegh. preferred *Vachellias* as host plants, while *Vachellia flava* (Forssk.) Kyal. & Boatwr. was the heavily infected host by *P. austroarabica*. *P. austroarabica* had a broad spectrum of host range (13 host plants), while *O. cernua* had a very narrow host range (only *Rumex nervosus* Vahl). Individual parasite and host species were markedly more abundant in the wet zone than in the low-altitude dry zone. Further research is needed to fully understand such distinctive groups of plants and their negative and positive ecological consequences on plant biodiversity and natural ecosystems.

Keywords: parasitic plants; host species; elevation; vegetation parameters; diversity; Saudi Arabia

Citation: Al-Robai, S.A. Ecology and Diversity of Angiosperm Parasites and Their Host Plants along Elevation Gradient in Al-Baha Region, Saudi Arabia. *Diversity* **2023**, *15*, 1065. https://doi.org/10.3390/d15101065

Academic Editors: Michael Wink and Ilias Travlos

Received: 6 September 2023
Revised: 28 September 2023
Accepted: 4 October 2023
Published: 7 October 2023

1. Introduction

Parasitic angiosperms are a diverse group of around 4500 species divided into 12 families and roughly 300 genera [1]. They are specialized plants that lose the capacity to photosynthesize entirely or partially and receive their organic and inorganic nutrients from the host plants through the haustoria. A parasitic plant's haustorium is a specialized structure that has a physiological connection with the host plants [2]. Although the number and diversity of these parasites vary depending on the biome and ecosystems, they are common elements in terrestrial environmental habitats worldwide [3]. Beyond having a bad impact on the host, their ecological roles are much more nuanced [4]. Some parasitic species can change the community competition patterns, facilitate the environmental cycling of nutrients [5], and modify diversity within communities [6]. It has been frequently shown that parasitic plants can inhibit the development and competitiveness of widespread plants, lower the rate of biomass formation in the ecosystem, and promote seed growth by creating openings for embryos to germinate [7,8]. This means that the overall harm caused by parasitic plants to their hosts could be changed into a benefit for wider plant communities [4]. Some of them play significant roles in controlling plant invasions and aiding in biodiversity restoration [9,10]. Furthermore, the haustorial link may aid in the flow of stress-responsive chemicals as well as potentially hazardous substances such as heavy metals [11,12]. However, a small number of parasitic plants are among the most devastating agricultural pests, costing billions of dollars in annual losses, food poverty, and ecological risks [13]. Mistletoe is an airborne parasitic plant that is pollinated and disseminated by birds and other visitors; it controls the natural structure of the plant groups throughout the local habitat [14].

Depending on their photosynthetic ability, they can be classified as holoparasites (photosynthetic) or hemiparasites (non-photosynthetic) [15] or as stem or root parasites according to where they attach [16]. Although the haustorium structure varies throughout parasitic plant families, it is a feature common to all species [16].

Orobanchaceae and the order Santalales have the highest number of parasitic plants [2], with Loranthaceae (76 genera, more than 1000 species) having most of them [17]. Loranthaceae is predominantly distributed in Asia, the Americas, Africa, and Australia, with certain species spreading in areas having mild temperatures in Europe and East Asia [18]. It is primarily composed of airborne parasitic plants but also includes three root parasitic species [18]. The plants in this family interact with insects, birds, and mammals and have a crucial role in the biological ecosystems in which they exist [14]. The family *Loranthaceae* has four genera, *Phragmanthera, Oncocalyx, Tapinanthus,* and *Plicosepalus*, all of which grow natively in Saudi Arabia, with six species belonging to these genera dispersed across the Kingdom's north, west, and south [19]. There have been few publications, most of which have focused on species in a relatively restricted area of the Al-Baha region [20–22]. These publications barely mention the distribution and host range of parasitic vascular plants in this region. A few studies [23,24] have been conducted, but they do not address the ecology and diversity of all parasitic plants and their host plants that are present throughout the country. Consequently, this work gives an in-depth investigation on the distribution and diversity of parasitic plants and their host range of flowering plants in the Al-Baha region.

2. Materials and Methods

2.1. Study Area

Al-Baha is a highland region in southwestern Saudi Arabia lying at the longitude 41/42 E and latitude 16/20 N with an elevation ranging from 1300 to 2450 m above sea level (Figure 1). It is bordered on the west by the Rocky Mountains and on the east by semi-arid mountains and has a broad diversity of habitats associated with different plant species. The majority of Al-Baha's region falls within the tropical and subtropical arid zones [20]. Summer temperatures range from 22 to 32 °C and winter temperatures range from 10 to 22 °C, and the rainfall in arid areas varies between 100 and 200 mm [25]. Alaqiq, Baljurashi, Al Mandaq, and Al Mikhwah areas receive annual precipitations of 142, 300, 316, and 200 mm, respectively [20].

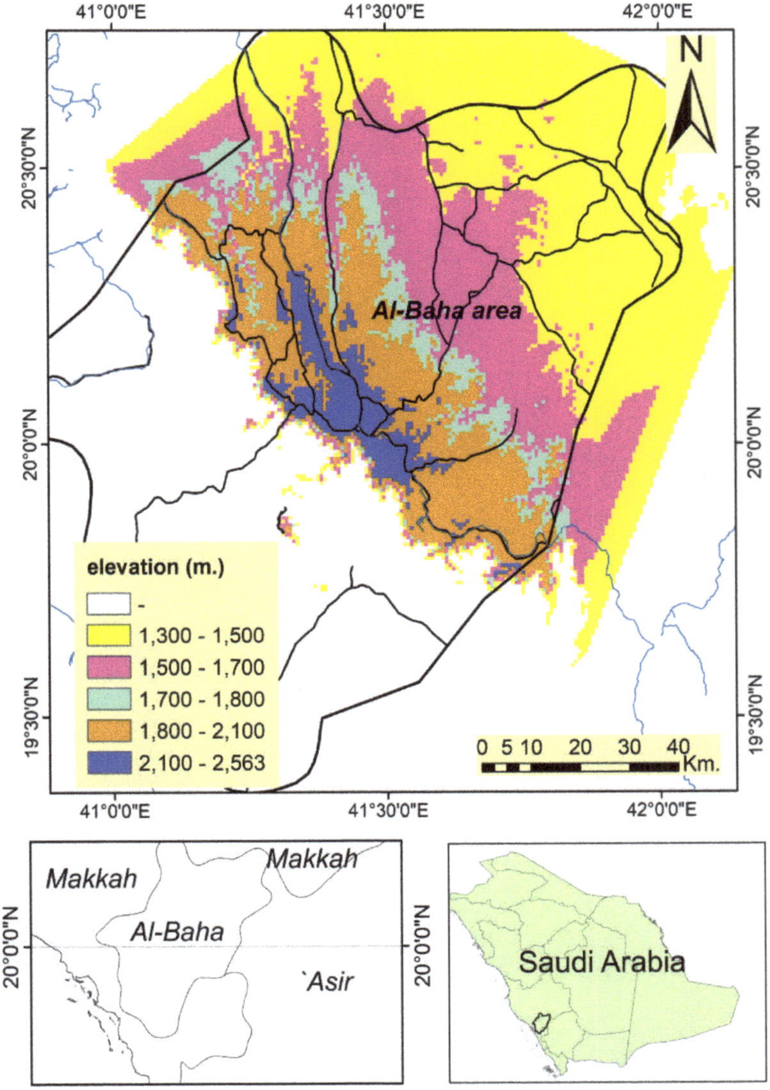

Figure 1. A map shows the locations of the study area.

2.2. Field Survey

Extensive field surveys (over 30 trips) were conducted to cover all the sites of the study region from April 2022 to March 2023. The region was divided into two zones, dry (low altitude) and wet (high altitude) zones. According to the elevation gradient, each zone was subdivided into two sites, dry zone (site 1: 1300–1500, site 2: 1501–1700 m.a.s.l.) and wet zone (site 3: 1800–2100, site 4: 2101–2400 m.a.s.l.). Further, each site was divided into eight stands with 10 quadrats (20 × 20 m). An overall 80 quadrat units were analyzed in each site. Plant samples were collected once, and there was no seasonal sampling.

2.3. Plant Collection and Species Identification

The examined plant taxa were identified and revised using different published volumes of Saudi Arabia flora [26–29]. Nomenclature was reviewed and updated using authenticated international data from the World Online [30]. Voucher specimen from each plant was deposited at Biology Department, Faculty of Science, Al-Baha University.

2.4. Floristic Analysis

Life form, habit, and life span categories were determined using the updated Raunkiaer [31] classification by Govaerts et al. [32]. Phytogeographical categories were recognized following Wickens [33] and Zohary [34].

2.5. Data Analysis

Quantitative measurements derived from the quadrat method were performed to determine the relative frequency, relative abundance, and relative density of each species in the surveyed region. From these parameters, the species importance value index (IVI) was calculated to assess the dominant species in the study region [35,36]. The importance value defines the importance of a species in the community structure or species composition in the study area [37].

$$\text{Density (D)} = \frac{\text{Total number of individuals of a species in all quadrats}}{\text{Total number of quadrats studied}} \times 100$$

$$\text{Frequency (F)} = \frac{\text{Total number of quadrats in which species occur}}{\text{Total number of quadrats studied}} \times 100$$

$$\text{Abundance (A)} = \frac{\text{Total number of individuals of a species in all quadrats}}{\text{Total number of quadrats in which species occur}} \times 100$$

$$\%\ \text{Relative density (RD)} = \frac{\text{Density of the species}}{\text{Total density of all species}} \times 100$$

$$\%\ \text{Relative frequency (RF)} = \frac{\text{Frequency of the species}}{\text{Total frequency of all species}} \times 100$$

$$\%\ \text{Relative abundance (RA)} = \frac{\text{Total number of individuals of a species in all quadrats}}{\text{Total abundance of all species}} \times 100$$

$$\text{Importance value index (IVI)} = \text{RD} + \text{RF} + \text{RA}$$

Microsoft Excel was utilized as a software package (16.0) to analyze data and construct the charts and histograms that were used to present the collected data.

3. Results

3.1. Floristic Composition of Parasitic Species

Eight species of parasitic angiosperm plants were recorded in the studied four sites, namely, *Phragmanthera austroarabica* A.G.Mill. & J.A.Nyberg, *Viscum schimperi* Engl., *Cuscuta campestris* Yunck., *Orobanche mutelii* F.W.Schultz, *Orobanche cernua* Loefl., *Loranthella deflersii* (Tiegh.) S.Blanco & C.E.Wetzel, *Plicosepalus acaciae* (Zucc.) Wiens & Polhill, and *Plicosepalus curviflorus* Tiegh. (Table A1 and Figure 2). These species belong to Loranthaceae (four species), Orobanchaceae (two species), Santalaceae (one species), and Convolvulaceae (one species). All parasitic species were found in the two surveyed dry and wet zones, except for *O. cernua* and *L. deflersii* which were reported in the wet zone only (Figure 3). The total number of individual parasitic species ranged between 392 and 405 in the wet zone and 172 and 190 in the dry zone. *P. austroarabica*, *P. curviflorus*, and *V. schimperi* were the most abundant parasitic species in the study area (Figure 4). According to the WFO [38], *C. campestris* and *O. mutelii* were exotic plants while the remaining species were indigenous plants.

Figure 2. Some parasitic species and their host plants. (**a**) *P. austroarabica* on *Calotropis procera*. (**b**) *P. curviflorus* on *Vachellia gerrardii*. (**c**) *C. campestris* on *N. glauca*. (**d**) *O. cernua* on *Rumex nervosus*.

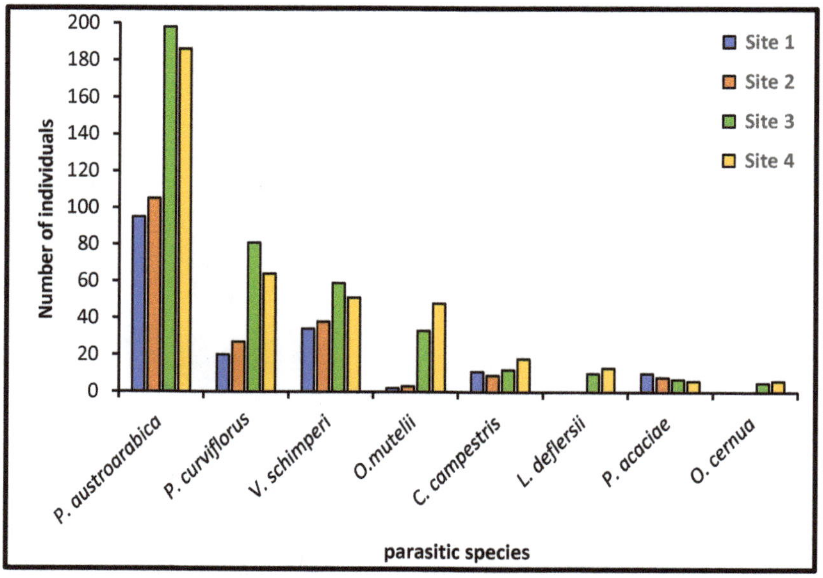

Figure 3. The number of individual parasitic species identified in each site of the study region.

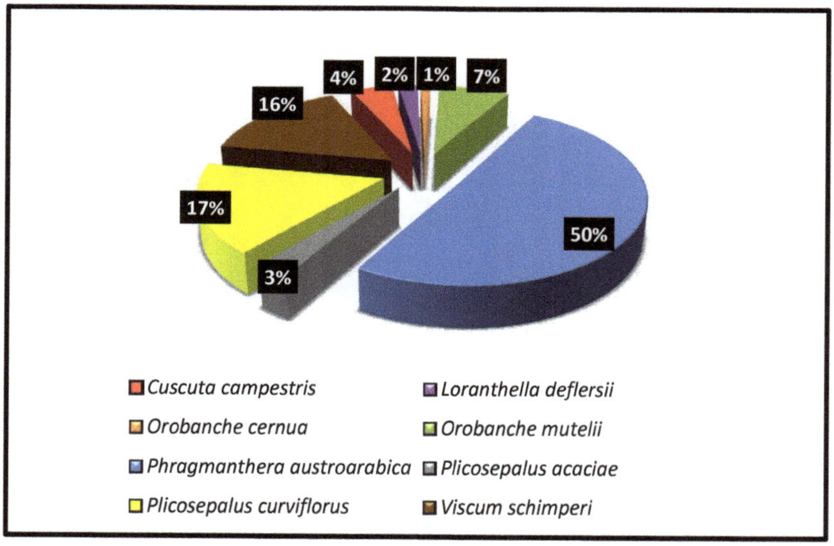

Figure 4. Percentage of the total number of individual parasitic species in the study region.

The identified angiosperm parasites can be categorized into four kinds: root hemiparasites, root holoparasites, stem hemiparasites, and stem holoparasites (Table A1). Most of the parasitic species belonged to stem hemiparasites (five species), followed by root holoparasites (two species), and stem holoparasites (one species). Of the total, five species (62.5%) were perennials/shrubs and three species were annual/herbs (37.5%).

The number of the host taxa that were attacked by parasitic species in the study area ranged between 1 and 13 species (Figure 5). *P. austroarabica* and *P. curviflorus* were found parasitizing on 13 and 5 of the host plants, respectively. *Vachellias* was the favored host for *P. austroarabica* and *P. curviflorus* with 83.4% and 73.5% infestations, respectively. *C. campestris* was found on *Nicotiana glauca* (2%), *Pluchea dioscoridis* (28%), and *Pulicaria undulata* (70%), while *L. deflersii* was reported on *Vachellia tortilis* subsp. *tortilis* (69.6%), *Tamarix senegalensis* (21.7%), and *Searsia retinorrhoea* (8.7%). The genus *Orobanche* was represented by two species, including *O. cernua* on only one host species (*Rumex nervosus*) and *O. mutelii* on *Bidens biternata* (22.1%) and *R. nervosus* (77.9%). *P. acaciae* was found parasitizing *Senegalia asak* (58.1%), *Vachellia tortilis* subsp *raddiana* (32.3), and *Tamarix aphylla* (9.7%). *V schimperi* was found attacking two different host species, *Ziziphus spina-christi* (92.9%) and *T. aphylla* (7.1%).

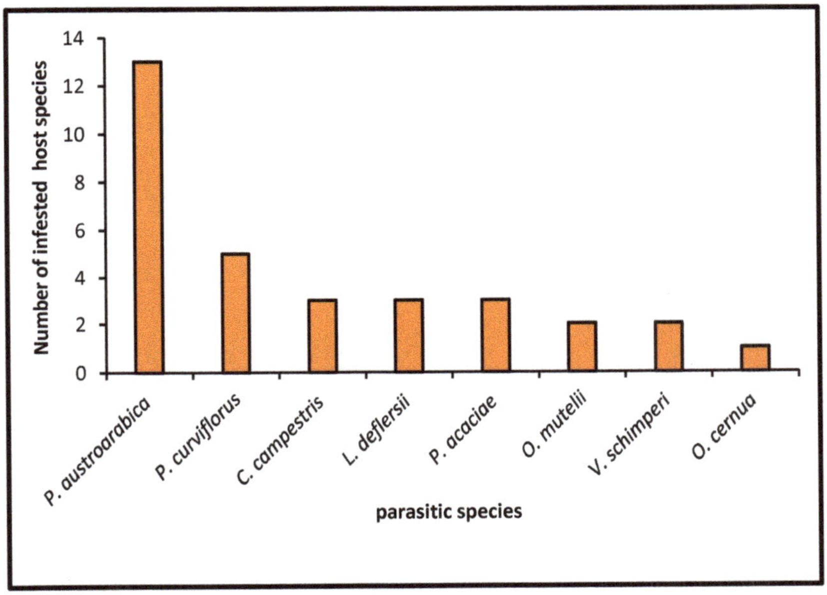

Figure 5. The number of host species infected by parasitic species in the study region.

3.2. Chorological Analysis of Parasitic Plants

The chorological analysis revealed that the identified parasitic species belonged to two major groups: biregional and monoregional (Table 1). Five species, accounting for 62.5% of the total number of species reported in the area, belonged to biregional taxa. Saharo-Arabian shared with Sudano-Zambezian had the highest share of species with three species (*L. deflersii*, *V. schimperi*, and *P. curviflorus*), followed by Saharo-Arabian /Irano-Turanian with two species. On the other hand, three species representing 37.5% of the parasitic species were monoregional. The three plants were native to the American (*C. campestris*), Saharo-Arabian (*P. austroarabica*), and Sudano-Zambezian (*P. acaciae*) regions. The mechanisms of seed dispersal for the eight parasitic plants are represented in Figure 6. Zoochory was the most prevalent seed dispersion method (six species), followed by anempchory (two species).

Table 1. The number of parasitic and host species and their relevant percentages (%) classified into different regional chorotypes in the study region.

Chorotype	Parasitic Species		Host Species	
	Number of Species	Percentage (%)	Number of Species	Percentage (%)
		Monoregional		
AM	1	12.5	-	-
SA-AR	1	12.5	-	-
SU-ZA	1	12.5	-	-
NEO	-	-	1	4.35
PAN	-	-	1	4.35
		Biregional		
SA-AR + SU-ZA	3	37.5	18	78.3
SA-AR + IR-TR	2	12.5	-	-
SU-ZA + IR-TR			1	4.35
		Pluriregional		
SA-AR + SU-ZA + ME	-	-	1	4.35
SA-AR + ME + TRO	-	-	1	4.35
Total	8	100	23	100

ME: Mediterranean; IR-TR: Irano-Turanian; SA-AR: Saharo-Arabian; SU-ZA: Sudano-Zambezian; AM: American; TRO: Tropical; NEO: Neotropical; PAN: Pantropical.

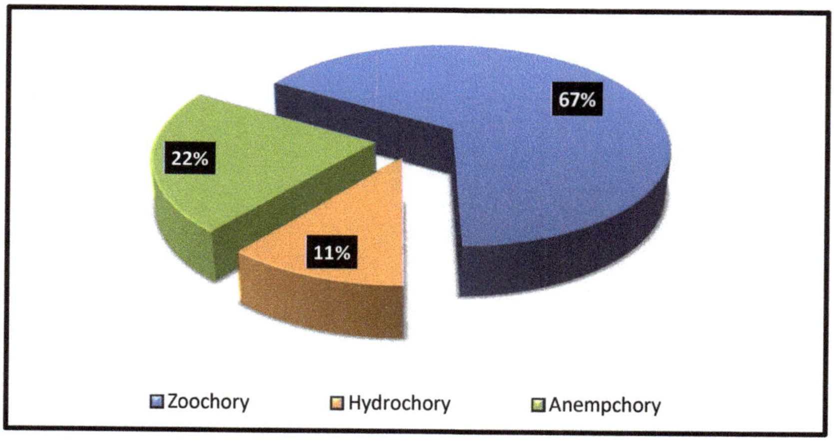

Figure 6. Dispersal mechanisms for parasitic species in the study region.

3.3. Diversity of Parasitic Species

The IVI values of the parasitic species were in the range of 107.28 to 10.21 (Table 2). The most important parasitic plants were *P. austroarabica*, *P. curviflorus*, and *V. schimperi*, with importance values of 107.28, 51.01 and 50.98, respectively, whereas *O. mutelii* was the least important species in the study region with an importance value of 10.21.

Table 2. Density (D), relative density (RD), frequency (F), relative frequency (RF), abundance (A), relative abundance (RA), and importance value index (IVI) for the parasitic species in the study region.

Parasite Species	D	RD	F	RF	A	RA	IVI
Cuscuta campestris Yunck.	0.156	4.28	10.94	8.38	1.43	8.36	21.02
Loranthella deflersii (Tiegh.) S.Blanco & C.E.Wetzel	0.072	1.98	6.56	5.02	1.1	6.43	13.43
Orobanche cernua Loefl.	0.034	0.93	2.81	2.15	1.22	7.13	10.21
Orobanche mutelii F.W.Schultz	0.269	7.38	19.06	14.59	1.41	8.25	30.22
Phragmanthera austroarabica A.G.Mill. & J.A.Nyberg	1.825	50.08	37.81	28.95	4.83	28.25	107.28
Plicosepalus acaciae (Zucc.) Wiens & Polhill	0.097	2.66	9.06	6.94	1.07	6.26	15.86
Plicosepalus curviflorus Tiegh.	0.6	16.47	29.69	22.73	2.02	11.81	51.01
Viscum schimperi Engl.	0.591	16.22	14.69	11.25	4.02	23.51	50.98
Total	3.644	100	130.62	100	17.1	100	300.01

3.4. Floristic Composition of Host Species

In the dry zone, 104 individuals representing 13 species and 5 families were found in site 1, whereas 102 individuals representing 13 species and 6 families were recorded in site 2 (Figures 7 and 8). In the wet zone, 148 individuals were recorded in site 3 representing 19 species and 10 families, whereas 164 individuals, 20 species, and 9 families were documented in site 4. The most represented host plant families in all the sites were Fabaceae, with 6, 6, 7, and 8 species in sites 1, 2, 3, and 4, respectively. Asteraceae was the second highest number with three species in each site. In each of the following families, only one species was identified: Apocynaceae, Barbeyaceae, Oleaceae, Polygonaceae, Rhamnaceae, Scrophulariaceae, and Solanaceae. In the surveyed sites, the perennial parasitic plants (91.3%) were considered the predominant type over the annual (8.7%) species (Table A2). Among the host species, 20 species were indigenous (87%) and three species were exotic (13%).

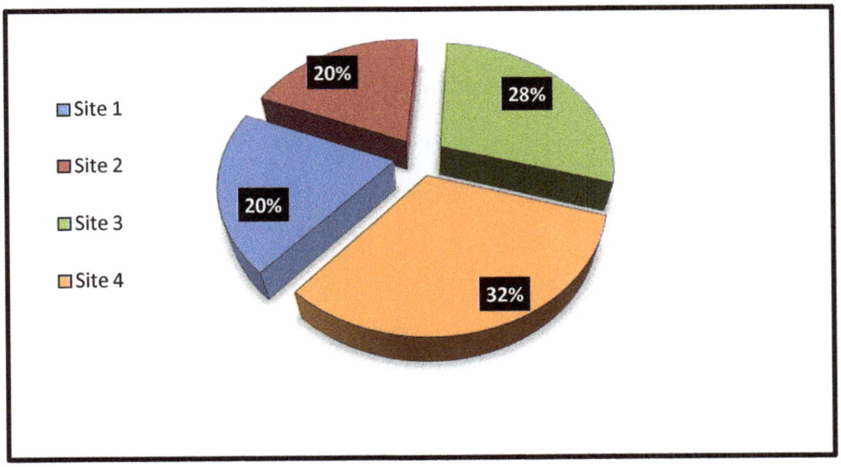

Figure 7. Percentage of the total number of individual host species at each site. Dry zone (site 1: 1300–1500, site 2: 1501–1700 m.a.s.l.); wet zone (site 3: 1800–2100, site 4: 2101–2400 m.a.s.l.).

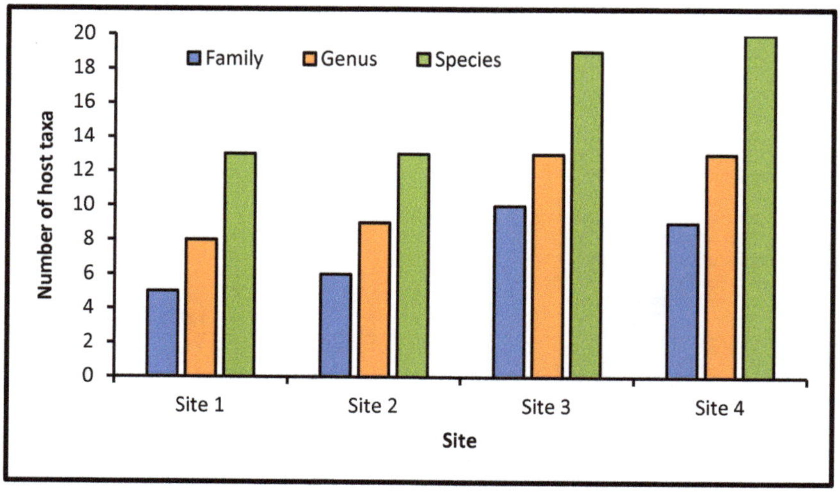

Figure 8. The number of families, genera, species, and total individuals of host plants in each site. Dry zone (site 1: 1300–1500, site 2: 1501–1700 m.a.s.l.); wet zone (site 3: 1800–2100, site 4: 2101–2400 m.a.s.l.).

Eighteen species were detected in each site of the study region (Table A2). *Vachellia origena*, *Barbeya oleoides*, *Buddleja polystachya*, and *Ficus carica* were found only in the wet zone, while *Olea europaea* subsp. *cuspidata* was absent in site 1. It also indicated that six host species were more affected by parasitic species (≥20%) and 17 species were less affected (<20%). The percentage of infested trees by parasitic species showed that *Vachellia flava* was the highest-infested tree (*P. austroarabica* = 47.5%), followed by *Z. spina-christi* (*V. schimperi* = 30.1%, *P. austroarabica* = 7.4%), *V. tortilis* subsp. *tortilis* (*L. deflersii* = 26.7%), *Vachellia gerrardii* (*P. austroarabica* = 19.8%, *P. curviflorus* = 5.8%), and *V. tortilis* subsp. *raddiana* (*P. austroarabica* = 13.8%, *P. acacia* = 5.3%, *P. curviflorus* = 5.3%).

The plant life form analysis demonstrated that three life forms of host species were reported in the four sites (Figure 9). Phanerophytes were the most common life form (19 species = 83%), followed by Therophytes (3 species = 13%), while Hemicryptophytes were represented by one species. The overall diversity of native trees was higher when compared with that of non-native tree species. Additionally, the number of perennial host plants in all sites were 21 species, whereas the annual plants were 2 species.

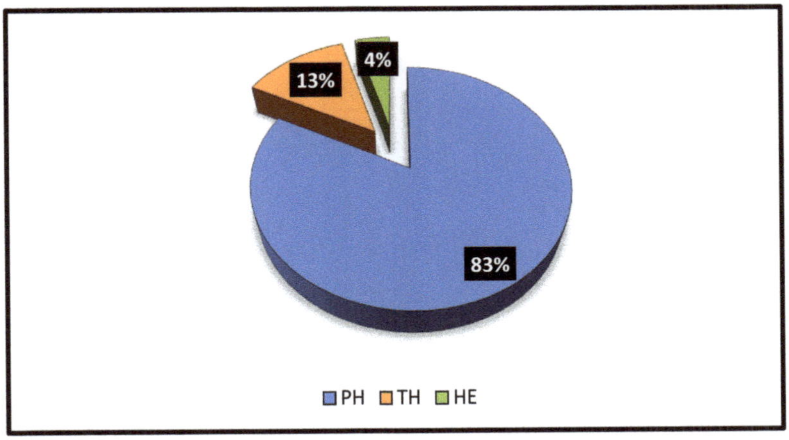

Figure 9. Life form categories of the host species in the study region. TH: Therophytes; PH: Phanerophytes; HE: Hemicryptophytes.

3.5. Chorological Analysis of Host Plants

The chorological analysis categorized the host species into three main categories, pluriregional, biregional, and monoregional. Most of the reported host plants (19 species = 83%) in the region belonged to the biregional group (Table 1). The biregional group had two chorotypes, with the Saharo-Arabian/Sudano-Zambezian chorotypes having the greatest number of species (18 species), while the Sudano-Zambezian/ Irano-Turanian included only one species, *T. aphylla*.

Only two pluriregional species were reported in the studied sites. Saharo-Arabian/ Sudano-Zambezian/Mediterranean was represented by one species (*T. senegalensis*), and Saharo-Arabian/Mediterranean/Tropical also was represented by one species (*Ficus palmata*). The monoregional elements were represented by two species. The detected monoregional species fall under two chorotypes: Neotropical (represented by one species, *B. biternata*) and Pantropical (represented by one species, *N. glauca*).

3.6. Diversity of Host Species

The calculated importance value indices (IVI) for the host species ranged between 32.44 and 2.31%. As shown in Table 3, seven plant species had the highest levels of IVI across the studied region. The dominant species were *N. glauca* (32.44%), *P. undulata* (28.24%), *S. asak* (27.14%), *Vachellia etbaica* (19.66%), *V. gerrardii* (19.09%), *O. europaea* subsp. *cuspidata* (18.59%), and *B. biternata* (18.21%). On the other hand, the least important species found in the study region were *V. tortilis* subsp. *tortilis* (4.68%), *Pistacia falcata* (4.36%), *T. senegalensis* (4.25%), *B. polystachya* (3.04%), and *F. carica* (2.31%). The importance value indices (IVI) for these species were 4.68%, 4.36%, 4.25%, 3.04%, and 2.31%, in order.

Table 3. Density (D), relative density (RD), frequency (F), relative frequency (RF), abundance (A), relative abundance (RA), and importance value index (IVI) for the host plant species in the study region.

Host Species	D	RD	F	RF	A	RA	IVI
Barbeya oleoides Schweinf.	0.28	1.61	13.13	3.2	2.14	2.22	7.03
Bidens biternata (Lour.) Merr. & Sherff	1.11	6.38	13.44	3.27	8.26	8.56	18.21
Buddleja polystachya Fresen.	0.07	0.4	4.06	0.99	1.62	1.68	3.07
Calotropis procera (Aiton) W.T.Aiton	0.49	2.82	19.06	4.64	2.59	2.68	10.14
Ficus carica L.	0.16	0.92	1.56	0.38	1	1.04	2.34
Ficus Palmata Forssk.	0.61	3.51	24.38	5.93	2.5	2.59	12.03
Nicotiana glauca Graham	2.17	12.47	13.44	3.27	16.12	16.71	32.45
Olea europaea subsp. *cuspidata* (Wall. & G.Don) Cif.	1.18	6.78	33.44	8.14	3.53	3.66	18.58
Pistacia falcata Becc. ex Martelli.	0.13	0.75	8.44	2.05	1.52	1.58	4.38
Pluchea dioscoridis (L.) DC.	0.36	2.07	8.44	2.05	4.25	4.4	8.52
Pulicaria undulata (L.) C.A.Mey.	1.79	10.29	12.5	3.04	14.35	14.87	28.2
Rumex nervosus Vahl	0.68	3.91	10	2.43	6.75	7	13.34
Senegalia asak (Forssk.) Kyal. & Boatwr.	1.76	10.11	56.56	13.76	3.12	3.23	27.1
Searsia retinorrhoea (Steud. ex Oliv.) Moffett	0.94	5.4	5.94	1.45	1.579	1.64	8.49
Tamarix aphylla (L.) H.Karst.	0.73	4.2	17.81	4.33	4.11	4.26	12.79
Tamarix senegalensis DC.	0.1	0.57	3.75	0.91	2.67	2.77	4.25
Vachellia etbaica (Schweinf.) Kyal. & Boatwr.	1.3	7.47	33.75	8.21	3.84	3.98	19.66
Vachellia flava (Forssk.) Kyal. & Boatwr.	0.25	1.44	11.56	2.81	2.16	2.24	6.49
Vachellia gerrardii (Benth.) P.J.H.Hurter	1.24	7.13	33.13	8.1	3.76	3.9	19.13
Vachellia origena (Hunde) Kyal. & Boatwr.	0.8	4.6	23.75	5.78	3.38	3.5	13.88
Vachellia tortilis subsp. *raddiana* (Savi) Kyal. & Boatwr.	0.48	2.76	22.5	5.48	2.11	2.19	10.43
Vachellia tortilis subsp. *tortilis*	0.09	0.52	2.81	0.68	3.33	3.45	4.65
Ziziphus spina-christi (L.) Willd.	0.68	3.91	37.5	9.13	1.8	1.87	14.91
Total	17.4	100.02	410.95	100	96.489	100	300

4. Discussion

Studies on the distribution and diversity of parasitic plants in natural plant communities are lacking for many regions of Saudi Arabia. To my knowledge, this is the first attempt to study the ecology and diversity patterns of these highly specialized plants in the Al-Baha region, in the southwest part of Saudi Arabia. Previous studies focused only on the general floristic compositions and the structure of plant communities in specific habitats. The current study recorded eight species of the parasitic plants from four families. There were 34 parasitic species belonging to eight families recorded in the flora of Saudi Arabia [39]. However, the number of parasitic plants in Saudi Arabia is much lower than that of Nepal (151) [40], Turkey (146), and China (678 species) [41]. This variation could be attributed to host availability and abiotic stress, which are the key factors affecting the distribution and abundance of parasitic plants or other ecological factors. Furthermore, the success of parasitic plants under adverse conditions is highly dependent on host selection [11].

In the southern Andes at high altitudes, Amico et al. [42] demonstrated that the Andean-Patagonian Forest is rich in parasitic mistletoes. This finding agreed with our result which showed that the high-altitude sites had high numbers of parasites as compared with the low-altitude sites. On the other hand, the abundance of parasitic plants at high-altitude sites could be related to the richness of host plants in these sites. However, Hechinger and Lafferty [43] revealed that high host diversity can assist the diversity of parasitic plants.

In the current study, the most dominant family of parasitic species was Loranthaceae with four species, followed by Orobanchaceae with two species. Stem hemiparasites were the dominant parasitic group with five species, while only two species belonging to root holoparasitic plants were detected in the studied sites. This result was consistent with the fact that the majority of parasitic species are hemiparasites [44]. Moreover, hemiparasitic plants had morphological characteristics with a wide range of host interaction [44], often parasitizing multiple plant species [45]. The diversity and distribution of both hemiparasitic and holoparasitic plants across the study region could be attributed to the effects of topography and climate factors, especially on the growth of seedlings and germination of parasitic species.

Drought stress is another factor that affects the growth and distribution of hemiparasitic and holoparasitic plants [11]. The deficiency of water availability negatively affects the development and the growth rate of seeds of the root parasites *Orobanche crenata* [46], *Striga hermonthica*, and *Alectra vogelii* [47].

Our results revealed that *P. austroarabica* and *P. curviflorus* parasitized 13 and 5 different host plants, respectively. *Vachellias* was the most favored host for *P. austroarabica* and *P. curviflorus* with 83.4% and 73.5% infestation, respectively. This might be due to the dominance of *Vachellias* trees, which were the highest-recorded host species in the study area. This finding was consistent with the results of Migahid [48], who stated that *P. austroarabica* and *P. curviflorus* are common parasites of *Vachellias* in Saudi Arabia. On the other hand, Amico et al. [42] reported 12 mistletoes that parasitize 43 species (from 23 families) out of 185 woody species in the Andean-Patagonian Forest. Moreover, eight of these mistletoes are specialists with restricted host range and the remaining are generalists. In the wet zone, *C. campestris* parasitized *N. glauca*, which is one of the most invasive species in Saudi Arabia. Previous studies in Chian indicated that the genus *Cuscuta* suppress the invasive plants *Ipomoea cairica, Mikania micrantha, Wedelia trilobata, Solidago canadensis, Bidens pilosa*, and *Humulus scandens* [49,50]. As a result, it is expected that the reported *C. campestris* in the wet sites will suppress the widespread *N. glauca*. Těšitel et al. [10] demonstrated that some native parasitic plant species could be used to repress plant invasions and help restore biodiversity. Thus, it is possible that the indigenous parasites identified in the study region could benefit this habitat by controlling invading plants.

The chorological analysis indicated that parasitic species from the Saharo-Arabian and Sudano-Zambezian regions dominated the region. According to Zohary [34], the vegetation of Saudi Arabia belongs to that of the Saharo-Arabian region. Abdel Khalik et al. [51] showed that the Saharo-Arabian and Sudano-Zambezian species were the best biomarkers of arid climate. Seed dispersal of parasitic species revealed that zoochory was the main dispersal mechanism for six species, whereas anemochory was the dispersion mode of both *Orobanche* species.

After feeding on mistletoe fruits, generalist birds regurgitate or defecate the sticky seeds which paste onto woody branches [52]. Birds dispersing mistletoe seeds demonstrate the great degree of coevolution between them [53], which also has an important function in pollination [52]. Other studies have reported that birds are the primary seed dispersers, with some seeds being dispersed by wind or hydraulic explosives [54,55].

As observed from the IVI analysis, *P. austroarabica* had the highest importance value (107.28) as compared with other parasitic species. Magray et al. [56] reported that the variation in the IVI of species might be caused by predominant environmental factors. Moreover, the difference in IVI among the sites may be due to the composition of plant species, human activities, and local ecological factors [57].

The number of host species, genera, and families varied across the surveyed sites. In the dry zone, 206 individuals from 14 species and 6 families were recorded, whereas 312 individuals from 22 species and 12 families were recorded in the wet zone. Vegetation analysis of the host species demonstrated that Fabaceae (seven species) and Asteraceae (three species) were the two top host species-rich families. These results were consonant with previous studies on the Saudi Arabian flora [58,59]. However, Fabaceae and Asteraceae were notified in the flora of the Mediterranean, North Africa, eastern Ethiopia, and northern Zambia [59]. Anacardiaceae, Moraceae, and Tamaricaceae each had only two species (8.7%), whereas seven families (30.43%) were represented by just one species. As compared with desert vegetation, the majority of plant species in Saudi Arabia are members of a few families and about 58% of the families were represented by a single species. Al Nafie [58] recorded that 24.2% of the families in Saudi Arabia's flora are represented by one species per family. In the surveyed region, the perennial types (91.3%) prevailed over the annuals (8.7%). The host plants included 20 indigenous (87%) and 3 (13%) exotic species.

In southwest China, Luo et al. [60] found only three hosts out of 88 species in the tropical forest that were parasitized by *Dendrophthoe pentandra*, which indicates that the abundance and host richness of species did not explain the frequency of infections at the sampling unit. Zhang et al. [41] reported that the abundance of parasitic plants in a particular site is usually determined by environmental (altitude, area, longitude, and latitude) and biological (dispersal vector and host availability) factors. Host plants' diversity and different kinds of habitats can also influence the spreading and density of parasitic species [61]. Further, the development and growth of parasitic species were closely correlated to the features of their hosts [62,63]. Roxburgh and Nicolson [64] demonstrated that the age of the host species was correlated with the number of mistletoes, so a high number of mistletoe clusters enhanced the probability that additional mistletoe seeds could germinate on that host tree, resulting in the growth of more mistletoe clusters.

The obtained results demonstrated that parasitic species infection varied among the tree species, with *V. flava* being the highest infested tree by *P. austroarabica*, followed by *Z. spina-christi* and *V. tortilis* subsp. *tortilis*. The woody parasites *P. acaciae* and *P. curviflorus* are widespread parasites of *Vachellia* [48]. Al-Rowaily et al. [24] reported a high infection incidence of *P. curviflorus* in different species of *Vachellia* genus in Saudi Arabia's southern and western areas, resulting in ecosystem degradation and loss of diversity and soil nutrients. Other studies demonstrated that mistletoe preferentially infects massive trees over small ones [64,65]. Therefore, the difference in infection rate could possibly be correlated to host size [66]. This can be linked to the fact that dispersal birds prefer larger trees for resting and feeding [67]. Further, some parasitic species preferred woody hosts, which may be consistent with the perennial life form and hemiparasitic nature [68,69]. On the other hand, some herbal species may work as bridging hosts to enable seedlings to live long enough to grow onto nearby shrubs or trees [70].

In the present investigation, three life forms were noticed, with Phanerophytes being the most prominent (19 species = 83%), followed by Therophytes (three species = 13%). This result agreed with the findings that were reported by Abbas et al. [59] and Elkordy et al. [71] in different regions of the Kingdom. The dominance of perennial species reflects the characterization of the vegetation in the studied region. This may be caused by low precipitation and a long dry season, which is not sufficient for the growth of annual species [72,73]. Further, perennial plants can acclimate to the extreme ecological conditions of the area.

According to the chorological analysis results, the biregional elements of the Saharo-Arabian/Sudano-Zambezian chorotypes had the most dominant share of host species by 18 species. Comparable findings were found in several investigations in Saudi Arabian flora [59,71,73].

Native plants showed higher levels of species richness and diversity when compared to non-native plants. The abiotic variables such as the amount of rainfall and the aridity of the area are the most affected factors on the dominance of perennial species [51,73,74]. This is a prominent characteristic of the vegetation of the Al-Baha region since perennial species might be more resistant to climatological change than annual species.

Based on the IVI analysis, the most important host species were *N. glauca*, *P. undulata*, *S. asak*, *V. etbaica*, *V. gerrardii*, *O. europaea* subsp. *cuspidata*, and *B. biternata*. Importance value indices can be applied in vegetation analysis to determine the order of species conservation. Species that have low importance values must be afforded higher conservation priority than species with high importance values [75].

The variations in the obtained results among the studied sites could be mainly due to the elevation gradient. However, Al-Robai et al. [21] revealed the effects of elevation gradient on structure and diversity of plant communities along the Alabna escarpment in the Al-Baha area. Further, Al-Namazi et al. [22] demonstrated that the high-attitude area of the Al-Baha region has a wide range of plant diversity.

5. Study Limitations

An important study limitation to consider is the inability to compare, monitor, and assess the distribution and diversity of parasitic plants in the surveyed region due to the lack of previous research on these species. The study focuses on the elevation gradient as the major element that may influence the distribution and diversity of parasite and host plants, whereas other environmental variables may have an impact on the structure and dynamics of these plants. It should be kept in mind that this study was conducted in a rather small and restricted location. Future research will expand on other habitats to cover more geographical regions across the country to fully examine the main factors that affect the diversity of parasitic plants and their hosts in the selected studied regions. This will be performed by employing more environmental parameters.

6. Conclusions

Vegetation analysis of density, abundance, and frequency revealed a clear variation of both parasite and host plants along the elevation gradient. Elevation, unique topography, and climatic conditions of the study region could possibly be responsible for these variations as well as other environmental factors. Eight parasites and 23 host species were recorded across the surveyed four sites. *Phragmanthera austroarabica* was widely spread throughout the sites and was reported in most of the studied sampling units. It represented half of the total individuals in the region and had the highest infection rate, followed by *P. curviflorus*. Half of the parasitic species (50%) found in the region belonged to Loranthaceae family. The highest alpha diversity of the reported hosts (20) was detected in site 4, while the lowest (13) was reported in sites 1 and 2. Fabaceae was the most common host plant family, and seven families were represented by only one host species. The most infested tree was *Vachellia flava*, followed by *Z. spina-christi*, and *V. tortilis* subsp. *raddiana* was the least infected species. Perennial, therophyte, and bioregional elements were the dominant host species in both the dry and wet zones. It is suggested that *Cuscuta campestris* can be exploited as a biological control tool to reduce the spread of *N. glauca* in the Al-Baha region. The majority of the recorded parasite and host plants were indigenous to Saudi Arabia. Further work at the local or broad geographical regions is recommended all over the country to better understand and document the diversity, structure, and dynamics of parasitic plants.

Funding: This research received no external funding.

Institutional Review Board Statement: Not applicable.

Data Availability Statement: The data presented in this study are available in this article.

Conflicts of Interest: The author declares no conflict of interest.

Appendix A

Table A1. List of parasitic species recorded in the studied region with their families, spatial distribution, kind of parasite, origin, habit, life span, chorotype, dispersal type, infested plant organ, hosts, and percentage of infestation.

Family/Parasite	Spatial Distribution	Kind of Parasite	Origin	Habit	Life Span	Chorotype	Dispersal Type	Infested Plant Organ	Host	INFC (%)
Convolvulaceae										
Cuscuta campestris Yunck.	1.2.3.4	Holop.	EXO	Herb	Ann.	AM	ZO + HY	Stem	*Nicotiana glauca* Graham	2
									Pluchea dioscoridis (L.) DC.	28
									Pulicaria undulata (L.) C.A.Mey.	70
Loranthaceae										
Loranthella deflersii (Tiegh.) S.Blanco & C.E.Wetzel	3.4	Hmip.	IND	Shrub	Per.	SA-AR + SU-ZA	ZO	Stem	*Searsia retinorrhoea* (Steud. ex Oliv.) Moffett	8.7
									Tamarix senegalensis DC.	21.7
									Vachellia tortilis subsp. *tortilis*	69.6
Phragmanthera austroarabica A.G.Mill. & J.A.Nyberg	1.2.3.4	Hmip.	IND	Shrub	Per.	SA-AR	ZO	Stem	*Barbeya oleoides* Schweinf.	2.6
									Buddleja polystachya Fresen.	0.17
									Calotropis procera (Aiton) W.T.Aiton	0.17
									Ficus carica L.	3.8
									Ficus Palmata Forssk.	0.17
									Olea europaea subsp. *cuspidata* (Wall. & G.Don) Cif.	3.1
									Pistacia falcata Becc. ex Martelli.	0.51
									Tamarix aphylla (L.) H.Karst.	2.6
									Vachellia flava (Forssk.) Kyal. & Boatwr.	11.5
									Vachellia gerrardii (Benth.) P.J.H.Hurter	33.7
									Vachellia origena (Hunde) Kyal. & Boatwr.	27.6
									Vachellia tortilis subsp. *raddiana* (Savi) Kyal. & Boatwr.	10.6
									Ziziphus spina-christi (L.) Willd.	3.6

Table A1. *Cont.*

Family/Parasite	Spatial Distribution	Kind of Parasite	Origin	Habit	Life Span	Chorotype	Dispersal Type	Infested Plant Organ	Host	INFC (%)
Plicosepalus acaciae (Zucc.) Wiens & Polhill	1.2.3.4	Hmip.	IND	Shrub	Per.	SU-ZA	ZO	Stem	*Senegalia asak* (Forssk.) Kyal. & Boatwr.	58.1
									Tamarix aphylla (L.) H.Karst.	9.7
									Vachellia tortilis subsp. *raddiana* (Savi) Kyal. & Boatwr.	32.3
Plicosepalus curviflorus Tiegh.	1.2.3.4	Hmip.	IND	Shrub	Per.	SA-AR + SU-ZA	ZO	Stem	*Senegalia asak* (Forssk.) Kyal. & Boatwr.	21.4
									Tamarix aphylla (L.) H.Karst.	5.2
									Vachellia etbaica (Schweinf.) Kyal. & Boatwr.	31.8
									Vachellia gerrardii (Benth.) P.J.H.Hurter	26.6
									Vachellia tortilis subsp. *raddiana* (Savi) Kyal. & Boatwr.	15.1
Orobanchaceae										
Orobanche mutelii F.W.Schultz	1.2.3.4	Holop.	IND EXO	Herb Herb	Ann. Ann.	SA-AR + IR-TR Med + IR-TR	AN AN	Root	*Bidens biternata* (Lour.) Merr. & Sherff	22.1
									Rumex nervosus Vahl	77.9
Orobanche cernua Loefl.	3.4	Holop.	IND	Herb	Ann.	SA-AR + IR-TR	AN	Root	*Rumex nervosus* Vahl	100
Santalaceae										
Viscum schimperi Engl.	1.2.3.4	Hmip.	IND	Shrub	Per.	SA-AR + SU-ZA	ZO	Stem	*Tamarix aphylla* (L.) H.Karst.	7.1
									Ziziphus spina-christi (L.) Willd.	92.9

Kinds of parasite abbreviations: Holop.: holoparasite; Hmip.: Hemiparasitic; INFC: infestation. Life span abbreviations: Per.: perennial; Ann.: annual. Origin abbreviations: IND: indigenous; EXO: exotic. Chorotype abbreviations: ME: Mediterranean; IR-TR: Irano-Turanian; SA-AR: Saharo-Arabian; SU-ZA: Sudano-Zambezian; AM: American. Dispersal type abbreviations: ZO: zoochory; HY: hydrochory; AN: anempchory.

Table A2. List of host species with their families, spatial distribution, life form, life span, origin, chorotypes, parasitic species, and the percentage of the infestation.

Family/Host	Spatial Distribution	Life Form	Life Span	Origin	Chorotype	Parasitic Species	Infestation (%)
Asteraceae							
Bidens biternata (Lour.) Merr. & Sherff	1.2.3.4	TH	Ann.	IND	NEO	*Orobanche mutelii* F.W.Schultz	2
Pluchea dioscoridis (L.) DC.	1.2.3.4	TH	Per.	IND	SA-AR + SU-ZA	*Cuscuta campestris* Yunck	11.3
Pulicaria undulata (L.) C.A.Mey.	1.2.3.4	HE	Per.	IND	SA-AR + SU-ZA	*Cuscuta campestris* Yunck	6.3
Anacardiaceae							
Pistacia falcata Becc. ex Martelli.	1.2.3.4	PH	Per.	IND	SA-AR + SU-ZA	*Phragmanthera austroarabica* A.G.Mill. & J.A.Nyberg	2.4
Searsia retinorrhoea (Steud. ex Oliv.) Moffett	1.2.3.4	PH	Per.	IND	SA-AR + SU-ZA	*Loranthella deflersii* (Tiegh.) S.Blanco & C.E.Wetzel	3.3
Apocynaceae							
Calotropis procera (Aiton) W.T.Aiton	1.2.3.4	PH	Per.	IND	SA-AR + SU-ZA	*Phragmanthera austroarabica* A.G.Mill. & J.A.Nyberg	0.6
Barbeyaceae							
Barbeya oleoides Schweinf.	3,4	PH	Per.	IND	SA-AR + SU-ZA	*Phragmanthera austroarabica* A.G.Mill. & J.A.Nyberg	2.2
Fabaceae							
Senegalia asak (Forssk.) Kyal. & Boatwr.	1.2.3.4	PH	Per.	IND	SA-AR + SU-ZA	*Plicosepalus acaciae* (Zucc.) Wiens & Polhill *Plicosepalus curviflorus* Tiegh.	2 3.4
Vachellia etbaica (Schweinf.) Kyal. & Boatwr.	1.2.3.4	PH	Per.	IND	SA-AR + SU-ZA	*Plicosepalus curviflorus* Tiegh.	8.2
Vachellia flava (Forssk.) Kyal. & Boatwr.	1.2.3.4	PH	Per.	IND	SA-AR + SU-ZA	*Phragmanthera austroarabica* A.G.Mill. & J.A.Nyberg	47.5
Vachellia gerrardii (Benth.) P.J.H.Hurter	1.2.3.4	PH	Per.	IND	SA-AR + SU-ZA	*Phragmanthera austroarabica* A.G.Mill. & J.A.Nyberg *Plicosepalus curviflorus* Tiegh.	19.8 5.8
Vachellia origena (Hunde) Kyal. & Boatwr.	3,4	PH	Per.	EXO	SA-AR + SU-ZA	*Phragmanthera austroarabica* A.G.Mill. & J.A.Nyberg	16
Vachellia tortilis subsp. *raddiana* (Savi) Kyal. & Boatwr.	1.2.3.4	PH	Per.	IND	SA-AR + SU-ZA	*Phragmanthera austroarabica* A.G.Mill. & J.A.Nyberg *Plicosepalus acaciae* (Zucc.) Wiens & Polhill *Plicosepalus curviflorus* Tiegh.	13.8 5.3 5.3
Vachellia tortilis subsp. *tortilis*	1.2.3.4	PH	Per.	IND	SA-AR + SU-ZA	*Loranthella deflersii* (Tiegh.) S.Blanco & C.E.Wetzel	26.7

Table A2. *Cont.*

Family/Host	Spatial Distribution	Life Form	Life Span	Origin	Chorotype	Parasitic Species	Infestation (%)
Moraceae							
Ficus carica L.	3,4	PH	Per.	EXO	SA-AR + SU-ZA	*Phragmanthera austroarabica* A.G.Mill. & J.A.Nyberg	20
Ficus Palmata Forssk.	1.2.3.4	PH	Per.	IND	SA-AR+ ME+TRO	*Phragmanthera austroarabica* A.G.Mill. & J.A.Nyberg	1.5
Oleaceae							
Olea europaea subsp. *cuspidata* (Wall. & G.Don) Cif.	2.3.4	PH	Per.	IND	SA-AR + SU-ZA	*Phragmanthera austroarabica* A.G.Mill. & J.A.Nyberg	2.9
Polygonaceae							
Rumex nervosus Vahl	1.2.3.4	TH	Ann.	IND	SA-AR + SU-ZA	*Orobanche mutelii* F.W.Schultz *Orobanche cernua* Loefl.	9.7 4.2
Rhamnaceae							
Ziziphus spina-christi (L.) Willd.	1.2.3.4	PH	Per.	IND	SA-AR + SU-ZA	*Phragmanthera austroarabica* A.G.Mill. & J.A.Nyberg *Viscum schimperi* Engl.	7.4 30.1
Scrophulariaceae							
Buddleja polystachya Fresen.	3,4	PH	Per.	IND	SA-AR + SU-ZA	*Phragmanthera austroarabica* A.G.Mill. & J.A.Nyberg	4.8
Solanaceae							
Nicotiana glauca Graham	1.2.3.4	PH	Per.	EXO	PAN	*Cuscuta campestris* Yunck	0.1
Tamaricaceae							
Tamarix aphylla (L.) H.Karst.	1.2.3.4	PH	Per.	IND	SU-ZA + IR-TR	*Phragmanthera austroarabica* A.G.Mill. & J.A.Nyberg *Viscum schimperi* Engl. *Plicosepalus acaciae* (Zucc.) Wiens & Polhill *Plicosepalus curviflorus* Tiegh.	3.8 5.1 1.3 4.3
Tamarix senegalensis DC.	1.2.3.4	PH	Per.	IND	SA-AR+SU-ZA+ME	*Loranthella deflersii* (Tiegh.) S.Blanco & C.E.Wetzel	15.6

Life form abbreviations: PH: Phanerophytes; TH: Therophytes; HE: Hemicryptophytes. Life span abbreviations: Per.: perennial; Ann.: annual. Origin abbreviations: IND: indigenous; EXO: exotic. Chorotype abbreviations: ME: Mediterranean; NEO: Neotropical; PAN: Pantropical; IR-TR: Irano-Turanian; SA-AR: Saharo-Arabian; SU-ZA: Sudano-Zambezian; TRO: Tropical.

References

1. Nickrent, D.L. Parasitic angiosperms: How often and how many? *Taxon* **2020**, *69*, 5–27. [CrossRef]
2. Al-Juhani, W.; Al Thagafi, N.T.; Al-Qthanin, R.N. Gene Losses and Plastome Degradation in the Hemiparasitic Species *Plicosepalus acaciae* and *Plicosepalus curviflorus*: Comparative Analyses and Phylogenetic Relationships among Santalales Members. *Plants* **2022**, *11*, 1869. [CrossRef] [PubMed]
3. Těšitel, J. Functional biology of parasitic plants: A review. *Plant Ecol. Evol.* **2016**, *149*, 5–20. [CrossRef]
4. Těšitel, J.; Li, A.R.; Knotková, K.; McLellan, R.; Bandaranayake, P.C.; Watson, D.M. The bright side of parasitic plants: What are they good for? *Plant Physiol.* **2021**, *185*, 1309–1324. [CrossRef]
5. Těšitel, J.; Mládek, J.; Fajmon, K.; Blažek, P.; Mudrák, O. Reversing expansion of *Calamagrostis epigejos* in a grassland biodiversity hotspot: Hemiparasitic *Rhinanthus major* does a better job than increased mowing intensity. *Appl. Veg. Sci.* **2018**, *21*, 104–112. [CrossRef]
6. Fibich, P.; Lepš, J.; Chytrý, M.; Těšitel, J. Root hemiparasitic plants are associated with high diversity in temperate grasslands. *J. Veg. Sci.* **2017**, *28*, 184–191. [CrossRef]
7. Demey, A.; Rüttingm, T.; Huygens, D.; Staelens, J.; Hermy, M.; Verheyen, K.; Boeckx, P. Hemiparasitic litter additions alter gross nitrogen turnover in temperate semi-natural grassland soils. *Soil Biol. Biochem.* **2014**, *68*, 419–428. [CrossRef]
8. Těšitel, J.; Mládek, J.; Horník, J.; Těšitelová, T.; Adamec, V.; Tichý, L. Suppressing competitive dominants and community restoration with native parasitic plants using the hemiparasitic *Rhinanthus alectorolophus* and the dominant grass *Calamagrostis epigejos*. *J. Appl. Ecol.* **2017**, *54*, 1487–1495. [CrossRef]
9. Wu, A.P.; Zhong, W.; Yuan, J.R.; Qi, L.Y.; Chen, F.L.; Liang, Y.S.; He, F.F.; Wang, Y.H. The factors affecting a native obligate parasite, *Cuscuta australis*, in selecting an exotic weed, *Humulus scandens*, as its host. *Sci. Rep.* **2019**, *9*, 3–10. [CrossRef]
10. Těšitel, J.; Cirocco, R.M.; Facelli, J.M.; Watling, J.R. Native parasitic plants: Biological control for plant invasions? *Appl. Veg. Sci.* **2020**, *23*, 464–469. [CrossRef]
11. Zagorchev, L.; Stöggl, W.; Teofanova, D.; Li, J.; Kranner, I. Plant parasites under pressure: Effects of abiotic stress on the interactions between parasitic plants and their hosts. *Int. J. Mol. Sci.* **2021**, *22*, 7418. [CrossRef] [PubMed]
12. Ashapkin, V.V.; Kutueva, L.I.; Aleksandrushkina, N.I.; Vanyushin, B.F.; Teofanova, D.R.; Zagorchev, L.I. Genomic and Epigenomic Mechanisms of the Interaction between Parasitic and Host Plants. *Int. J. Mol. Sci.* **2023**, *24*, 2647. [CrossRef] [PubMed]
13. Albanova, I.A.; Zagorchev, L.I.; Teofanova, D.R.; Odjakova, M.K.; Kutueva, L.I.; Ashapkin, V.V. Host resistance to parasitic plants—Current knowledge and future perspectives. *Plants* **2023**, *12*, 1447. [CrossRef] [PubMed]
14. Watson, D.M. Mistletoe—A keystone resource in forests and woodlands worldwide. *Annu. Rev. Ecol. Syst.* **2001**, *32*, 219–249. [CrossRef]
15. Shin, H.W.; Lee, N.S. Understanding plastome evolution in hemiparasitic Santalales: Complete chloroplast genomes of three species, *Dendrotrophe varians, Helixanthera parasitica,* and *Macrosolen cochinchinensis*. *PLoS ONE* **2018**, *13*, e0200293. [CrossRef]
16. Yoshida, S.; Cui, S.; Ichihashi, Y.; Shirasu, K. The haustorium, a specialized invasive organ in parasitic plants. *Annu. Rev. Plant Biol.* **2016**, *67*, 643–667. [CrossRef]
17. Su, H.J.; Liang, S.l.; Nickrent, D.L. Plastome variation and phylogeny of *Taxillus* (Loranthaceae). *PLoS ONE* **2021**, *16*, 0256345. [CrossRef]
18. Vidal-Russell, R.; Nickrent, D.L. Evolutionary relationship in the showy mistletoe family (Loranthaceae). *Am. J. Bot.* **2008**, *95*, 1015–1029. [CrossRef]
19. Noman, O.M.; Mothana, R.A.; Al-Rehaily, A.J.; Nasr, F.A.; Khaled, J.M.; Alajmi, M.F.; Al-Said, M.S. Phytochemical analysis and anti-diabetic, anti-inflammatory and antioxidant activities of *Loranthus acaciae* Zucc. grown in Saudi Arabia. *Saudi Pharm. J.* **2019**, *27*, 724–730. [CrossRef]
20. Al-Aklabi, A.; Al-Khulaidi, A.W.; Hussain, A.; Al-Sagheer, N. Main vegetation types and plant species diversity along an altitudinal gradient of Al Baha region, Saudi Arabia. *Saudi J. Biol. Sci.* **2016**, *23*, 687–697. [CrossRef]
21. Al-Robai, S.A.; Mohamed, H.A.; Ahmed, A.A.; Al-Khulaidi, A.W.A. Effects of elevation gradients and soil components on the vegetation density and species diversity of Alabna escarpment, southwestern Saudi Arabia. *Acta Ecol. Sin.* **2019**, *39*, 202–211. [CrossRef]
22. Al-Namazi, A.A.; Al-Khulaidi, A.W.A.; Algarni, S.; Al-Sagheer, N.A. Natural plant species inventory of hotspot areas in Arabian Peninsula: Southwest Al-Baha region, Saudi Arabia. *Saudi J. Biol. Sci.* **2021**, *28*, 3309–3324. [CrossRef] [PubMed]
23. Abulfaith, H.A.; Emara, H.A. Altitudinal distribution of the hemiparasitic Loranthaceae in southwestern Saudi Arabia. *Biotropica* **1988**, *20*, 81–83. [CrossRef]
24. Al-Rowaily, S.L.; Al-Nomari, G.S.; Assaeed, A.M.; Facelli, J.M.; Dar, B.M.; El-Bana, M.I.; Abd-ElGawad, A.M. Infection by *Plicosepalus curviflorus* mistletoe affects the nutritional elements of *Acacia* species and soil nutrient recycling in an arid rangeland. *Plant Ecol.* **2020**, *221*, 1017–1028. [CrossRef]
25. Ibrahim, M.M. Study of cystic echinococcosis in slaughtered animals in Al Baha region, Saudi Arabia: Interaction between some biotic and abiotic factors. *Acta Trop.* **2010**, *113*, 26–33. [CrossRef] [PubMed]
26. Chaudhary, S.A. *Flora of the Kingdom of Saudi Arabia*; Ministry of Agriculture and Water: Riyadh, Saudi Arabia, 1999; Volume 1.
27. Chaudhary, S.A. *Flora of the Kingdom of Saudi Arabia*; Ministry of Agriculture and Water: Riyadh, Saudi Arabia, 2000; Volume 2.
28. Chaudhary, S.A. *Flora of the Kingdom of Saudi Arabia*; Ministry of Agriculture and Water: Riyadh, Saudi Arabia, 2001; Volume 3.

29. Collenette, S. *Wild Flowers of Saudi Arabia*; National Commission for Wild Life Conservation and Development: Riyadh, Saudi Arabia, 1999.
30. POWO (Plants of the World Online). Facilitated by Royal Botanic Gardens, Kew Botanical Garden. 2023. Available online: https://powo.science.kew.org/ (accessed on 3 June 2023).
31. Raunkiaer, C. *The Life Forms of Plants and Statistical Plant Geography*; Clarendon Press: Oxford, UK, 1934.
32. Govaerts, R.; Frodin, D.G.; Radcliffe-Smith, A.; Carter, S. *World Checklist and Bibliography of Euphorbiaceae (with Pandaceae)*; Royal Botanic Gardens, Kew: London, UK, 2000.
33. Wickens, G. The Flora of Jebel Marra (Sudan Republic) and Its Geographical Affinities. *Kew Bull. Addit. Ser.* **1978**, *144*, 496. [CrossRef]
34. Zohary, M. *Geobotanical Foundations of the Middle East*; Gustav Fischer: Stuttgart, Germany, 1973.
35. Phillips, E.A. *Methods of Vegetation Study*; Holt, Rhinehart and Winston, Inc.: New York, NY, USA, 1959.
36. Mueller-Dombois, D.; Ellenberg, H. *Aims and Methods of Vegetation Ecology*; John Willey and Sons, Inc.: New York, NY, USA, 1974.
37. Isa, N.; Razak, S.A.; Abdullah, R.; Khan, M.N.; Hamzah, S.N.; Kaplan, A.; Dossou-Yovo, H.O.; Ali, B.; Razzaq, A.; Wahab, S.; et al. Relationship between the Floristic Composition and Soil Characteristics of a Tropical Rainforest (TRF). *Forests* **2023**, *14*, 306. [CrossRef]
38. WFO (World Flora Online). Facilitated by Royal Botanic Gardens, Kew and Missouri Botanical Garden. 2023. Available online: http://http://www.worldfloraonline.org/ (accessed on 3 June 2023).
39. PFSA (Parasitic flora of Saudi Arabia). Cite maintained by Jacob Thomas, Herbarium, Department of Botany and Microbiology, College of Science, King Saud University, Riyadh, KSA. 2023. Available online: https://www.plantdiversityofsaudiarabia.info/index.htm (accessed on 3 June 2023).
40. O'Neill, A.R.; Rana, S.K. An ethnobotanical analysis of parasitic plants (Parijibi) in the Nepal Himalaya. *J. Ethnobiol. Ethnomed.* **2016**, *12*, 14–28. [CrossRef]
41. Zhang, G.; Li, Q.; Sun, S. Diversity and distribution of parasitic angiosperms in China. *Ecol. Evol.* **2018**, *8*, 4378–4386. [CrossRef]
42. Amico, G.C.; Nickrent, D.L.; Vidal-Russell, R. Macroscale analysis of mistletoe host ranges in the Andean-Patagonian forest. *Plant Biol.* **2019**, *21*, 150–156. [CrossRef]
43. Hechinger, R.F.; Lafferty, K.D. Host diversity begets parasite diversity: Bird final hosts and trematodes in snail intermediate hosts. *Proc. R. Soc. B Biol. Sci.* **2005**, *272*, 1059–1066. [CrossRef] [PubMed]
44. Twyford, A.D. Parasitic plants. *Curr. Biol.* **2018**, *28*, R857–R859. [CrossRef]
45. Brown, M.R.; Moore, P.G.P.; Twyford, A.D. Performance of generalist hemiparasitic Euphrasia across a phylogenetically diverse host spectrum. *New Phytol.* **2021**, *232*, 2165–2174. [CrossRef] [PubMed]
46. Moral, J.; Lozano-Baena, M.D.; Rubiales, D. Temperature and water stress during conditioning and incubation phase affecting *Orobanche crenata* seed germination and radicle growth. *Front. Plant Sci.* **2015**, *6*, 408. [CrossRef] [PubMed]
47. Dawoud, D.A.; Sauerborn, J. Impact of drought stress and temperature on the parasitic weeds *Striga hermonthica* and *Alectra vogelii* in their early growth stages. *Exp. Agric.* **1994**, *30*, 249–257. [CrossRef]
48. Migahid, A.M. *Flora of Saudi Arabia*, 4th ed.; King Saud University Press: Riyadh, Saudi Arabia, 1996; Volume 2.
49. Yu, H.; Liu, J.; He, W.M.; Miao, S.L.; Dong, M. *Cuscuta australis* restrains three exotic invasive plants and benefits native species. *Biol. Invasions* **2011**, *13*, 747–756. [CrossRef]
50. Li, J.; Jin, Z.; Song, W. Do native parasitic plants cause more damage to exotic invasive hosts than native non-invasive hosts? An implication for biocontrol. *PLoS ONE* **2012**, *7*, e34577. [CrossRef]
51. Abdel Khalik, K.; Al-Gohary, I.; Al-Sodany, Y. Floristic composition and vegetation: Environmental relationships of Wadi Fatimah, Mecca, Saudi Arabia. *Arid Land Res. Manag.* **2017**, *31*, 316–334. [CrossRef]
52. Muche, M.; Muasya, A.M.; Tsegay, B.A. Biology and resource acquisition of mistletoes, and the defense responses of host plants. *Ecol. Process.* **2022**, *11*, 24. [CrossRef]
53. Medel, R.; Vergara, E.; Silva, A.; Kalin-Arroyo, M. Effects of vector behavior and host resistance on mistletoe aggregation. *Ecology* **2004**, *85*, 120–126. [CrossRef]
54. Okubamichael, D.Y.; Grifths, M.E.; Ward, D. Host specifcity, nutrient and water dynamics of the mistletoe *Viscum rotundifolium* and its potential host species in the Kalahari of South Africa. *J. Arid Environ.* **2011**, *75*, 898–902. [CrossRef]
55. Rist, L.; Shaanker, R.U.; Ghazoul, J. The spatial distribution of mistletoe in a Southern Indian tropical forest at multiple scales. *Biotropica* **2011**, *43*, 50–57. [CrossRef]
56. Magray, J.A.; Wani, B.A.; Islam, T.; Ganie, A.H.; Nawchoo, I.A. Phyto-ecological analysis of *Phytolacca acinosa* Roxb. assemblages in Kashmir Himalaya, India. *Front. For. Glob. Change* **2022**, *5*, 976902. [CrossRef]
57. Rashid, K.; Rashid, S.; Islam, T.; Ganie, A.H.; Nawchoo, I.A.; Ahmad Khuroo, A. Vegetation and soil ecology of threatened Himalayan Trillium habitats in Kashmir Himalaya. *Nord. J. Bot.* **2023**, *2023*, e03925. [CrossRef]
58. AlNafie, A.H. Phytogeography of Saudi Arabia. *Saudi J. Biol. Sci.* **2008**, *15*, 159–176.
59. Abbas, A.M.; Al-Kahtani, M.A.; Alfaifi, M.Y.; Elbehairi, S.E.I.; Badry, M.O. Floristic diversity and phytogeography of Jabal Fayfa: A subtropical dry zone, south-west Saudi Arabia. *Diversity* **2020**, *12*, 345. [CrossRef]
60. Luo, Y.; Sui, Y.; Gan, J.; Zhang, L. Host compatibility interacts with seed dispersal to determine small-scale distribution of a mistletoe in Xishuangbanna, Southwest China. *J. Plant Ecol.* **2016**, *9*, 77–86. [CrossRef]

61. Ma, R.; Miao, N.; Zhang, H.; Tao, W.; Mao, K.; Moermond, T.C. Generalist mistletoes and their hosts and potential hosts in an urban area in southwest China. *Urban For. Urban Green.* **2020**, *53*, 126717. [CrossRef]
62. Sargent, S. Seed fate in a tropical mistletoe: The importance of host twig size. *Funct. Ecol.* **1995**, *9*, 197–204. [CrossRef]
63. Guerra, T.J.; Pizo, M.A.; Silva, W.R. Host specificity and aggregation for a widespread mistletoe in Campo Rupestre vegetation. *Flora* **2017**, *238*, 148–154. [CrossRef]
64. Roxburgh, L.; Nicolson, S.W. Differential dispersal and survival of an African mistletoe: Does host size matter? *Plant Ecol.* **2008**, *195*, 21–31. [CrossRef]
65. de Buen, L.L.; Ornelas, J.F.; García-Franco, J.G. Mistletoe infection of trees located at fragmented forest edges in the cloud forests of Central Veracruz, Mexico. *For. Ecol. Manag.* **2002**, *164*, 293–302. [CrossRef]
66. Ndagurwa, H.G.; Mundy, P.J.; Dube, J.S.; Mlambo, D. Patterns of mistletoe infection in four *Acacia* species in a semi-arid southern African savanna. *J. Trop. Ecol.* **2012**, *28*, 523–526. [CrossRef]
67. Aukema, J.E.; Martínez del Rio, C. Where does a fruit-eating bird deposit mistletoe seeds? Seed deposition patterns and an experiment. *Ecology* **2002**, *83*, 3489–3496. [CrossRef]
68. Kokubugata, G.; Nakamura, K.; Forster, P.I.; Wilson, G.W.; Holland, A.E.; Hirayama, Y.; Yokota, M. *Cassytha pubescens* and *C. glabella* (Lauraceae) are not disjunctly distributed between Australia and the Ryukyu Archipelago of Japan—Evidence from morphological and molecular data. *Aust. Syst. Bot.* **2012**, *25*, 364–373. [CrossRef]
69. Zhang, C.; Ma, H.; Sanchez-Puerta, M.V.; Li, L.; Xiao, J.; Liu, Z.; Ci, X.; Li, J. Horizontal gene transfer has impacted cox1 gene evolution in *Cassytha filiformis*. *J. Mol. Evol.* **2020**, *88*, 361–371. [CrossRef]
70. Zhang, H.; Florentine, S.; Tennakoon, K.U. The angiosperm stem hemiparasitic genus *Cassytha* (Lauraceae) and its host interactions: A review. *Front. Plant Sci.* **2022**, *13*, 864110. [CrossRef]
71. Elkordy, A.; Nour, I.H.; Ellmouni, F.Y.; Al Shaye, N.A.; Al-Bakre, D.A.; EL-Banhawy, A. Floristic Diversity of Jabal Al-Ward, Southwest Tabuk Region, Kingdom of Saudi Arabia. *Agronomy* **2022**, *12*, 2626. [CrossRef]
72. Erfanzadeh, R.; Pétillon, J.; Maelfait, J.P.; Hoffmann, M. Environmental Determinism versus Biotic Stochasticity in the Appearance of Plant Species in Salt-Marsh Succession. *Plant Ecol. Evol.* **2010**, *143*, 43–50. [CrossRef]
73. Abdel Khalik, K.; El-Sheikh, M.; El-Aidarous, A. Floristic Diversity and Vegetation Analysis of Wadi Al-Noman, Mecca, Saudi Arabia. *Turk. J. Bot.* **2013**, *37*, 894–907. [CrossRef]
74. Vico, G.; Brunsell, N.A. Tradeoffs between water requirements and yield stability in annual vs. perennial crops. *Adv. Water Resour.* **2018**, *112*, 189–202. [CrossRef]
75. Zegeye, H.; Teketay, D.; Kelbessa, E. Diversity, regeneration status and socio-economic importance of the vegetation in the islands of Lake Ziway, south-central Ethiopia. *Flora* **2006**, *201*, 483–498. [CrossRef]

diversity

MDPI

Article

Ecology and Diversity of Weed Communities in the Northern Andes under Different Anthropogenic Pressures

Yessica P. Duque [1], Carlos E. Giraldo-Sánchez [2,*], Mario A. Quijano-Abril [3] and Jose M. Rojas [3,4]

1 Facultad de Ciencias Agropecuarias, Universidad Católica de Oriente, Rionegro 054048, Colombia; yessica.duque8329@uco.net.co
2 Grupo de Investigación de Sanidad Vegetal, Universidad Católica de Oriente, Rionegro 054048, Colombia
3 Grupo de Investigación Estudios Florísticos, Herbario Universidad Católica de Oriente, Rionegro 054048, Colombia; maquijano@uco.edu.co (M.A.Q.-A.); miguel.rojas4@udea.edu.co (J.M.R.)
4 Instituto de Biología, Facultad de Ciencias Exactas y Naturales, Universidad de Antioquia, Medellín 050001, Colombia
* Correspondence: cegiral0@gmail.com or csanchez@uco.edu.co

Abstract: Weeds can have both positive and negative effects on agricultural environments. However, despite the growing interest in the ecology of weed communities in agricultural areas, a few studies have been carried out in the northern region of the Andes of Colombia, where urban and agricultural expansion have generated highly disturbed scenarios. The aim of this study was to analyze the diversity of vegetation and weed seed banks in three agricultural production systems and a forest ecosystem in the northern Andes of Colombia. Hill numbers were used to compare diversity, Beta diversity to assess changes in composition, and range—abundance–dominance curves at different sites. Likewise, indicator species were analyzed to find species associations to each system. The results revealed differences in the composition of weeds between the forest ecosystem and the agricultural production systems, with higher equitability in the forest ecosystem and higher dominance in agricultural systems. Significant differentiation was observed among the dominant species within each agricultural system, particularly highlighting those species considered pests due to their unique life history traits. These traits confer them with a greater advantage in the face of various anthropogenic selection pressures. These findings highlight the impact of anthropogenic disturbances on the ecological dynamics of weed communities in different ecosystems, which should be considered when planning integrated weed management techniques.

Keywords: soil seed banks; surface vegetation; composition; dominance; weeds; forest ecosystems; agricultural production systems

Citation: Duque, Y.P.; Giraldo-Sánchez, C.E.; Quijano-Abril, M.A.; Rojas, J.M. Ecology and Diversity of Weed Communities in the Northern Andes under Different Anthropogenic Pressures. *Diversity* **2023**, *15*, 936. https://doi.org/10.3390/d15080936

Academic Editors: Ilias Travlos and Mario A. Pagnotta

Received: 29 June 2023
Revised: 12 August 2023
Accepted: 14 August 2023
Published: 17 August 2023

1. Introduction

Weeds are adventitious plants that grow in crops without being sown intentionally but can play a crucial role in these systems [1]. These plants have short life cycles, produce abundant seeds, and form seed banks in the soil that ensure their persistence over time in various ecosystems [2,3]. Due to their nature, they can positively or negatively affect agricultural environments, either through soil conservation [4], associated beneficial fauna and allelopathic effects on crops [5], or due to the intense competition exerted by some species for resources such as nutrients, water, and sunlight [6,7]. Some studies suggest a high correlation between severe soil disturbances, with an increase in annual weed communities [8,9]. However, although tillage stimulates the germination of weed seeds dormant in the soil, the effect caused by this and other cultural practices on seed banks depends on the weed species and the interactions with the environment [10–12]. Thus, the diversity of weed species that comprise the seed banks could fluctuate over time and with different crop rotation systems [13]. In this way, understanding the dynamics of weed communities subjected to different anthropic pressures in different agricultural systems could help design more efficient and environmentally friendly management strategies.

Biodiversity and ecological dynamics of weed communities in agricultural production systems have been the subject of recent studies due to the negative impact that human activities have on these highly modified environments [12]. Authors have suggested that the composition of weed communities in agriculture shows a wide variation related to the different types of crops and the historical uses of the soil [2]. For example, in a study that evaluated seed bank dynamics for five years under a corn–soybean rotation system, seed density in the soil decreased by almost 90% during the first year under a productive system based on corn, and the trend was maintained by rotating with soybeans during the following years [2]. Thus, understanding the diversity variability of weed communities in areas under different anthropogenic pressures, including various crops and associated tasks, is essential to identify problematic species in agricultural production systems and developing more efficient and sustainable management strategies [14]. However, even though in recent years there has been a growing interest in understanding the ecology and diversity of weed communities in agricultural areas with high anthropogenic pressure, little has been studied in the northern region of the Andes, where urban and agricultural expansion have generated highly disturbed scenarios [15].

In the northern part of the Colombian Andes, agriculture, livestock, and urbanization have historically caused significant impacts on plant cover, generating drastic changes in soils and promoting the colonization of weeds, many of them invasive [15,16]. For example, a study carried out in a production system of roses under greenhouse conditions in the Sabana de Bogotá area to evaluate the diversity of weed species in cultivated fields registered 46 species, of which 2, *Cardamine hirsuta* and *Pennisetum clandestinum*, showed a marked dominance with 67% of the total plant cover [17]. Similarly, in a peach orchard in the same region, a low species diversity was found, with a high dominance of *Oxalis corniculata* reaching a 68% coverage [18]. However, some weed species in this Andean area have become a serious problem in forested areas without agricultural pressure, as is the case of the invasive species *Thunbergia alata*, in which an average density of 493 seeds/m^2 has been documented, with a viability of 100% stored in the soil [19]. This suggests that the various aspects of the life histories of weeds could be related to the structuring of communities according to anthropogenic dynamics that alter natural ecosystems, both in cover and in seed banks in the soil.

Understanding the diversity of species that comprise seed banks can provide valuable information on the relationships of weeds with their environment [20] and the effect of transforming natural areas into different agricultural contexts. Such information could help minimize the use of herbicides and agricultural inputs, contributing to the conservation and health of ecosystems [21]. In this sense, the aim of this study was to analyze the diversity of vegetation and weed seed banks in four areas of the northern Andes of Colombia under different anthropogenic pressures (little intervened forest, and vegetable, avocado, and livestock farming), with the hypothesis that species diversity changes according to the type of intervention, and that some dominant species can be identified as specific indicators of each system. Consistent with this, the following questions are answered: (a) Can differences in the diversity (Alpha) of the weed communities of the seed banks between the different agricultural production systems and the forest ecosystem be observed? (b) Are there differences in weed species composition in the soil seed banks between the agricultural production systems and the forest ecosystem (Beta diversity)? (c) Which weed species in the seed banks of the soil manage to manifest themselves in its surface vegetation and can potentially be considered "weeds"? (d) Is a higher dominance of weed species observed in the ecosystems most affected by human activity? (e) Which weed species could be considered indicators of each evaluated system?

This work provides novel information related to the composition of the weed communities in the seed banks of little-studied areas with different anthropic impacts. It also provides basic knowledge on patterns of diversity and ecology of different weed species associated with anthropized ecosystems in one of the most diverse regions in the world and with high rates of endemism, as is the northern area of the Andean mountain range in Colombia [22]. These findings may contribute significantly to integrated weed management plans, prioritizing those that show a marked dominance and that could be considered specific to each productive system evaluated.

2. Materials and Methods

2.1. Study Area

The study was carried out in the highlands of the Oriente antioqueño area located in the central mountain range (Cordillera) of the Andes, to the southeast of the department of Antioquia, Colombia. In this region, the low montane very humid forest life zone predominates [23] with average temperatures between 18 °C and 21 °C, elevations from 2100 to 2400 m a.s.l. and annual rainfall between 1500 and 4000 mm. This is one of the areas that shows the highest agricultural suitability in the department, being considered a substantial agricultural breadbasket [24], where different production systems have been established, including vegetables, flowers, fruit trees, coffee, sugarcane, and livestock, among others [25].

The sampling was carried out in four sites, three of them corresponding to agricultural production systems farming vegetables (FV), avocado (A), and livestock (L), and a forest fragment identified as a low secondary vegetation cover (VC) [26] (Figure 1).

The forest ecosystem (VC) is in the initial stages of a secondary succession since it has undergone various anthropic intervention phenomena due to deforestation and urban expansion. Its vegetation type is mainly shrubby and herbaceous, with an irregular canopy and occasional trees and vines, with heights of less than five meters [26]. It is dominated by some tree species such as *Cecropia peltata*, *Andesanthus lepidotus*, *Cavendishia pubescens*, *Vismia baccifera*, and *Myrsine guianensis*, and shows an area of 7000 m^2. The livestock productive system (L) corresponds to extensive livestock farming for milk production, an area dominated by the Kikuyu grass *Pennisetum clandestinum* used for animal feed. The avocado production system (A) is dedicated to the production of export-type avocados, with an extension of 200,000 m^2. Finally, the vegetable production system (FV) includes open fields dedicated to cultivating lettuce, potato, and tomato in an area of 10,000 m^2. For weed management in the production systems, the non-selective agrochemical Paraquat, which contains the chemical molecule 1,1′-dimethyl-4,4′-bipyridinium ion dichloride, applied every two months, is used in the livestock area (L) and avocado crop (A). For vegetable crops (FV), selective herbicides with the active ingredient Oxyfluorfen, with the chemical molecule 2-chloro-1-(3-ethoxy-4-nitrophenoxy)-4-(trifluoromethyl)benzene, are used with a frequency of application of every two months.

2.2. Sampling Design

A specific sampling was carried out for each production system, including registering soil seed banks and their surface vegetation. In VC and L, the samples were collected haphazardly in a zigzag pattern; in the A, the planted tree area was sampled, and in the FV, samples were collected on the production furrows in areas where weeds were present. In each productive system, an area of 5000 m^2 was delimited, establishing five plots of 0.25 m^2. Each plot was subdivided into 25 quadrats of 0.1 m. Twenty soil subsamples were extracted from each plot collected with the help of an auger at a depth of 10 cm according to the recommendations of Buhler et al. [2]. The subsamples were grouped, homogenized, and stored in hermetically sealed plastic bags for later analysis, following the methodology described in Bigwood and Inouye [27]. The number of species found in the surface vegetation was recorded in each plot, including the presence or absence of the species without considering their abundances.

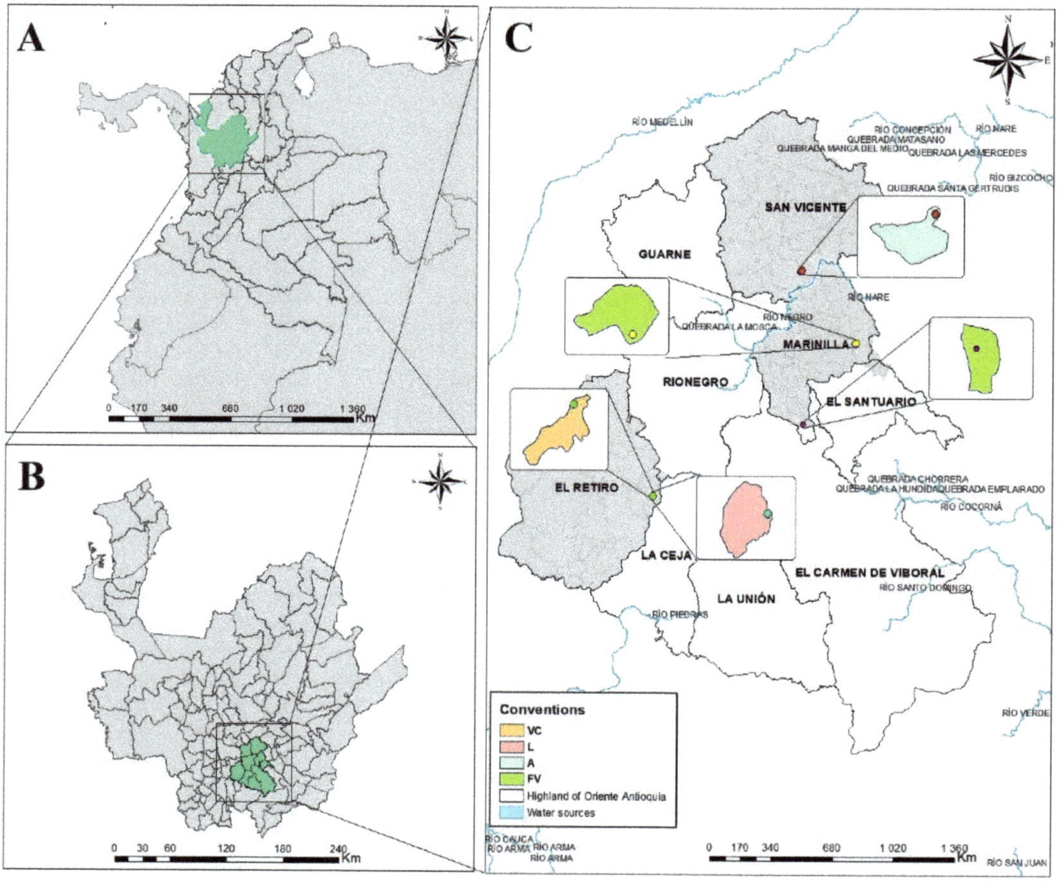

Figure 1. Study area. (**A**) Geographical location of the northern Andes, Colombia. (**B**) Location of the department of Antioquia, indicating the highlands of the Oriente antioqueño area. (**C**) Spatial location of the agricultural production systems sampled. VC = low secondary vegetation area, L = livestock production system area, A = avocado production system area, and FV = vegetable production system area.

The seedling emergence method [28] was used to determine the species that comprise the soil seed banks in each production system. In this method, the homogenized material of the soil samples is arranged in 50 × 30 cm germination trays covered with plastic lids to prevent the interference of other weed seeds that may be present in the environment. The trays were left for four months while the seeds germinated. During this time, they were supplied with surface irrigation every 15 days until reaching field capacity, according to the requirements of each tray. All the experiments were carried out under greenhouse conditions with a mean temperature of 26.39 ± 4.85 °C and a relative humidity of 60.83 ± 10.79%. Every 15 days during the four months, the emerged seedlings were counted and recorded, and their taxonomic identification was carried out through comparisons between specimens stored in the reference collection of the herbarium of Universidad Católica de Oriente (HUCO). In addition, taxonomic identification guides from the list of vascular plants of Antioquia [29], the illustrated guide to identifying weed plants of the Marengo agricultural center (CAM) [30], and the guide of frequent weeds in coffee cultivation in Colombia [31] were used.

2.3. Data Analysis

Only zero-order (q0) Alpha diversity was calculated for the surface vegetation in the four productive systems assessed and used to compare the weed species found in the soil seed bank with their surface vegetation. Likewise, the Alpha diversity of the soil seed banks of each productive system was computed using the numbers of species equivalents [32,33]. For Order 0 diversity or richness (q0), the number of species found in the soil seed banks was considered excluding abundance. Order 1 diversity, calculated as the exponential of the entropy of the Shannon index (q1), analyzes the diversity weighted by the relative frequency of each species in the community. Finally, Order 2 diversity calculated as the inverse of the Gini–Simpson index (q2) considers the most abundant or dominant species in the community, discarding rare species. From the data obtained, the confidence intervals were calculated [34]. At the same time, the completeness of the seed banks sampling in each productive system was evaluated using the sample coverage (*Cm*) proposed by Chao and Jost [34] and based on the following equations:

$$E(Cm) = \sum_{i=1}^{S} pi[1-(1-pi)^m] = 1-\sum_{i=1}^{S} pi(1-pi)^m, m > 0, \quad (1)$$

where

Cm = sampling coverage
S = total number of species sampled
pi = relative abundance of the *i*th species
m = sample size.

This value indicates the proportion of the total community represented by the weed species found in each study area. When its value approaches 100%, the complete sample is considered concerning the sampling effort made [34]. The estimation was performed in the iNEXT web application [34]. In addition, rank–abundance curves were established for each production system to compare the dominance of the weed species in the soil seed banks [35].

The Beta diversity was calculated for three orders (β0, β1, and β2) following the multiplicative partition of diversity [36] to evaluate the change in the composition of weed species, seed banks, and between each production system. The Baselga partitioned Beta index was also calculated [37] to estimate the proportion of turnover and nesting in differentiating communities between sites. Additionally, comparisons were made between the composition of the species found in the soil seed bank with the superficial vegetation in each of the productive systems using Whittaker's complementarity index [38], which exclusively contemplates matrices of presence and absence of the species between the compared units. The above is performed to estimate the representation of the seed bank in the coverage of each site. Finally, a species indicator analysis was performed to estimate the weed species best associated with each production system using the indicator value method (*IndVal*) according to Dufrene and Legendre [39] and calculated as follows:

$$\boldsymbol{IndVal} = specificity \times Fidelity \times 100, \quad (2)$$

where

Specificity = *Nindij*/*Nindi*
Fidelity = *Ntrapij*/*Ntrapj*
Nindij = the average number of individuals of species *i* in type *j* habitat
Nindi = the sum of the average number of individuals of species *i* in all habitat types
Ntrapij = the number of individuals in habitat *j* where species *i* is present
Ntrapj = the total number of individuals in habitat *j*.

The species with high IndVal (higher than 50%) were considered the best indicators of the productive systems, while the species with low percentages (less than 25%) were not considered indicators [39]. Statistical significances (*p*-values) were estimated using 9999 random permutations of sites between groups. The data were analyzed in the statistical software Past version 4.13 [40].

3. Results

3.1. Species Richness

From the soil seed banks, a total of 3332 weed seedlings were recorded emerging from the four productive systems, belonging to 38 species, 36 genera, and 21 botanical families. Asteraceae was the most representative family with 12 species (31.57%) of the total number of emerged seedlings, followed by Cyperaceae and Polygonaceae with three species each (7.89%), and Caryophyllaceae with two species (5.26%). The other families were only represented by a single species (Table 1). The highest seedling emergence in the soil seed banks was constituted by genera *Cardamine* (723), *Trixella* (636), *Verbena* (380), *Polygonum* (375), *Oxalis* (200), *Cyperus* (277), and *Gnaphalium* (179). The productive system with the highest number of individuals was L with 1872, followed, with a much lower number, by A with 534, FV with 525, and VC with 401, i.e., the lowest number of individuals. The species predominating in the soil seed banks of the four productive systems evaluated were *Cardamine hirsuta* with 282 seedlings and *Polygonum nepalense* with 136 individuals in FV. Conversely, *Trixella arvensis, Oxalis corniculata*, and *Cyperus odoratus* dominated in L with a total number of 605, 152, and 83 seedlings, respectively (Table 1).

Table 1. Weed species abundance found in the soil seed banks of four productive systems evaluated in the Oriente antioqueño region, northern Andes, Colombia.

Family	Species	VC	L	A	FV
Amaranthaceae	*Amaranthus viridis* L.				3 *
Apiaceae	*Centella asiatica* (L.) Urb	7 *	6		
Araliaceae	*Hydrocotyle umbellata* L.	6 *			
Asteraceae	*Ageratum conyzoides* L.	20 *		12 *	
	Erechtites valerianaefolia C.E.C. Fisch	5 *	22 *	1 *	
	Porcellites radicata (L.) Cass.	15 *	7 *	48 *	
	Jaegeria hirta (Lag.) Less.	3 *	6 *	5 *	
	Gnaphalium americanum Mill.		110 *	60 *	9 *
	Sonchus oleraceus L.		18 *		
	Artemisia vulgaris L.		5		20 *
	Conyza bonariensis (L.) Cronquist		1 *	9 *	
	Emilia sonchifolia (L.) DC.			6 *	
	Galinsoga quadriradiata Ruiz and Pav.				13 *
	Senecio vulgaris L.				3 *
	Acmella oppositifolia (Lam.) R.K. Jansen				1 *
Brassicaceae	*Cardamine hirsuta* L.	2	263	176	282
Caryophyllaceae	*Stellaria media* (L.) Vill.	46	1	8 *	2 *
	Drymaria villosa Schltdl. and Cham			15 *	
Commelinaceae	*Commelina diffusa* Burm. f.	12 *	2 *		1 *
Convolvulaceae	*Ipomoea purpurea* (L.) Roth				1 *

Table 1. *Cont.*

Family	Species	VC	L	A	FV
Cyperaceae	*Cyperus odoratus* L.	67	83	48 *	12 *
	Cyperus rotundus L.	51	2	14 *	
	Kyllinga erecta Schumach.	42	3		
Fabaceae	*Mimosa albida* Humb. and Bonpl. ex Willd	3 *			
Melastomataceae	*Chaetogastra kingii* (Wurdack) P.J.F. Guim. and Michelang.	14 *			3 *
Lythraceae	*Cuphea carthagenensis* (Jacq.) J.F. Macbr.	20 *	4 *		
Iridaceae	*Sisyrinchium micranthum* Cav.	27 *	10 *	4 *	
Lamiaceae	*Trixella arvensis* (L.) Fourr.	1	605 *	4 *	26 *
Phyllanthaceae	*Phyllanthus niruri* L.	2 *			
Rubiaceae	*Richardia scabra* L.	32 *	5	20 *	
Oxalidaceae	*Oxalis corniculata* L.	1	152 *	39 *	8 *
Poaceae	*Bromus* sp. L.	16 *			
	Paspalum paniculatum L.		11 *	50 *	5 *
Polygonaceae	*Polygonum nepalense* Meisn.	4 *	125 *	9 *	136 *
	Polygonum segetum Kunth		1 *		
	Rumex crispus L.		49 *		
Plantaginaceae	*Plantago major* L.	5 *	1 *	6 *	
Verbenaceae	*Verbena litoralis* Kunth		380 *		
N° individual per site		401	1872	534	525
N° species per site		23	25	19	16

VC = low secondary vegetation area, L = livestock production system area, A = avocado production system area, and FV = vegetables production system area. * Species registered both in the seed bank and in the surface vegetation.

The plots of the four-soil seed bank production systems exhibited variability in terms of both richness and abundance. Regarding richness, the weed communities displayed no significant differences in the number of species. However, when it came to abundance, notable variations were observed in relation to the number of individuals present in each of the production system plots (Table 2).

Table 2. Average richness and abundance with standard deviations (in brackets) across sampled plots in four production systems of the Oriente antioqueño region, Northern Andes, Colombia.

	L	VC	A	FV
Richness	12.8 (±3.49)	12.2 (±5.67)	8.2 (±2.16)	7.2 (±2.77)
Abundance	374.4 (±148.79)	80.2 (±56.82)	106.8 (±41.49)	105 (±63.65)

The total diversity observed in the soil seed banks of the four productive systems was 38 species, with a sampling coverage of 0.999, indicating a representative sampling. The richness and diversity of the seed banks differed between the productive systems; the diversity of L showed the highest richness value of the soil seed banks (q0) with 25 species, followed by VC with 23 and A with 19. On the other hand, FV registered the lowest richness value with 16 species (Table 3). Even though L obtained a higher richness (q0), the weed community of the seed banks was more equitable in VC since its (q1) and (q2) diversity values were higher compared to those of the other systems. Furthermore, regarding the diversity weighted by the most abundant species (q2), VC registered a higher value of

11.07 equivalent species once again, followed by A with 6.40, L with 5.47, and FV with 2.77 (Figure 2).

Table 3. Alpha diversity values for the weed species registered in four productive systems sampled in the Oriente antioqueño region, northern Andes, Colombia.

Diversity		VC	L	A	FV	Total
Observed	q0	23 [18.89; 27.11]	25 [19.16; 30.84]	19 [17.59; 20.41]	16 [8.37; 23.63]	38 [32; 43]
	q1	14.11 [13.03; 15.19]	7.75 [7.37; 8.13]	9.89 [9.10; 10.68]	4.30 [3.84; 4.76]	13.40 [12.87; 13.94]
	q2	11.07 [9.80; 12.35]	5.47 [5.16; 5.77]	6.40 [5.48; 7.33]	2.77 [2.50; 3.04]	8.58 [8.18; 8.97]
Estimated	q0	23.86 [18.42; 29.30]	27.53 [18.54; 36.51]	19.00 [17.78; 20.22]	18.19 [8.52; 27.86]	40.19 [32.80; 47.58]
	q1	14.39 [13.28;1 5.49]	7.79 [7.40; 8.18]	10.01 [9.13; 10.88]	4.35 [3.88; 4.81]	13.46 [12.92; 14.00]
	q2	11.21 [9.89; 12.54]	5.47 [5.17; 5.78]	6.44 [5.55; 7.33]	2.77 [2.50; 3.04]	8.59 [8.19; 8.98]
Sample coverage		0.995	0.998	1	0.994	0.999

VC = low secondary vegetation area, L = livestock production system area, A = avocado production system area, and FV = vegetables production system area. Note: The numbers inside the [ ] indicate the 95% confidence interval.

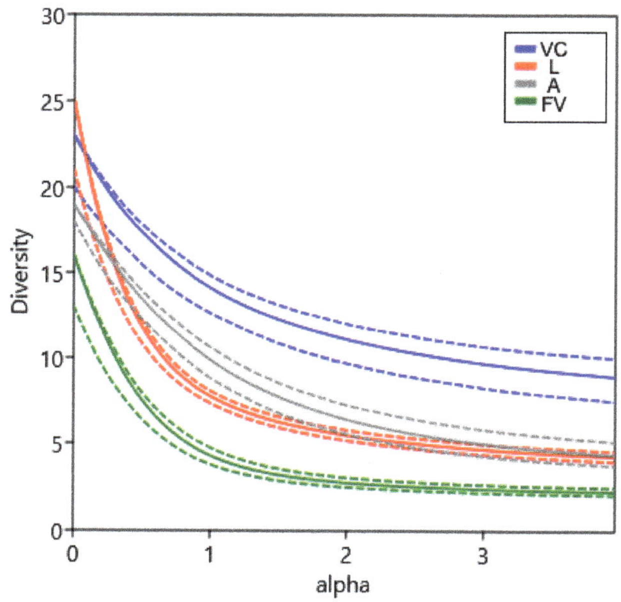

Figure 2. Alpha diversity profiles based on the number of equivalent species from the soil seed banks of weeds from the four production systems sampled in the Oriente antioqueño region, northern Andes, Colombia. VC = low secondary vegetation area, L = livestock production system area, A = avocado production system area, and FV = vegetables production system area. Alpha indicates order of diversity q0, q1 and q2. Dashed lines indicate confidence intervals (95%).

The range–abundance graph revealed fluctuations in the dominance of weed species from the seed banks in each production system. In VC, no species showed a clear dominance. On the other hand, in the L productive system, a higher dominance of species was demonstrated, with a total of six weed species with values higher than 100 individuals (*Trixella arvensis* with 605 individuals, *Verbena litoralis* with 380, *Cardamine hirsuta* with 263, *Oxalis corniculata* with 152, *Polygonum nepalense* with 125 and *Gnaphalium americanum* with 110 individuals). In A, a single species, *Cardamine hirsute*, dominated with 176 individuals, while in FV, two dominant species, *Cardamine hirsuta* with 282 and *Polygonum nepalense* with 136 individuals, were found (Figure 3).

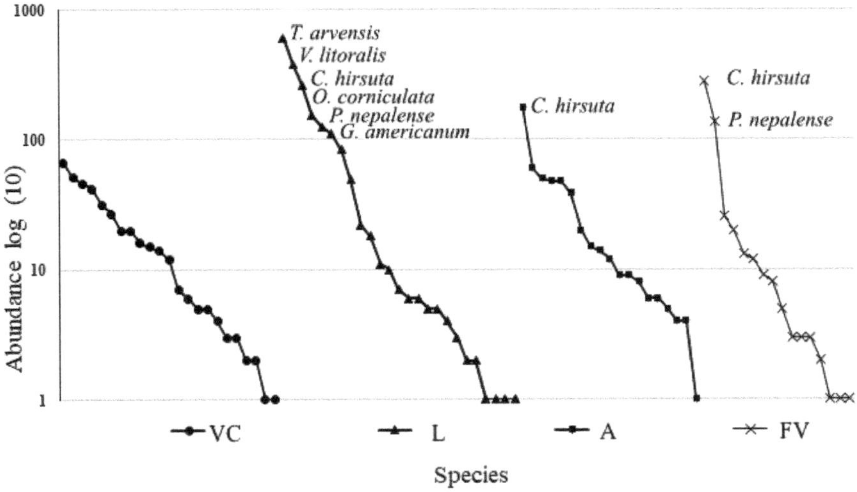

Figure 3. Weed species range–abundance curves in the seed banks of the sampled production systems. The names of species with abundances higher than 100 individuals are provided. VC = low secondary vegetation area, L = livestock production system area, A = avocado production system area, and FV = vegetable production system area.

3.2. Species Composition

3.2.1. Seed Bank Composition

Richness (β0) between the comparisons of the VC and the L and A productive systems showed a low dissimilarity in the composition of weed species of the seed banks since they showed values of higher than 50% similarity, represented by [0.29; 0.33; 0.27]. In addition, when the VC, L, and A are compared with FV, the richness (β0) tends to increase the dissimilarity of the species. Regarding (β1), the comparison between VC and FV showed a high dissimilarity, represented by a 79% variation between the species of these systems. Regarding (β2), when comparing the forest ecosystem (VC) with the agricultural systems, there was a high dissimilarity of species, registering values of higher than 50% of difference, represented by [0.84; 0.60; 0.91] (Table 4). In addition, the variations in the composition of the weed communities were mainly caused by the high turnover values of the species ($\beta C\text{-}_{bal}$), reflected in the paired comparison between the forest ecosystem with the agricultural production systems, recording values of [0.690; 0.655; 0.935] (Table 5).

Table 4. Beta diversity of the soil seed banks of weeds in multiplicative partition among the four productive systems evaluated in the Oriente antioqueño region, northern Andes, Colombia.

	VC × L	VC × A	VC × FV	L × A	L × FV	A × FV
β0	0.29	0.33	0.58	0.27	0.51	0.54
β1	0.41	0.46	0.79	0.24	0.21	0.30
β2	0.84	0.60	0.91	0.43	0.41	0.15

VC = low secondary vegetation area, L = livestock production system area, A = avocado production system area, and FV = vegetables production system area.

3.2.2. Composition of the Seed Banks and Their Surface Vegetation

The Beta diversity in the four productive systems showed a high similarity between the composition of weed species that comprise the soil seed banks concerning their surface vegetation, showing values higher than 50%, corresponding to the 76.19% of similarity in VC, 77.27% in L, 93.75% in FV, and 100% in A. Further, the species found in the soil seed banks were also registered in the superficial vegetation of this productive system.

Table 5. Beta partitioned diversity of the soil seed banks of weeds among the four productive systems evaluated in the Oriente antioqueño region, northern Andes, Colombia.

		β.Bray.bal (Turnover)			
		VC	L	A	FV
β.Bray.gra (nesting)	VC	0	0.69	0.655	0.935
	L	0.2	0	0.299	0.133
	A	0.048	0.389	0	0.571
	FV	0.008	0.487	0.003	0

VC = low secondary vegetation area, L = livestock area, A = avocado production system area, FV = vegetables production system area.

3.2.3. Indicator Species

In the VC, three weed species were registered as indicators of the productive system, with IndVal values higher than 60%. These species were *Cyperus rotundus* (60.9%; $p = 0.0048$) and *Stellaria media* (64.56%; $p = 0.0053$), and the species that had the highest representativeness was *Kyllinga erecta* (74.67%; $p = 0.0024$). L registered the following indicator species: *Oxalis corniculata* (76%; $p = 0.001$), *Verbena litoralis* and *Sonchus oleraceus* (80%; $p = 0.0021$), *Trixella arvensis* (95.13%; $p = 0.0003$) and *Rumex crispus* with 100% ($p = 0.0003$). We recorded *Conyza bonariensis* as an indicator species but with a value lower than 60% (54%; $p = 0.0186$). Lastly, in FV, *Galinsoga quadriradiata* (80%; $p = 0.0013$) was the only indicator species for the system (Figure 4).

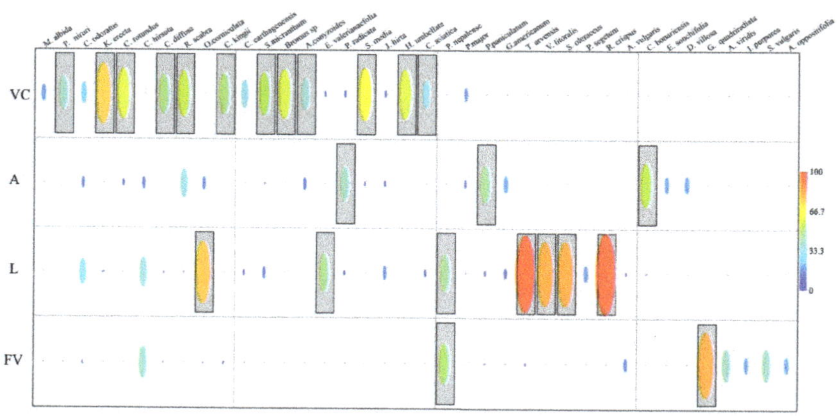

Figure 4. Indicator analysis of weed species of the four production systems sampled in the Oriente antioqueño region, northern Andes, Colombia. VC = low secondary vegetation area, A = avocado production system area, L = livestock productive system area, and FV = vegetables productive system area. Yellow, orange, and red colors correspond to InVal values higher than 60%.

4. Discussion

Human activities exerted on ecosystems can have an impact on the ecological dynamics of weed communities, as well as on the composition and dominance of species [41,42]. Among these activities, agriculture and deforestation have been identified as the leading causes of impacts on natural areas [42,43]. Although several studies have documented variations in the diversity of weed communities related to disturbances [12,14,43–45], in the Andes, there is a shortage of research that evaluates the effect of anthropogenic disturbance associated with different agricultural practices on weed vegetation cover [17,18]. In this work, the taxonomic diversity of weed communities in one of the areas of greatest anthropogenic pressure, the northern Andes in Colombia, was studied by comparing the surface cover and the soil seed banks in three intensive agricultural production systems and a slightly intervened forest ecosystem.

Our results indicate that, despite finding no significant differences in weed richness between the compared areas (Alpha q0), variations were observed in the composition of the communities, mainly reflected in the high turnover values (βC_{-bal}) and minor nesting (βC_{-gra}). In addition, the diversity profiles were less equitable in the agricultural production systems compared to the forested area, suggesting a higher dominance of species considered "weeds", which have a negative economic impact on these production systems. Additionally, the second-order Beta diversity (β^2) for the seed banks [0.15; 0.91] suggests that weed communities under different anthropic pressures have similar structuring patterns in which dominance increases, but differences in dominant species are probably the result of contrasting life histories that allow their response in different ways to selection pressures generated by human activity.

Several studies have documented the differences in the composition and abundance of weed species between production systems and forest ecosystems; they attribute these changes to variations in the growth habits of the species and the agricultural management supplied to the weeds to control their growth [12,46–49]. Likewise, previous studies have shown that the composition of weed communities in soil seed banks is influenced by human activities [2,43]. Similarly, recent research suggests that differences in weed species composition between natural ecosystems and agricultural production system areas are subject to constant anthropogenic disturbances caused by tillage, with higher dominance of weed species in agricultural systems that have experienced major interventions [12,43,49–51]. Thus, the disturbance frequencies governed by tillage in production systems generate notorious changes in the weed community, where highly disturbed environments tend to be simpler and less stable in the abundance of weed populations [12,43,49–51]. For example, as a disturbance in soil seed banks increases in corn, soybean, and oat cropping, the dominance of weed species capable of adapting to these areas tends to increase [52,53]. Similarly, it has been observed that the weed communities in the forest cover do not reflect dominance in the most conserved areas, in contrast to the most intervened areas such as cropping fields since the species of the conserved ecosystems are less abundant and competitive [42].

However, cases have also been reported where, in agricultural production system areas, the abundance, diversity, and uniformity of the weed community in the seed banks tend to increase as the disturbances caused by soil tillage decrease [49,54]. Gurber and Claupein [55] recorded a higher abundance of weed species in more conserved areas compared to those highly disturbed. These authors relate their results to the high capacity of weeds to produce a large number of seeds that persist in the soil forming seed banks, also suggesting that the physical and chemical characteristics of the soil can influence the diversity of weeds in less disturbed ecosystems.

Sharp [56] argued that the differences in the life cycles of the weed species (annual vs. perennial) may influence the dominance of the species. It has been documented that environments with high anthropogenic disturbances favor the growth of weeds with annual cycles, which have the capacity to grow rapidly when tillage is interrupted, reaching reproductive maturity in a short time [8,9,12]. This allows the species produce a large number of seeds in a single season, increasing their ability to disperse and colonize open areas [12,57–59]. In addition, it has been indicated that the notable abundance of weed species in cultivated areas and pastures may be related to reproductive strategies and seed dispersal mechanisms, allowing the species the expansion to new habitats and quick colonization of those disturbed areas [12,42]. These studies support the current results since, in the agricultural production systems, there was a marked dominance and abundance of weed species with annual cycles, including *Trixella arvensis, Verbena litoralis, Cardamine hirsuta, Polygonum nepalense,* and *Gnaphalium Americanum.* The only exception was *Oxalis corniculata.* The species found in the current study as dominant, which could be considered weeds in agricultural production systems, show certain characteristics that allow them the domination of agricultural areas. For example, it has been reported that the high abundance of the species *Cardamine hirsuta* in horticultural crops could be related to its ability to easily adapt to disturbed environments and grow in open habitats with higher availability of direct light [17]. Likewise, its reproductive capacity offers it advantages to dominate, since it is a species with a short cycle (annual) and a high germination potential; it has been recorded to produce approximately 5000 seeds with germination percentages higher than 90% [60,61]. In addition, this species has self-dispersal mechanisms, favoring its dominance in anthropized environments [17,60,62]. Similarly, it is argued that the dominance of *Oxalis corniculata* in crop fields is related to its polymorphic reproduction since it is a species capable of easily reproducing both by seeds and vegetatively [63,64]. Conversely, *Polygonum nepalense* has been reported as a weed in productive systems of the Colombian Andes, and its dominance is due to its reproductive traits through the production of a large number of seeds (approximately 27,900/m^2); the seeds have the ability to survive for long periods in the soil and form seed banks, in addition to having a wide range of adaptation to disturbed ecosystems [17,65,66].

Other findings correspond to the fact that the diversity of soil seed banks is higher in the less disturbed ecosystem. These findings are reflected in high values of first- and second-order Alpha diversity (q1 and q2) in the forest ecosystem, presenting a more equitable and homogeneous behavior in terms of its species, and are supported by previous research. For example, Mitja and Miranda [42] found similar results, indicating that forest covers with some degree of conservation may show a higher diversity of weed species, and the diversity in this type of habitat may be related to the stability dynamics of forest ecosystems [49]. In line with these findings, in the current study, the weed community in the soil seed banks and the surface vegetation of the forested area were found not to show competitiveness characteristics. These results support the idea that less intervened ecosystems offer favorable conditions for the coexistence of multiple weed species in an equilibrium [49,67].

On the other hand, the weed communities in the soil seed banks in the agricultural production systems showed a lower first- and second-order Alpha diversity compared to the forest ecosystem. These results agree with previous investigations, indicating that the low diversity of weeds in soil seed banks is a consequence of the high disturbance pressure exerted by man on agricultural systems for their control [49,68], where intensive management practices can favor the growth and establishment of some weed species while restricting the development of others [49,67,68]. Therefore, agricultural management, including selection pressures in production systems, could be a contributing factor to the low weed diversity observed in these environments [12,41,49,51,57,67].

It is crucial to recognize that while the findings imply that the sampling effort for each production system was sufficient and that the observed and expected diversity are in alignment, logistical constraints tied to material collection limited our ability to conduct sampling at only one site within each production system for this study. Despite this constraint, species estimation indicates a marginal increase in just two additional species with doubled sampling efforts. As such, the decision to expand the number of plots at this specific site seems to offer minimal impact. However, an exception arises in the context of the vegetable production system, wherein further sampling could potentially unveil a modest rise in newly identified species. This circumstance suggests that future research endeavors should consider this system as a potential focal point for deeper exploration.

The agricultural production systems assessed have historically experienced various types of agricultural management to control weed populations. Among these management practices, the application of herbicides and overgrazing stand out in livestock areas. In avocado and vegetable production systems, soil tillage practices are highlighted. Some studies have reported that the diversity of weeds in soil seed banks and their surface vegetation is influenced by the type of management used in each production system, suggesting that the use of herbicides, soil tillage, or grazing result in communities of different weeds [8,12,41,43,49,52,67,69]. However, although agricultural management practices in production systems can reduce the diversity of weeds, it has been observed that the abundance of these species does not always decrease and may even tend to increase the dominance of a few species that have the capacity to easily compete with other weed species [43–45,70], agreeing with the results found in the current study.

The results suggest that understanding the variability of the diversity of weed communities in soil seed banks and their surface vegetation in areas under different anthropogenic pressures in the northern Colombian Andes provides valuable information on the ecological dynamics of weeds species, which could facilitate the identification of those that really represent a problem in agricultural production systems. In addition, a clear understanding of how different agricultural management practices interact to condition weed communities is a key component in the development of integrated weed management programs focused on agricultural efficiency and environmental sustainability based on ecological approaches that promote the biodiversity of ecosystems [42,45,46,54,71–75].

5. Conclusions

The composition of the weed communities in soil seed banks and the superficial vegetation differed between the little-intervened forest ecosystem and the agricultural production systems assessed. A greater equitability was observed in the forest ecosystem, while there was a higher dominance of some species in the agricultural systems. Likewise, the highest differentiation occurred in the species composition between sites and the dominant species in the agricultural systems, those with a negative impact on agricultural production standing out probably due to their life history traits that could make them more successful in the face of different anthropic selection pressures in each productive system. Thus, understanding the dynamics of weed communities subjected to various anthropic pressures in different production systems could help design more efficient and environmentally friendly weed management strategies.

Author Contributions: Conceptualization, Y.P.D. and C.E.G.-S.; methodology, Y.P.D., C.E.G.-S. and J.M.R.; Software, C.E.G.-S.; validation, C.E.G.-S. and M.A.Q.-A.; formal analysis, C.E.G.-S.; research, Y.P.D.; Resources, data curation, C.E.G.-S.; writing—original draft preparation, Y.P.D. and J.M.R.; writing—proofreading and editing, Y.P.D., C.E.G.-S., M.A.Q.-A. and J.M.R.; display, C.E.G.-S.; supervision, C.E.G.-S.; project management, Y.P.D. and C.E.G.-S.; fund acquisition, M.A.Q.-A. All authors have read and agreed to the published version of the manuscript.

Funding: This research was funded by the "Dirección de Investigación Desarrollo e Innovación, Universidad Católica de Oriente (UCO)", Project code: 202207.

Institutional Review Board Statement: On behalf of UniversidadCatólica de oriente, ANLA approved the biological collection of plant in the study area (permit number 1726, of 2019).

Data Availability Statement: Data are available from the corresponding author upon request.

Acknowledgments: We express our gratitude to Wilmar Sánchez for his support in the fieldwork and Daniela Alzate for the technical support in preparing the different maps. We also thank the owners of the properties where the study was carried out and the herbarium of the Universidad Católica de Oriente (HUCO) for their support throughout the entire research process.

Conflicts of Interest: The authors declare no conflict of interest.

References

1. Monaco, T.J.; Weller, S.G.; Ashton, F.M. *Weed Science: Principles and Practices*, 4th ed.; Wiley-Blackwell: Hoboken, NJ, USA, 2002; ISBN 978-0-471-37051-2.
2. Buhler, D.D.; Kohler, K.A.; Thompson, R.L. Weed Seed Bank Dynamics During a Five-Year Crop Rotation. *Weed Technol.* **2001**, *15*, 170–176. [CrossRef]
3. Baker, H. The Evolution of Weeds. *Annu. Rev. Ecol. Syst.* **1974**, *5*, 1–24. [CrossRef]
4. Williams, C. Potential Valuable Ecological Functions of Nonindigenous Plants. In *Assessment and Management of Plant Invasions*; Luken, J., Thieret, J., Eds.; Springer: New York, NY, USA, 1997; pp. 26–34.
5. Jordan, N.; Vatovec, C. Agroecological Benefits from Weeds. In *Weed Biology and Management*; Inderjit, Ed.; Springer: Dordrecht, The Netherlands, 2004; pp. 137–158.
6. Gideon, K.M.; Samuel, O.A.; Joseph, S.; Marius, M.; Emmanuel, H. Effect of Linear View Approach of Weed Management in Agro-Ecosystem: A Review. *Afr. J. Agric. Res.* **2021**, *17*, 238–246. [CrossRef]

7. Zimbahl, R.L. *Fundamentals of Weed Science*; Project Manager: Denver, CO, USA, 2018; ISBN 978-0-12-811143-7.
8. Thomas, A.G.; Derksen, D.A.; Blackshaw, R.E.; Van Acker, R.C.; Légère, A.; Watson, P.R.; Turnbull, G.C. A Multistudy Approach to Understanding Weed Population Shifts in Medium- to Long-Term Tillage Systems. *Weed Sci.* **2004**, *52*, 874–880. [CrossRef]
9. Sheley, R.L.; James, J.; Rinella, M.-M.; Blumenthal, D.; DiTomaso, J. *Invasive Plant Management on Anticipated Conservation Benefits: A Scientific Assessment*; USDA Natural Resources Conservation Service: Lawrence, KS, USA, 2011; pp. 291–336.
10. Buhler, D.D.; Mester, T.C.; Kohler, K.A. The Effect of Maize Residues and Tillage on Emergence of Setaria Faberi, Abutilon Theophrasti, Amaranthus Retroflexus and Chenopodium Album. *Weed Res.* **1996**, *36*, 153–165. [CrossRef]
11. Farooq, M.; Flower, K.C.; Jabran, K.; Wahid, A.; Siddique, K.H.M. Crop Yield and Weed Management in Rainfed Conservation Agriculture. *Soil Tillage Res.* **2011**, *117*, 172–183. [CrossRef]
12. Nichols, V.; Verhulst, N.; Cox, R.; Govaerts, B. Weed Dynamics and Conservation Agriculture Principles: A Review. *Field Crops Res.* **2015**, *183*, 56–68. [CrossRef]
13. Liebman, M.; Dyck, E. Crop Rotation and Intercropping Strategies for Weed Management. *Ecol. Appl.* **2011**, *3*, 92–122. [CrossRef]
14. Travlos, I.S.; Cheimona, N.; Roussis, I.; Bilalis, D.J. Weed-Species Abundance and Diversity Indices in Relation to Tillage Systems and Fertilization. *Front. Environ. Sci.* **2018**, *6*, 11. [CrossRef]
15. Raczka, M.F.; Mosblech, N.A.; Giosan, L.; Valencia, B.G.; Folcik, A.M.; Kingston, M.; Baskin, S.; Bush, M.B. A Human Role in Andean Megafaunal Extinction? *Quat. Sci. Rev.* **2019**, *205*, 154–165. [CrossRef]
16. Correa Ayram, C.A.; Etter, A.; Díaz-Timoté, J.J.; Rodríguez Buriticá, S.; Ramírez, W.; Corzo, G. Spatiotemporal Evaluation of the Human Footprint in Colombia: Four Decades of Anthropic Impact in Highly Biodiverse Ecosystems. *Ecol. Indic.* **2020**, *117*, 106630. [CrossRef]
17. Plaza, G.; Quintana, D.; Aponte, L.; Chaves, B. Characterization of the Weed Community of a Rose Greenhouse Production System in the Bogota Plateau. *Agron. Colomb.* **2009**, *27*, 385–394.
18. Moreno-Preciado, O.E.; Balaguera-Lopez, H.E. *Caracterización de La Comunidad de Malezas y Su Diversidad en una Modelación Estadística en un Cultivo de Duraznero (Prunus persica (L.) Batsch.)*; Revista UDCA Actualidad & Divulgación Científica: Bogotá, Colombia, 2021; Volume 24, pp. 1–10. [CrossRef]
19. Quijano-abril, M.A.; Castaño-López, M.; de los, Á.; Marín-Henao, D.; Sánchez-Gómez, D.; Rojas-Villa, J.M.; Sierra-Escobar, J. Functional Traits of the Invasive Species Thunbergia Alata (Acanthaceae) and Its Importance in the Adaptation to Andean Forests. *Acta Bot. Mex.* **2021**, *128*, 1–23. [CrossRef]
20. Buhler, D.D.; Hartzler, R.G.; Forcella, F. Weed Seed Bank Dynamics: Implications to Weed Management. *J. Crop Prod.* **1998**, *1*, 145–168. [CrossRef]
21. Chauhan, B.S.; Johnson, D.E. The Role of Seed Ecology in Improving Weed Management Strategies in the Tropics. *Adv. Agron.* **2010**, *105*, 221–262. [CrossRef]
22. Myers, N.; Mittermeler, R.A.; Mittermeler, C.G.; Da Fonseca, G.A.B.; Kent, J. Biodiversity Hotspots for Conservation Priorities. *Nature* **2000**, *403*, 853–858. [CrossRef]
23. Holdridge, L.R. *Life Zone Ecology*; Tropical Science Center: San José, CR, USA, 1967; ISBN B0007JDYF2.
24. Zapata, D.; Barrera, M.; Gómez, R.; Naranjo, L. *Plan de Crecimiento Verde y Desarrollo Compatible Con el Clima en el Oriente Antioqueño*; CORNARE: Cali, Colombia, 2017.
25. Dirección de Competitividad y Desarrollo Empresarial. *(CCOC) Concepto Economico del Oriente Antioqueño 2018*; Cámara de Comercio del Oriente Antioqueño: Rionegro, Colombia, 2017; pp. 1–94.
26. IDEAM. *Leyenda Nacional de Coberturas de La Tierra. Metodología CORINE Land Cover Adaptada Para Colombia, Escala 1:100,000*; Martinez-Ardila, N.J., Murcia-Garcia, U.G., Eds.; Instituto de Hidrología, Meteorología y Estudios Ambientales: Bogotá, Colombia, 2010; ISBN 9789588067292.
27. Bigwood, D.W.; Inouye, D.W. Spatial Pattern Analysis of Seed Banks: An Improved Method and Optimized Sampling. *Ecology* **1998**, *69*, 497–507. [CrossRef]
28. Montenegro, A.; Ávila, Y.; Mendivelso, H.; Vargas, O. Potential of the Seed Bank in the Regeneration of Plant Communities at Jaboque Wetland, Bogotá, Colombia. *Caldasia* **2006**, *28*, 285–306.
29. Idárraga-Piedrahíta, Á.; Ortiz, R.D.C.; Callejas Posada, R.; Merello, M. *Flora de Antioquia: Catálogo de Las Plantas Vasculares. Vol II. Listado de Las Plantas Vasculares del Departamento de Antioquia*; Universidad de Antioquia: Medellin, Colombia, 2011.
30. Lorena Gámez, A.; Rojas, L.M. *Guía Ilustrada de Plantas Arvenses del Centro Agropecuario Marengo (CAM)*; Universidad Nacional de Colombia (Bogotá): Bogotá, Colombia, 2018; ISBN 978-958-783-393-5.
31. Salazar Gutiérrez, L.F. *Arvenses Frecuentes en el Cultivo de Café En Colombia*; Cenicafé: Manizales, Colombia, 2021; ISBN 9789588490489.
32. Hill, M.O. Diversity and Evenness: A Unifying Notation and Its Consequences. *Ecology* **1973**, *54*, 427–432. [CrossRef]
33. Jost, L. Entropy and Diversity. *Oikos* **2006**, *113*, 363–375. [CrossRef]
34. Chao, A.; Chiu, C.-H.; Hsieh, T.C. Proposing a Resolution to Debates on Diversity Partitioning. *Ecology* **2012**, *93*, 2037–2051. [CrossRef]
35. Magurran, A.E. *Ecological Diversity and Its Measurement*; Springer: Dordrecht, The Netherlands, 1988.
36. Jost, L. Partitioning Diversity into Independent Alpha Beta Concepts. *Ecology* **2007**, *88*, 2427–2439. [CrossRef]
37. Baselga, A. Separating the Two Components of Abundance-Based Dissimilarity: Balanced Changes in Abundance vs. Abundance Gradients. *Methods Ecol. Evol.* **2013**, *4*, 552–557. [CrossRef]

38. Whittaker, R.H. Vegetation of the Siskiyou Mountains, Oregon and California. *Ecol. Monogr.* **1960**, *30*, 279–338. [CrossRef]
39. Dufrêne, M.; Legendre, P. Species Assemblages and Indicator Species: The Need for a Flexible Asymmetrical Approach. *Ecol. Monogr.* **1997**, *67*, 345–366.
40. Hammer, Ø.; Harper, D.A.T.; Ryan, P.D. Paleontological Statistics Software Package for Education and Data Analysis. *Palaeontol. Electron.* **2001**, *4*, 9.
41. Wilson, W.L.; Abernethy, V.J.; Murphy, K.J.; Adam, A.; Mccracken, D.I.; Downie, I.S.; Foster, G.N.; Furness, R.W.; Waterhouse, A.; Ribera, I. Prediction of Plant Diversity Response to Land-Use Change on Scottish Agricultural Land. *Agric. Ecosyst. Environ.* **2003**, *94*, 249–263. [CrossRef]
42. Mitja, D.; Miranda, I.S. Weed Community Dynamics in Two Pastures Grown after Clearing Brazilian Amazonian Rainforest. *Weed Res.* **2010**, *50*, 163–173. [CrossRef]
43. Barroso, J.; Miller, Z.J.; Lehnhoff, E.A.; Hatfield, P.G.; Menalled, F.D. Impacts of Cropping System and Management Practices on the Assembly of Weed Communities. *Weed Res.* **2015**, *55*, 426–435. [CrossRef]
44. Dorado, J.; Lo, C. The Effect of Tillage System and Use of a Paraplow on Weed Flora in a Semiarid Soil from Central Spain. *Weed Res.* **2006**, *46*, 424–431. [CrossRef]
45. Armengot, L.; José-maría, L. Weed Harrowing in Organically Grown Cereal Crops Avoids Yield Losses without Reducing Weed Diversity. *Agron. Sustain. Dev.* **2012**, *33*, 405–411. [CrossRef]
46. Shrestha, A.; Knezevic, S.Z.; Roy, R.C.; Ball-Coelho, B.R.; Swanton, C.J. Effect of Tillage, Cover Crop and Crop Rotation on the Composition of Weed Flora in a Sandy Soil. *Weed Res.* **2002**, *42*, 76–87. [CrossRef]
47. Bàrberi, P.; Silvestri, N.; Bonari, E. Weed Communities of Winter Wheat as Influenced by Input Level and Rotation. *Weed Res.* **1997**, *37*, 301–313. [CrossRef]
48. Swanton, C.J.; Clements, D.R.; Derksen, D.A. Weed Succession under Conservation Tillage: A Hierarchical Framework for Research and Management. *Weed Technol.* **1993**, *7*, 286–297. [CrossRef]
49. Légère, A.; Stevenson, F.C.; Benoit, D.L. Diversity and Assembly of Weed Communities: Contrasting Responses across Cropping Systems. *Weed Res.* **2005**, *45*, 303–315. [CrossRef]
50. Anderson, R.L. A Multi-Tactic Approach to Manage Weed Population Dynamics in Crop Rotations. *Agron. J.* **2005**, *97*, 1579–1583. [CrossRef]
51. Buhler, D.D.; Stoltenberg, D.E.; Becker, R.L.; Gunsolus, J.L. Perennial Weed Populations after 14 Years of Variable Tillage and Cropping Practices. *Weed Sci.* **1994**, *42*, 205–209. [CrossRef]
52. Cardina, J.; Herms, C.P.; Doohan, D.J. Crop Rotation and Tillage System Effects on Weed Seedbanks. *Weed Sci.* **2002**, *50*, 448–460. [CrossRef]
53. Sosnoskie, L.M.; Cardina, J.; Herms, C.P.; Kleinhenz, M. Weed Seedbank Community Composition in a 35-Year-Old Tillage and Rotation Experiment. *HortScience* **2019**, *39*, 845C. [CrossRef]
54. Santín-Montanyá, M.I.; Martín-Lammerding, D.; Walter, I.; Zambrana, E.; Tenorio, J.L. Effects of Tillage, Crop Systems and Fertilization on Weed Abundance and Diversity in 4-Year Dry Land Winter Wheat. *Eur. J. Agron.* **2013**, *48*, 43–49. [CrossRef]
55. Gruber, S.; Claupein, W. Effect of Tillage Intensity on Weed Infestation in Organic Farming. *Soil Tillage Res.* **2009**, *105*, 104–111. [CrossRef]
56. Sharp, D. A Phytosociological Study of Weed Communities on the Southwestern Coastal Plain of North Carolina. *Vegetatio* **1976**, *31*, 103–119. [CrossRef]
57. Tuesca, D.; Puricelli, E.; Papa, J.C. A Long-Term Study of Weed Flora Shifts in Different Tillage Systems. *Weed Res.* **2001**, *41*, 369–382. [CrossRef]
58. Taa, A.; Tanner, D.; Bennie, A.T.P. Effects of Stubble Management, Tillage and Cropping Sequence on Wheat Production in the South-Eastern Highlands of Ethiopia. *Soil Tillage Res.* **2004**, *76*, 69–82. [CrossRef]
59. Heijting, S.; Van Der Werf, W.; Kropff, M.J. Seed Dispersal by Forage Harvester and Rigid-Tine Cultivator in Maize. *Weed Res.* **2009**, *49*, 153–163. [CrossRef]
60. Kudoh, H.; Nakayama, M.; Lihová, J.; Marhold, K. Does Invasion Involve Alternation of Germination Requirements? A Comparative Study between Native and Introduced Strains of an Annual Brassicaceae, Cardamine Hirsuta. *Ecol. Res.* **2007**, *22*, 869–875. [CrossRef]
61. Pedraza, M.; Plaza, G.A. Reconocimiento y Caracterización Ecológica de La Flora Arvense Asociada Al Cultivo de Uchuva. *Agron. Colomb.* **2007**, *25*, 306–313.
62. Yatsu, Y.; Kachi, N.; Kudoh, H. Ecological Distribution and Phenology of an Invasive Species, *Cardamine hirsuta* L., and Its Native Counterpart, Cardamine Flexuosa with, in Central Japan. *Plant Species Biol.* **2003**, *18*, 35–42. [CrossRef]
63. Robertson, K.R. The Oxalidaceae in the Southeastern United States. *J. Arnold Arbor.* **1975**, *56*, 223–239. [CrossRef]
64. Holt, J.S. Factors Affecting Germination in Greenhouse-Produced Seeds of Oxalis Corniculata, a Perennial Weed. *Am. J. Bot.* **1987**, *74*, 429–436. [CrossRef]
65. Gómez-Aristizábal, A.; Rivera-Posada, H. *Descripción de Malezas en las Plantaciones de Café*; Federación Nacional de Cafeteros de Colombia, Subgerencia General Técnica, Centro Nacional de Investigaciones de Café: Chinchiná, Colombia, 1987.
66. Roberts, H.A.; Stokes, F.G. Studies on the Weeds of Vegetable Crops. VI. Seed Populations of Soil Under Commercial Cropping. *J. Appl. Ecol.* **1966**, *3*, 181. [CrossRef]
67. Storkey, J.; Neve, P. What Good Is Weed Diversity? *Weed Res.* **2018**, *58*, 239–243. [CrossRef] [PubMed]

68. Köllmann, P.; Waldhardt, R. Farming Intensity Affects Soil Seedbank Composition and Spontaneous Vegetation of Arable Weeds. *Diversity* **2022**, *14*, 111. [CrossRef]
69. Ludvíková, V.; Pavl, V.; Pavl, L.; Gaisler, J.; Hejcman, M. Sward-Height Patches under Intensive and Extensive Grazing Density in an Agrostis Capillaris Grassland. *Folia Geobot.* **2015**, *50*, 219–228. [CrossRef]
70. Gazoulis, I.; Kanatas, P.; Antonopoulos, N. Cultural Practices and Mechanical Weed Control for the Management of a Low-Diversity Weed Community in Spinach. *Diversity* **2021**, *13*, 616. [CrossRef]
71. Chauhan, B.S.; Johnson, D.E. Ecological Studies on Cyperus Difformis, Cyperus Iria and Fimbristylis Miliacea: Three Troublesome Annual Sedge Weeds of Rice. *Ann. Appl. Biol.* **2009**, *155*, 103–112. [CrossRef]
72. Gaba, S.; Perronne, R.; Fried, G.; Gardarin, A.; Bretagnolle, F.; Biju-Duval, L.; Colbach, N.; Cordeau, S.; Fernández-Aparicio, M.; Gauvrit, C.; et al. Response and Effect Traits of Arable Weeds in Agro-Ecosystems: A Review of Current Knowledge. *Weed Res.* **2017**, *57*, 123–147. [CrossRef]
73. Romero, A.; Chamorro, L.; Sans, F.X. Weed Diversity in Crop Edges and Inner Fields of Organic and Conventional Dryland Winter Cereal Crops in NE Spain. *Agric. Ecosyst. Environ.* **2008**, *124*, 97–104. [CrossRef]
74. Tscharntke, T.; Klein, A.M.; Kruess, A.; Steffan-Dewenter, I.; Thies, C. Landscape Perspectives on Agricultural Intensification and Biodiversity—Ecosystem Service Management. *Ecol. Lett.* **2005**, *8*, 857–874. [CrossRef]
75. Riemens, M.M.; Groeneveld, R.M.; Kropff, M.J.; Lotz, L.A.; Renes, R.J.; Sukkel, W.; van der Weide, R.Y. Linking Farmer Weed Management Behavior with Weed Pressure: More than Just Technology. *Weed Sci.* **2010**, *58*, 490–496. [CrossRef]

MDPI
St. Alban-Anlage 66
4052 Basel
Switzerland
www.mdpi.com

Diversity Editorial Office
E-mail: diversity@mdpi.com
www.mdpi.com/journal/diversity

www.ingramcontent.com/pod-product-compliance
Lightning Source LLC
Chambersburg PA
CBHW061614210926
51850CB00009B/2319